The SAGE Companion to the City

EDITED BY
Tim Hall, Phil Hubbard and John Rennie Short

Los Angeles • London • New Delhi • Singapore

First published 2008

SAGE Publications Ltd
1 Oliver's Yard
55 City Road
London EC1Y 1SP

SAGE Publications Inc.
2455 Teller Road
Thousand Oaks, California 91320

SAGE Publications India Pvt Ltd
B 1/I 1 Mohan Cooperative Industrial Area
Mathura Road
New Delhi 110 044

SAGE Publications Asia-Pacific Pte Ltd
33 Pekin Street #02-01
Far East Square
Singapore 048763

Library of Congress Control Number: 2007930357

British Library Cataloguing in Publication data

A catalogue record for this book is available from the British Library

ISBN 978-1-4129-0206-9
ISBN 978-1-4129-0207-6 (pbk)

Typeset by C&M Digitals (P) Ltd., Chennai, India
Printed in India at Replika Press Pvt Ltd
Printed on paper from sustainable resources

CONTENTS

LIST OF TABLES AND FIGURES

Tables

ACKNOWLEDGEMENTS

This book represents a collaborative project. We are especially grateful to all the contributors; they responded to our requests and mercifully failed to comment on the length of time the project took to complete. The brief was simple and ambitious: bring together a list of active scholars to write accessible, informative pieces on aspects of the contemporary city. In practice of course, even the best-laid plans rarely unfold so easily or seamlessly. We acknowledge the patience of the contributors and publishers. We would also like to thank Robert Rojek who initiated the project, got the editors together and then gave us time and space to develop the project. Parts of Chapter One were previously published in Chapter 4 of *Landscapes: Ways of Imagining the World* by Winchester, Kong and Dunn (2003). They are reproduced here with permission from the publishers. Figure 2.2 is reproduced with permission of Bristol Record Office; Figure 13.1 is reproduced with kind permission of Max Ellis, the Junkyard.co.uk. All other photographs are copyright the authors or have been sourced from copyright-free websites/archives.

LIST OF CONTRIBUTORS

Stuart C. Aitken, Department of Geography, San Diego State University, USA.

David Bell, Department of Geography, Leeds University, UK.

Lisa Benton-Short, Department of Geography, The George Washington University, Seattle, USA.

Angus Cameron, Department of Geography, Leicester University, UK.

Colin Chant, Social Sciences, Open University, UK.

Jon Coaffee, School of Environment and Development, Manchester, UK.

Gerald Daly, Environmental Studies, York University, Toronto, Canada.

Joyce Davidson, Department of Geography, Queens University, Ontario, Canada.

John R. Gold, Department of Geography, Oxford Brookes University, UK.

Steve Herbert, Department of Geography/Law, Societies and Justice Program, University of Washington, USA.

Sarah Holloway, Department of Geography, Loughborough University, UK.

Keith Lilley, School of Geography, Archaeology and Palaeoecology, Queens University Belfast, UK.

Mark Jayne, Department of Geography, School of Environment and Development, University of Manchester, UK.

Yeong-Hyun Kim, Department of Geography, Ohio University, USA.

Lily Kong, Department of Geography, National University of Singapore.

Alan Latham, Department of Geography, University College London, UK.

Derek McCormack, School of Geography, University of Oxford, UK.

Don McNeill, University of Western Sydney, Australia.

Malcolm Miles, School of Art and Performance, University of Plymouth, UK.

David Murakami Wood, School of Architecture, Planning and Landscape, Newcastle University, UK.

Andy C. Pratt, Cities Research Centre, and Department of Geography and Environment, London School of Economics, UK.

Marie Price, Department of Geography, The George Washington University, Seattle, USA.

Mick Smith, Department of Environmental Studies, Queens University, Ontario, Canada.

Gill Valentine, Department of Geography, Leeds University, UK.

Kevin Ward, Department of Geography, School of Environment and Development, University of Manchester, UK.

David Wilson, Department of Geography, University of Illinois at Urbana-Champaign, USA.

INTRODUCTION

Tim Hall, Phil Hubbard and John Rennie Short

We are in the midst of the Third Urban Revolution. The first began over 6,000 years ago and saw the first cities in Mesopotamia. These new cities were less the result of an agricultural surplus and more the reflections of concentrated social power that organized sophisticated irrigation schemes and vast building projects. The First Urban Revolution, independently experienced in Africa, Asia and the Americas, ushered wrenching social changes, new ways of doing things and new ways of experiencing, seeing and representing the world. The Second Urban Revolution began in the eighteenth century with the linkage between urbanization and industrialization that inaugurated the creation of the industrial city and unleashed unparalleled rates of urban growth. From 1800 to 2000 urban growth has been one of the most significant features of global demographic change.

Like these previous episodes of city-building, the Third Urban Revolution is a complex phenomenon that began in the last quarter of the twentieth century. It is marked by a major redistribution of economic activities following a putative global shift as manufacturing declines in importance in the West and new centres of industrial production emerge elsewhere. This is mirrored in the global growth of services – especially advanced producer services – which have become the cutting-edge of rapid urban economic development. Consequently, urban landscapes have become revalorized and devalorized at an often bewildering pace: central cities have characteristically become sites of new urban spectacle; inner cities are pockmarked by sites of gentrified renaissance as well as rampant poverty and criminality; inner suburbs show the first inklings of decline; ex-urban development continues apace as gated communities and mixed-use developments sprawl into the former countryside. Urban growth seems inexorable around the world, just as urban decline seems unavoidable throughout the globe. Moreover, the city is the setting, context and platform for new forms of identity construction, with cities long-structured around masculine production making way for spaces where new forms of identity can be forged in a kaleidoscope of (re)invented and (re)discovered identities. Cities are thus associated with diasporic and hybrid identities in which ethnicity and race categories become blurred; sexual identities whose coordinates disturb established notions of sex and gender; and urban subcultures for whom the streets throw up a range of lifestyle possibilities oriented on urban consumption and style. Accordingly, the city remains a furnace of individual creativity and innovation, with music, art and culture, literature, fashion, media, science and technology indelibly associated with the contemporary city.

The city accordingly serves as the eye of a veritable hurricane of economic change and social dislocation. At times, it certainly feels like we are in a hurricane: witness violent riots (such as the riots in Paris's suburbs in the autumn of 2005 or the Sydney beach riots earlier that year), spectacular acts of terrorism (9/11 in New York or 7/7 in London) or the steady drumbeat of civic disobedience (such as the anti-corruption sentiment in urban China and in Budapest, Hungary). Yet, for most of us, most of the time, the city seems a quite banal space, a scene of calm activity as people get on with their everyday lives, working, playing and loving in ways that embody and articulate – as well as resist and silence – dominant narratives and broader socio-economic forces. The fact that cities are so rarely the focus for insurrection, rebellion and disobedience is quite remarkable given the sheer diversity of life which congregates within them, and it is this capacity of the city to function in the face of complexity and contradiction that poses perhaps the key urban question of our (global) times.

Hence, we are in the throes of a revolution that we are only just beginning to see, name and theorize. The new lexicon which has emerged to describe cities – for example, as 'postmodern', 'global', 'intransitive', 'networked', 'hybrid' – offers some purchase on the rich complexity and deep contradictions of the Third Urban Revolution, but much remains to be said and done before we can make any sense of the new forms of urbanism which characterize the twenty-first century. Luckily, urban studies can draw upon a fertile tradition of scholarship that has sought to delineate and describe the city, and there is within this literature a rich legacy of concepts and theories that provide a springboard for exploring the geographies, histories, economies and socialities of the contemporary city. This book thus offers a number of 'cuts' through the contemporary city, considering how different aspects of city life have been conceptualized, quantified and qualified by generations of scholars so as to identify specific themes and languages which appear to offer us the basis for constructing an urban theory fit for contemporary times. Each chapter thus looks back at a body of work, dissecting it to draw out a number of key ideas that hold relevance in the contemporary content. *The Sage Companion to the City* thus represents a forward-looking collection designed to put down some signposts as to where urban studies may be heading – yet in doing so also offers a critical reflection on where it has been.

Placing urban studies

Though scholars have reflected on the role of cities since the First Urban Revolution, it was not until the rapid urbanization and industrialization of the Second Urban Revolution that the city began to be taken seriously as an object of study. The dramatic growth of cities, propelled by and organized around production, brought together individuals from disparate backgrounds in ways never experienced before. New building methods, urban technologies and innovations in transport rapidly transformed people's relationships with their surroundings – and one another – to the extent that *urbanism* began to be defined as a distinctive way of life. The sheer pace of change, and the need to develop a mental sensibility that could deal with the experience of being surrounded by strangers led some of the leading sociologists of the day (e.g. Durkheim, Simmel, Weber) to identify new phenomena that were innately urban (such as the adoption of a blasé outlook, indifference to strangers, a preoccupation with appearances and a dissolution of kinship ties). Furthermore, stark social and economic juxtapositions (e.g. over-crowded working-class terraces nestling alongside the spectacular residences of the *nouveau riche* bourgeoisie) raised new questions about the inequalities wrought by urbanization, with some notable commentators (not least Marx and Engels) rallying against the sheer inhumanity and inhospitable nature of city life.

It was from this foment that urban studies began to emerge as a distinctive disciplinary endeavour, albeit one that sat uneasily across the natural and social sciences. Individually and collectively, however, historians, sociologists, economists and political theorists began to note the new social spaces emerging in cities as traditional communities based on blood and kinship began to be replaced by more functional (and *organic*) forms of sociality. They noted the formation of ethnic enclaves, 'skid row' areas, zones in transition as well as elite residential tracts and affluent urban estates. More practically minded writers in the fields of architecture and town planning contributed to understandings of these new urban landscapes, making suggestions as to how cities could be modernized to enhance them aesthetically and socially. Such ideas clearly chimed with debates in geography, where environmental determinism was a popular perspective within a discipline largely preoccupied with regional description. Yet it was the Chicago School of Sociology (under Robert Park and Ernest Burgess) that was to give urban studies its most visible articulation, with a slew of exhaustive urban ethnographies and pioneering studies completed up to the 1940s. Mapping the extraordinary diversity of life within North American cities, the Chicago School identified the city as a unique organism whose life cycles demanded to be studied, noting the human adaptations occurring as the city itself reorganized.

One of the legacies of the Chicago School – the notion of teleological models describing the distinctive social areas and sectors of the city – was subsequently to inspire geographers and sociologists to develop theories of the city predicated on notions that land values decreased with distance from the city's most accessible point (i.e. the centre). Refinements of this notion over time led to increasingly sophisticated attempts to model the city, with innovations in computation and statistics allowing the development of models offering a better approximation of the city's form and function. Geography's new attempt to rebrand itself as a spatial science in the 1950s and 1960s (with the associated borrowing of ideas from mathematics, economics and even physics) witnessed ever more elegant and predicative models of urban land-use. Moreover, at the same time that geographers were shedding light on the internal dynamics of cities, economic geographers were developing Christaller's central place theory to speculate as to the way inter-city relations bequeathed national urban systems characterized by specific distributions of cities and people across space. Through engagements with economic theories of bid-rent and profit maximization, urban studies thus began to develop sophisticated ideas about the role of cities in organizing production and consumption across different nation-states.

The 1960s, however, also bought new urban phenomena to the fore, with the worsening 'inner-city' problem in US and European cities flagging up the stark racialization of the city. Cities also became the focus of anti-war and anti-nuclear protests, and, in the face of such social unrest, questions began to be raised about the social relevance of urban studies and the contribution of academics to alleviating urban poverty and inner-city decay. In this context, a new generation of scholars sought to develop a radical critique of capitalist urbanism, developing Marxist theories of the city that emphasized the active role of the city in producing and sustaining capitalism by assuaging class conflict and aiding capital accumulation. In turn, it was suggested that the city could become a site from where the oppressed could challenge the banality of everyday urbanism and overthrow existing orders: the idea that the city could be turned against the powerful became a strong motif. Boundaries between activist and academic thus became blurred, and some urbanists became pivotal in encouraging insurrection (the French-based Marxists Henri Lefebvre, Manuel Castells and Guy Debord, for example, were directly implicated in the student riots of 1968).

This attempt to situate urban studies within an explicitly politicized theoretical context thus brought urban scholars into dialogue with political scientists, and generated some powerful critiques of urban governors. The 'new' urban sociology of the period hence injected a political urgency into urban studies, and stimulated varied attempts to theorize the role of the city in mediating capital–labour conflicts at scales from the individual household to the entire urban-region. Yet at the same time, many urban researchers rejected structural or Marxist readings to focus on people's more or less rational decision-making processes, foregrounding questions of perception, choice and behaviour as they impinged on housing choice. People-centred theories of how cities are made through agency rather than structure thus posed a challenge to radical Marxist thinking, while feminist writing on the gendering of cities ultimately suggested class might be just one of many factors which determine the shape and structure of the urban landscape. Questions of identity and difference were thus a preoccupation for many, with issues of race, sexuality and culture becoming more important as a 'cultural turn' became evident across the social sciences in the 1980s and 1990s. Related to this was the notion that modern, industrial cities had been superseded by a post-industrial city, a more flexible, complex and divided city than its predecessor, with the ordered and production-based logic of the industrial era giving way to a more invidious mode of social control based on people's role as consumer-citizens. The result is a 'post-modern' city of different ethnic enclaves, consumer niches and taste communities, spun out across a decentred landscape where the boundaries between city and country are often hard to discern.

The apparent complexity of the contemporary city, and the obvious limitations of existing urban theories to explain its forms, has hence challenged any notion that urban studies has progressed towards a more complete or better understanding of how cities work. New ideas about complexity and contingency thus abound, with some post-structural theorists insisting that we remain sensitive to the particularity of each and every urban event, and avoid simplifying cities. Though difficult to define, post-structuralism's emphasis on questions of language, representation and power points to a different way of understanding the production of space, involving the entwining of immaterial and material forces. Notably, many of the key proponents of post-structural thought – Foucault, Derrida, Deleuze, Irigaray, Baudrillard – sought to offer accounts open to the messiness of life. Foucault, for example, developed a critique which destabilized the authority of the scholar and posed important questions about the power of disciplinary (and disciplined) accounts of the social world. Critical of the totalizing discourse characteristic of social science, Foucault argued for the recovery of *subjugated* urban knowledges (those disqualified or dismissed by the powerful and authoritative).

In the wake of such Foucauldian critique, it has been difficult for urban researchers to argue that they have a privileged gaze, or offer urban models that hold in each and every instance. Anxieties about common academic tropes of urban representation – maps, models, statistics – have hence ushered in experiments in new forms of city writing, with 'views from above' being joined by a diverse range of embodied 'views from below' as urban scholars explore the potential of street poetry, art and performance to speak to the experiences of different urban dwellers. Moreover, the materiality of the city itself is now frequently understood to encompass representations of the city which have a force and life of their own. Metaphors of the city as text abound, with 'readings' of cities based on a range of different sources and sites (including the urban landscape itself, which has been considered ripe for iconographic and semiotic deconstruction). The whole notion of what a city is – a dense, heterogeneous collection of people – has also been revised as notions about the agency of cities are widened to encompass the role of new technologies and media (not least the Internet). For many commentators, virtual cities are now as valid an object for study as 'real cities'.

Urban studies has hence undergone a number of broad 'sea changes' in the last 150 years, with new ideas, theories and approaches emerging at specific moments, shaking scholars out of any complacency that they have answered the 'urban question'. Yet throughout, urban studies appears to have remained fixated on a number of world cities – London, Paris, New York, Berlin, Los Angeles – to the detriment of studies of smaller and more 'ordinary' cities (as well as those beyond the West). Equally, urban studies has tended to be associated with academics working within higher education, the majority of whom are white, middle-class and heterosexual, and consequently locked into specific ways of 'viewing' the world. Though becoming more open to ideas of difference over time, the result is that urban studies has often failed to capture the sheer diversity and excessive liveliness of cities. It is this that constitutes one of the principal challenges that lies ahead as scholars seek to further their understanding of cities after the Third Urban Revolution.

Urban prospects

How are cities to develop? What are the prospects for hope or abjection in the future? Urban studies has long had one eye on the future of cities, typically outlining how conditions and processes of the present might be shaped in the future. A number of commentators in this collection offer their views, hopes and fears for the cities of the future and in doing so often suggest critiques of the present. Predicting the future is difficult and can often go spectacularly awry. Before exploring some brief thoughts on the future of the city it is worth reflecting on the purpose of looking forward in these ways.

Historically, urban studies has been shot through with concerns about the inequalities (social, economic, political or otherwise) within and between cities and the impacts of these on the lives of individuals and communities. As this introduction points out, the 'relevance' or 'applicability' of urban studies has varied through time but there has been a strong tradition of developing knowledges that can be used to make urban life better. There are many examples of scholars from this tradition becoming active in the improvement of cities, whether through activism (as protestors, advocates or squatters) or through professional practice (as planners, architects, advisers). The main reason that we, as urban scholars, might wish to look to the future of cities, then, is to hope that we might be able to make them better places. This raises crucial questions of the various channels through which this might be achieved.

Urban scholars might shape the cities of the future in at least three ways. First, education is not simply a process that students go through to make them more employable (although this is how it is increasingly seen by a number of central governments and, indeed, universities). Rather education is something that can develop citizens and encourage them to become active in the shaping of their lives and the lives of their communities (however the latter might be defined). Education raises awareness and prompts enquiry that can ripple out beyond the classroom walls long after the assignment deadline date. We live on an urban planet, and urban studies should recognize its important role in equipping students to become responsible citizens and life-long learners. Of course, students are not passive recipients of urban knowledge, and develop their own ideas about how they might live in a more sustainable and social responsible manner. Yet the role of urban scholars in opening the eyes of students – and wider society – to the possibilities of the city should never be underestimated.

Second, as we have noted, urban studies has a long tradition of direct engagement, either through activism or practice. This is a tradition that has not died out. A number of scholars have left the classroom and taken to the streets either through protest or through community activism of various kinds. Despite the many constraints, this activity endures and flourishes. There are many urban studies courses around the world, for example, that seek to take students, and academics, out of the familiar comfort zones of the classroom and the library and to place them at the heart of real cities and real urban problems. Many urban scholars are involved in grassroots activism, mobilization and protest; others have taken roles as councillors, politicians or formed pressure groups, agitating for change through more formal processes.

Finally, urban studies scholars produce knowledge that is not purely 'academic'. Rather, it is knowledge that is 'useful' and might be used by those outside the academy who are involved in developing, managing and running cities. It is here, though, that the prospects for the influence of urban studies scholars seem less encouraging. The dialogues between policy-makers, practitioners and urban studies scholars seem particularly barren at the moment, with the former not recognizing the worth of the latter's work and the latter probably being guilty of not seeking sustained and meaningful engagement with the former. Engaging with policy-makers and practitioners is a challenge that urban studies scholars should be prepared to meet to ensure the vitality of the discipline(s) into the future.

And what of cities themselves? What are their prospects? There are two ways that we might consider this. The first is to look back and ask what are the aspects of the city that have endured and what are the prospects that these might endure or change in the future? Poverty and inequality are two aspects that seem inevitable aspects of cities and city lives. It is difficult to imagine that they will vanish, or reduce significantly, in the future; indeed, most commentators suggest they seem to be getting worse. Environmental catastrophes, such as Hurricane Katrina's impact on New Orleans, brutally expose the inequalities of the city, given that it is the weakest and poorest who suffer most in times of urban crises. Yet perhaps such instances also provide an impetus for reorganizing the goods and bads of urban society between different racial, class and religious communities. No doubt patterns of inequality will shift as cities continue to change and develop. Whether they can ever be wished out of existence is another thing entirely given the capitalist city seems to thrive on inequalities.

Hence, one way of thinking about cities is to provide a dystopian reading of its inequalities and expose its pernicious social divides. Much urban writing – especially from the left – is of this ilk, and is fiercely critical of the city produced by capitalism, especially in its current neo-liberal guise. Yet there is also a tradition of urban writing that is more optimistic, and finds the seeds of change in a variety of everyday spaces and urban rituals of inhabitation. Such writing argues that the best place to think about the future of cities is not perched atop a skyscraper looking down (aping the 'plannner's eye view' of the city), but at street level, engaging with the everyday life of cities. The second way we might usefully think about city futures is therefore to look around us at emergent trends in everyday urbanism and imagine how these might be nutured and blossom into new urban formations.

Even so, such visions of the future of cities need to be tempered with the post-structuralist's wariness of over-arching grand visions or theories. Throughout this book the authors raise this concern and cite examples of the failures of past universal urban theories. Cities are hugely diverse entities, and are always more complex than the theories we develop about them.

The structure of this book

As we have outlined above, *The Sage Companion to the City* is intended to be more than a retrospective wallow in the archives of urban studies. Looking forward, and working through some of the themes outlined above, this volume includes contributions by a selection of those working at the leading-edge of urban scholarship in the disciplines of geography, history, sociology and public policy. Eschewing a chronological or theoretically-structured approach, each chapter instead demonstrates how urban studies has engaged with a particular *theme* (or set of themes) that is at the heart of debates surrounding urbanism and urbanization. Each chapter accordingly showcases enduring concerns and more recent departures in urban studies, and includes extracts from both classic and lesser known texts to demonstrate the variety of theoretical and methodological approaches that have been adopted by scholars in urban studies. As such, each chapter provides a taster of an urban literature that is incredibly rich and varied, and the book is designed to inspire the reader to explore this varied literature.

Inevitably, there are many silences and gaps here, as urban studies is a broad cross-disciplinary endeavour that includes practitioners as well as academics, and our choice of contributors and chapters reflects our own preoccupations as Anglo-American geographers. However, the chapters are arranged into a number of sections, each of which covers particular themes current in urban studies and highlights the range of ways in which leading figures have responded to the challenge of studying these particular facets of urbanization. While there is some overlap and spillage between sections, each hopefully offers coherent reflections on a set of debates within urban studies. The first, *Histories and Ideologies*, explores some of the vexing issues surrounding the changing role of cities over time, not least their role as centres of religious, productive, scientific and cultural life. Questions of historicity are also addressed here through reflections on the role of memory in cities, and a particular focus on the city as a palimpsest on which successive generations have imposed their identities and ideologies.

In Section Two, *Economies and Inequalities*, we consider the work cities perform as well as the work that is carried out in cities. As generations of urban scholars have noted, cities often appear to be organized according to the imperatives of production and consumption, bequeathing spaces of investment and disinvestment which condemn some to a life of poverty or disadvantage. The contrast between the street spaces of informal labour and the corporate citadels of international finance is one clear manifestation of this, but so too is the segregation of consumer spaces catering for different 'taste communities'. Questions of class and capital remain crucial, of course, but a consideration of how Marxist theories of capital accumulation are played out in globalizing and neo-liberalizing cities is a preoccupation for urban scholars – and a major theme in this section.

Section Three explores *Communities and Contestation*. The dissolution of kinship ties and the erosion of community was a much-noted phenomenon in early urban sociology, with the rise of the individual seen as an integral part of urbanization. Yet converse theories of residential clustering, association and cooperation have been posited, with scholars noting the positive role that neighbourhood formation has in promoting the political claims of marginalized groups such as ethnic minorities or gay and lesbian communities. On the one hand, agglomeration and residential clustering can create cities of segregation typified by mistrust, resentment and fear. Finding oneself in urban space is hence a process fraught with contradiction. In the section, therefore, a prime concern is with how the everyday task of getting by and getting along can reinforce dominant senses of who the city belongs to, where and how. On the other hand, it is by considering such questions of inhabitation that we begin to sense how the city is pregnant with possibility, a melting pot of different subjectivities and identities.

Our final section provides something of a summation. *Order and Disorder* is a section that captures the ambivalence of urban space and the maelstrom of change that typifies post-millennial cities. Attempts to impose an order on this obvious complexity are associated with the state and the law, and often tied into notions of criminality and immorality (that which is considered unfitting or 'out of place'). Yet ordering is not always oppressive, and often intervention is underpinned by a utopian vision or dream of what the city might be. Thinking through the play of forces that ultimately produces the 'urban order' allows us to reflect on both the city that has been and the city that might be. As our final contribution in this section stresses, we need to recognize that the city evokes nightmares of loss and disappearance deep-rooted in our collective psyche. Urban studies is perhaps also haunted by the legacies of previous times: in this section contributions unpick some of these legacies to pose some provocative questions about the types of city we want – as well as the types of urban scholar we wish to be.

Taken together, we hope the contributions to this volume provide a useful roadmap for those embarking on their first foray into urban studies. We also hope more veteran urbanists will take sustenance from what we have to offer. Of course, some will go away disappointed, and note the absence of commentary on Third World cities, or those in the post-socialist world. Others will lament the lack of space devoted to issues of urban design, planning and architecture, which also contribute to the rich tapestry of urban theory. Notwithstanding such predictable critiques, we offer this volume as a resource that will hopefully stimulate and provoke readers to develop their own take on the nature of cities.

SECTION ONE
HISTORIES AND IDEOLOGIES

One of the key debates in urban studies revolves around questions of city formation. Ultimately, the question of where the first cities emerged may never be resolved – despite a growing mountain of archaeological evidence. Yet even if our intellectual curiosity about where urbanism began will never be sated, exploring processes of urban formation will remain an important theme in urban studies because it allows us to explore vital questions of what cities are for. For example, contrasting ideas that cities emerged for defensive reasons, for the purposes of trade or as centres of cultural significance encourage us to reflect on the multiple roles that cities serve, and to consider the changing role of cities in pre-industrial, industrial and post-industrial societies. Moreover, the wide array of research completed by urban historians on changing urban forms has shed light on the way city space is produced through particular combinations of charisma and context – that is to say, it has demonstrated that architects, planners and politicians may shape the city, but not in circumstances of their own choosing. Considering how particular imperatives – religious, economic, political – have shaped the urban landscape in different times and places is thus a major theme in urban studies.

Within the literature about urban transformation, much energy has been expended exploring how the city has been modelled and remodelled so as to reflect dominant ideas and ideologies. In specific contexts, the rich and powerful have sought to impose their identity on city space, constructing spectacular and iconographic buildings that symbolize their authority.

In our first chapter, Lily Kong therefore focuses on the ways in which cities can be read as betraying the locus of power in society. Noting specific ways in which urban landscapes may impose the will of the powerful on the less affluent or advantaged, Lily Kong dwells on the social production of urban space. The idea of dominant or *hegemonic* values are made to appear natural is crucial here, as is the idea that capitalism produces landscapes which legitimate and celebrate particular forms of production and consumption.

Often, however, the city has been shaped so as to reflect dominant religious ideas and values, with the layout of streets, plots and buildings reflecting certain ideas about religious and moral order. In Chapter 2, Keith Lilley explores the influence of religious values on the urban landscape, stressing that the city has often been constructed so as to mirror 'god-given' cosmological orders. In other senses, too, the city has become a key site in the celebration (and occasionally, contestation) of religious values, with highly-charged sacred landscapes often providing a focus to urban life. Current debates around the place of religion and faith in multicultural societies are thus intimately bound into discussions of city space, given that sacred sites frequently serve some faith groups but not others. In this sense, while many landscapes reflect the dominant religious and moral values in society, there are also some associated with marginalized, residual and emerging religious groups.

What is clearly evident, therefore, is that the urban landscape is constantly evolving so as to reflect different imperatives and social mores. While in some contexts these continue to reflect religious values, faith in religion has often been superseded by a faith in science and technology to deliver certainty, order and 'the good life'. As Colin Chant describes, science and technology have become increasingly important influences on urban form, with new transport technologies allowing the expansion of cities both outwards and upwards. Often embedded in the urban fabric – to the extent that their importance is forgotten – infrastructures of wires, pipes and cables also create intricate networks that which have allowed for the production of more 'modern' and commodious cities. Far from determining the shape of cities, however, Chant shows that technology needs to be considered as entwined in complex processes of city-making, and requires to be theorized as caught up in networks whose agency is never easily discerned.

New technologies have also been examined for their role in allowing architects and planners to envisage new ways of organizing city life. In particular, the possibilities of constructing using new materials (glass, steel, concrete) was to inspire a generation of architects and planners in the early twentieth century to explore a new modern style which was fit to the 'machine age'. The ideas of this international 'Modern' movement are examined by John Gold, who excavates the way these imaginaries resonated with wider anxieties about the deleterious nature of industrial cities. As he shows, Modernism was to have little influence on the form of cities until the political will to rebuild cities in a modern idiom emerged after the Second World War. Noting the compromises often made in the process of modern city design, Gold's chapter underlines that modernization is always highly ambivalent: every attempt to order the city bequeaths contrary disorders, with city 'improvement' having highly iniquitous effects.

In the final chapter in this section, the focus is on another dimension of the history of cities: the way that past city orders are never obliterated, but remain apparent in sites and spaces of memorialization. Whose history and memories are celebrated and whose are repressed is always a matter of considerable contention, and nowhere is this more true than in the urban landscape, where statuary and memorials are an obvious testament to particular histories, but not others. Lisa Benton-Short's chapter provides a neat synopsis of such contested geographies of memorialization. In the final analysis, such readings of place, memory and conflict are of interest not just in and of themselves, but because they remind us that the city is a site that is always in high tension.

1 POWER AND PRESTIGE

Lily Kong

This chapter

○ Considers the idea that power is inscribed in the urban landscape in a variety of ways

○ Shows how the design, use and symbolism of urban space works to reproduce certain social, economic and political values

○ Concludes that while the city is infused with power, it is also a site of resistance and opposition

---

Cities and power

Cities are the medium and outcome of power: that is to say, they are the means by which power is expressed, and at the same time, the result of power and influence. This is evident in many senses. For example, cities express the power of the state and capital, but also of various social and civic groups, and indeed, individuals. States have the capacity to determine the shape of cities through government planning departments and planning legislatures, while capital has the capacity to shape cities through ownership of land and resources. Social and civic groups can express their power and ideologies as well, most commonly through influencing the use and occupation of urban space. Sometimes, particularly powerful individuals, including megalomaniac ones, express their power and control over cities through monuments and shrines.

Power most commonly extends to the shaping of urban form, through urban design and layout, as well as the form of buildings, but it is also the case that the use of urban spaces constitutes an expression of power in itself. For example, not only are urban segregation and the architecture of buildings illustrations of the exercise of power, so too are the appropriation of public spaces by human groups for their own use. Such appropriation is evident in events such as the open demonstrations of June 1989 in Tiananmen Square, China, where students called for democratic reforms, and the peaceful demonstrations in the streets of Taipei on 26 March 2005 to protest against the passing of the anti-secession law in China. Both represent a takeover of public space by ordinary citizens to express the power of 'the people'. Sometimes, such expressions achieve the desired goal – at other times, not.

The expression of power is also evident in a range of cityscapes at a variety of geographical scales. At one end, witness the plan and layout of entire cities, such as the design and construction of capital cities Canberra and Brasilia from scratch. At the other end of the scale, the body in the city may also be an inscription of power, such as through the patriarchal denial of access rights to Thai temples for menstruating women. In between, power may be expressed through specific streetscapes, landmarks, buildings, and monuments through their construction, demolition, redesign and use.

Power, therefore, takes many forms, ranging from open command (e.g. the openly racist planning frameworks of apartheid South Africa) to hegemonic control through the use of persuasive ideological strategies (e.g. the (re)naming of streets, often in post-colonial cities, to purvey certain ideologies of the powerful). Generally, the latter tends to be more effective as those subject to control do not recognize it and, indeed, may embrace the ideologies of the powerful as their own. This hegemonic role relies on the naturalization of ideological systems, made possible in cities through built environments which make the socially-constructed appear to be the natural order of things.

Having briefly introduced various aspects of power in the city, including who exercises power, how and where, I will now turn to elaborate on key conceptual ideas that help explicate power and prestige in the city. These are 'power', 'ideology', 'hegemony' and 'landscape'. Following this, I will illustrate how state, capital and individuals can influence, shape and, indeed, define cities. This is not to suggest that they act in mutually exclusive ways. Rather, there are instances when they intersect and reinforce one another. For example, a particularly strong (and sometimes megalomaniac) personality who is also a powerful political leader may stamp his/her imprint on the cityscape so that it is not always easy to tease apart when the influence is personal and when it is official. The same may be said of a particularly powerful capitalist. Hence, 'state' and 'capital' will be used as the key organizing frameworks for discussion below, with evidence of individual power and influence integrated with the exercise of state and capital power. I will begin with the power of the state, followed by the power of capital, giving due cognisance to the role of the individual in each section. I will end with a discussion of the intersections of state and capital.

Power, ideology, hegemony and landscape

Power is manifested through its 'transformative capacity'. This means the ability to 'intervene in a given set of events so as in some ways to alter them' (Giddens,

1987: 7). Power may be allocative, extending to control over material facilities, or authoritative, extending to control over human activities. While power may be applied via overt force, it is the 'institutional mediation of power' (Giddens, 1987: 9) or the silent 'repetition of institutionalized practices' which occurs more frequently on an everyday basis. In this respect, the concepts of 'ideology' and 'hegemony' are central to understanding how power is expressed and exercised in real everyday ways in cities.

By 'ideology', I refer to 'a system of signification which facilitates the pursuit of particular interests' and sustains specific 'relations of domination' within society (Thompson, 1981: 147). This is to be understood alongside Gramsci's (1973) notion of 'hegemony' as the means by which domination and rule are achieved. Hegemony does not involve controls which are clearly recognizable as constraints in the traditional coercive sense. Instead, hegemonic control involves a set of values which the majority is persuaded to adopt. So as to persuade the majority, these values are portrayed as 'natural' and 'common-sense'. This is 'ideological hegemony'. The most successful ruling group is that which attains power through ideological hegemony rather than coercion. When hegemonic control is successful, the social order endorsed by the elite is, at the same time, the social order the masses desire.

One of the key ways in which power and prestige can be expressed, maintained and, indeed, enhanced, is through the control and manipulation of landscapes, and there is no better exemplification of a cultural landscape than the city, which bears the imprint of human vision, desire and struggle, while shaping human behaviour and action. The city as landscape is 'imprecise and ambiguous', according to Cosgrove (1984: 13) (see Extract 1.1). It embodies and reflects the negotiation of power between the dominant and subordinated in society (Anderson, 1992: 28). On the one hand, cities as landscapes are social constructions of the powerful – planners, architects, administrators, politicians, property owners, developers – intent on advancing state ideology or consumer capitalism. On the other, cities as landscapes are also 'multicoded spaces' which, through everyday use, are constantly reinterpreted through everyday practices of 'reading' and 'writing' different languages in the built environment (Goss, 1988: 398).

Extract 1.1: From Cosgrove, D. (1984) *Social Formation and Symbolic Landscape*, London: Croom Helm, pp. 15–16.

...It is in the origins of landscape as a way of seeing the world that we discover its links to broader historical structures and processes and are able to locate landscape study within a progressive debate about society and culture. ... [L]andscape represents an historically specific way of experiencing the world developed by, and meaningful to, certain social groups. Landscape ... is an ideological concept. It represents a way in which certain classes of people have signified themselves and their world through their imagined relationship with nature, and

(Continued)

through which they have underlined and communicated their own social role and that of others with respect to external nature. Geography until very recently has adopted the landscape idea in an unexamined way, implicitly accepting many of its ideological assumptions. Consequently, it has not placed the landscape concept within an adequate form of historical or social explanation (Relph, 1981). To do so requires not so much a redefinition of landscape as an examination of geography's own purposes in studying landscape, a critical recognition of the contexts in which the landscape idea has intellectually evolved and a sensitivity to the range and subtlety of human creativity in making and experiencing the environment.

In this chapter, the focus is primarily on cities as impositions of the predilections and prejudices of a powerful state, of capital, and individuals. Indeed, cities exemplify how ideologies are produced and reproduced, through the design and use of its built landscapes, as well as the symbolism of these landscapes.

Power and prestige in the city: the role of the state

The city offers clear evidence of state control over landscapes. Planning laws and other legal and fiscal devices are the most common ways in which the state shapes the city's built and natural environments. Such authoritative authorship of cityscapes is sometimes the outcome of collective vision and ideology, such as through the work of state planning departments or shared official racial ideology, but at other times may be the outcome of more individual visions, whether through the megalomaniac tendencies of political leaders or the artistry and vision of influential architects.

Colonial cities exemplify the power of the colonial state, and this is evident in diverse contexts. Anthony King's work on the impact of Western industrial colonialism on non-Western cities is undoubtedly the seminal work in this respect. King offered insights into how the power structure of colonialism was reinforced by the creation of a segregated city with a colonial sector and an indigenous sector (See Extract 1.2).

Extract 1.2: From King, A.D. (1976) *Colonial Urban Development: Culture, Social Power and Enviroment*, Boston: Routledge and Kegan Paul, pp. 39–40.

...The dominance-dependent relationship [in the colonial city] can be seen at the city level. Here, for a variety of reasons other than those already discussed, the indigenous and colonial parts of the city were kept apart. ... This was either explicit and legally enforced through the creation of distinct areas (or 'reserves') for different racial groups with separate (and unequal) facilities, or it was implicit (as in twentieth-century 'imperial' India), with residential areas so characterized by cultural characteristics or economic deterrents (for example, the cost of land and housing) as to effectively prevent residential infiltration except by those willing and able to adopt the attributes and life-style of the colonial inhabitants. ...

The segregation of areas performed numerous functions, the first of which was to minimize contact between colonial and colonized populations (Balandier, 1951: 47). For the colonial community they acted as instruments of control, both of those outside as well as those within their boundaries. They helped the group to maintain its own self-identity, essential in the performance of its role within the colonial social and political system. They provided a culturally familiar and easily recognizable environment which – like dressing for dinner – was a formal, visible symbol providing psychological and emotional security in a world of uncertain events.

Segregation of the indigenous population provided ease of control in the supervision of 'native affairs'. It was economically useful in cutting down the total area subject to maintenance and development. The colonial environment offered a model for emulation by members of the indigenous society. Segregation was also an essential element in preserving the existing social structure, where residential separation in environments, differing widely in levels of amenity and environmental quality, simply reflected existing social relationships. ...

The colonial city was a 'container' of cultural pluralism but one where one particular cultural section had the monopoly of political power. ... The extensive spatial provision within the colonial settlement area, as well as the spatial division between it and the indigenous settlement, are to be accounted for not simply in terms of cultural differences but in terms of the distribution of power. Only this can explain why labour and urban amenities were available in the spacious, cultivated areas in the colonial settlement, but not in the indigenous town.

King's colonial cities reflect a broadly shared ideology among many colonial rulers that translated into control over the city and its people, so that the patterns King and others observed were evident in a number of cities. At the same time, it is not difficult to identify cityscapes which bear the imprint of more individual power, facilitated either by the individual's powerful position in government, or by state support and development of the individual's vision. I discuss two cases below. The first illustrates the establishment of monuments as a conscious and unmediated projection of power by Indonesian Old Order leader Sukarno. The second illustrates how the greening of the Singapore city is a direct expression of the power of Singapore's longest-serving Prime Minister (now Minister Mentor) Lee Kuan Yew.

Gerald Macdonald (1995: 273) draws attention to how '[s]cholars have characterized the Guided Democracy as a period of frenetic symbolic activity on the part of the President. Sukarno manipulated symbols as an expression of his personal power ...'. Sukarno was the nationalist leader who declared Indonesian independence on 17 August 1945, and who became President in the newly formed Republic of Indonesia in 1950. The period dubbed 'Guided Democracy' was one when Sukarno abolished the system of parliamentary democracy and 'ruled through a process of consultation with the major political forces in the country – the Communist, Islamic, and Nationalist political parties, and the army' (Macdonald, 1995: 273). Sukarno's political philosophy extended into his treatment of the cityscape.

One example of the interplay of personal and state expression of power is in the construction and symbolic meaning of *Monas* [*Monumen Nasional*] or the National Monument. This is located in the middle of a central square in downtown Jakarta known as Medan Merdeka. The monument is dedicated to Indonesia's war of independence. Modelled after the Washington Monument (ironically, given the anti-West rhetoric that Sukarno led), the monument is over 100 metres high, made of marble and topped by 32 kilogrammes of gold. Its richness of material in a city sharply divided by social class is a clear expression of the strong desire to display a nationalistic sentiment, richly. This emphasis on the power of the national state to make its own landscapes to express the break of the Indonesian Republic from its pre-colonial past is also matched by a more personal symbolism of power. Sukarno is known to have referred to the monument's phallic symbolism as representing his and Indonesia's virility (Macdonald, 1995: 278–84). Indeed, Macdonald (1995: 278) writes that 'even today, cab drivers delight in informing tourists that the monument is "Sukarno's last erection"'.

For a second example, I draw from the experience of Singapore. One dimension of Singapore's contemporary cityscape that is clearly and definitely attributable to the personal vision of Minister Mentor Lee Kuan Yew is the greening of Singapore. In the 1950s, Singapore was characterized by overcrowding in the central area, a proliferation of slums and squatter settlements, poor sanitary conditions and infrastructure. In the 1960s, major efforts were expended to provide affordable public housing, to redevelop the central area, and to provide adequate infrastructure. Simultaneously, in anticipation of a harsher built-up urban environment, a greening programme was introduced to turn Singapore into a 'Garden City'. This was to be achieved through the large-scale planting of trees and shrubs all over the island, which was given greatest impetus in 1963 with the then Prime Minister Lee Kuan Yew's introduction of Tree-Planting Day. Suitable trees and shrubs were planted along highways and roads, in public gardens, open spaces, parks, recreational grounds and approaches to public buildings. Subsequently, plants were also introduced to camouflage concrete structures in order to soften the harshness, and in particular creepers and climbers were trained on to retaining walls, lampposts, flyovers and bridges. As the programme progressed and Singapore attained a reputation as a 'green' city, the aim of 'colouring' the island to create an aesthetically pleasant environment received further attention. Specifically, the Parks and Recreation Department increasingly introduced a variety of ornamental trees and shrubs brought from other countries and more kaleidoscopic colours in the choice of vegetation. Simultaneously, the neighbourhood parks created in different parts of the city capitalized on the aesthetic quality of nature (Kong and Yeoh, 1996).

These wide-ranging efforts were introduced for three main reasons. First, it was recognized that nature can contribute to a salubrious environment which ensures people's health. Second, nature forms part of an aesthetic landscape, beautifying the ugliness of slum settlements of the past and softening the harsh built-up landscape of the new Singapore. Third, nature would provide a setting for recreation, with the construction of facilities like parks for leisurely activities and the harnessing of various waterscapes for sports (Kong and Yeoh, 1996).

While the greening efforts in Singapore are quite singular in their own right from a global perspective, what is also significant about this initiative is the personal impetus and attention given to it by the then Prime Minister Lee Kuan Yew. It stems from a personal vision, coupled with the power and will of a forceful leader who himself recognized the need for a sustained effort to transform an unhygienic, slum-ridden city into one with enviable public hygiene, affordable housing, and verdant tropical environs. Thus, Lee (2000: 201) writes in his memoirs: 'To achieve First World standards in a Third World region, we set out to transform Singapore into a tropical garden city.' He purposefully went about setting up a department 'dedicated to the care of trees after they had been planted' and 'met all senior officers of the government and statutory boards to involve them in the "clean and green" movement' (Lee, 2000: 201). His ideology was evident as follows:

> We planted millions of trees, palms and shrubs. Greening raised the morale of people and gave them pride in their surroundings. We taught them to care for and not vandalise the trees. We did not differentiate between middle-class and working-class areas. The British had superior white enclaves in Tanglin and around Government House that were neater, cleaner and greener than the 'native' areas. That would have been politically disastrous for an elected government. (Lee, 2000: 202)

The will translated into a conquering of nature, in one sense. While 'nature did not favour us with luscious green grass as it has New Zealand and Ireland', the then Prime Minister invited an Australian plant expert and a New Zealand soil expert to study Singapore's soil conditions. They advised that in Singapore, regular fertilizer application (preferably compost which would not wash away too easily, and lime to counter Singapore's acidic soil) would be necessary. After a successful trial, schools, sports fields and stadiums were thus treated, and 'gradually, the whole city greened up' (Lee, 2000: 202–3). He did not stop there, as he further records:

> Because our own suitable varieties of trees, shrubs and creepers were limited, I sent research teams to visit botanical gardens, public parks and arboreta in the tropical and subtropical zones to select new varieties from countries with a similar climate in Asia, Africa, the Caribbean and Central America. ... Our botanists

brought back 8,000 different varieties and got some 2,000 to grow in Singapore. They propagated the successful sturdy ones and added variety to our greenery. (Lee, 2000: 204)

Colonial cities, city monuments and garden cities – these three examples illustrate the power of the state or individuals within the state machinery, who subscribe to particular ideologies and/or have a particular vision (of grandeur and personal symbolism), and who translate them into urban form. In the context of colonial cities, the defining ideologies were colonial and racist in essence. In the context of garden cities, a green ideology prevailed, though it was one rooted mainly in the belief that a green environment contributed to good health (and hence productive workers), an aesthetic environment and recreational opportunities (for human consumption), rather than one anchored in environmental ethics. In the example of Indonesian monuments in Jakarta, the impetus was the desire to express a president's personal power, managed through state apparatus. These three examples are distinct enough to illustrate variations in the ways in which power transforms cityscapes, but sufficiently similar to illustrate the important role of the state in such transformations.

Power and prestige in the city: the role of capital

Having addressed the role of the state, I turn now to the role of capital. For Karl Marx, capital and power are inseparable, and power is defined in terms of control over the means of production. Capital rewrites landscapes in conspicuous ways, for example, through towering skyscrapers that dominate skylines. The power of capital to influence and shape the city must, however, be understood alongside the influence of the design professions, and the conflicts between the two must be acknowledged and analyzed. Three examples are discussed below to illustrate the power of capital to shape the city, namely, capital's role in commissioning towering skyscrapers that changed New York's nineteenth-century skyline; the commodification of urban public spaces in late twentieth-century Hong Kong; and the power of shopping malls to change landscapes and shape human behaviour. These examples are chosen to illustrate the different levels at which capital may exert its power – first, at the level of the skyline, viewed from the horizon; second, at the level of public ground space, in terms of use access; and third, in terms of interior space and its influence on human behaviour.

New York's commercial landscape in the latter half of the nineteenth century provides excellent examples of capital's power to effect change, and how this sometimes runs against the aesthetic judgement of the design profession. Domosh (1989) detailed how the city's emerging mercantile and entrepreneurial class commissioned many of New York's skyscrapers at that time, seeking to communicate prestige to potential customers. Prominent structures were preferred, as they symbolically expressed their new wealth while serving as a form of advertising.

Further, they were to bestow cultural legitimacy on those who commissioned them. Citing the historian Frederic Cople Jaher, Domosh (1989: 34) argues that New Yorkers, more than Bostonians, commissioned buildings that were ornate and towering. The New Yorkers were 'more inventive, their firms had shorter lives, and they were greater credit risks. Because they lacked stability and cohesiveness, New York's commercial elites wanted physical expression of their power' (Domosh, 1989: 34). This was particularly true in the case of the highly competitive, relatively new industries of newspaper publishing and life insurance, which relied on reaching out widely to an urban audience. Many of these were also owned and run by magnates who were keen not only to advertise, but to assert their corporate egos, and to affirm their cultural legitimacy as arbiters of art and good taste. They therefore commissioned tall, imposing structures and sought to insert their own visions of aesthetics and cultural value, such as the Woolworths building, completed in 1913 (see Figure 1.1).

Yet, the power of capital did not always confer cultural legitimacy, for the kind of conspicuous building merchants desired compromised the architectural beauty in the eyes of designers. They argued that buildings constructed mainly for profit (usually in haste) did not result in a work of art. Further, the ornamentation that clients desired in order to achieve prominence ran against the grain of contemporary aesthetic evaluations (Domosh, 1989: 36–7). Thus, instead of conferring cultural legitimacy, these magnates lost cultural respect among the designer class, so that commercial power and cultural power did not always coincide.

While the example of New York illustrates the power of capital to shape skylines in obvious and very material ways, closer to ground level, the power of capital may also be effected through the commodification of public urban space, and sometimes in less prominent but nevertheless real ways. Cuthbert (1995; Cuthbert and McKinnell, 1997) discussed the commodification of public urban space using his concept of 'ambiguous spaces', those seemingly public spaces which are nevertheless owned and subject to control and surveillance by corporate powers. Increasingly, he argued, control of social space has been transferred from the public to the private sector in three ways. First, large new building complexes belonging to banks, insurance and property companies, multinational corporations and the like, are encouraged to create and donate 'public space' at ground or podium level, and in turn, are rewarded with plot ratio benefits. Ironically, however, once constructed, these spaces are given back to private ownership, in that use of the space is controlled privately. Second, pedestrian movement is channelled through corporate space. Third, large shopping centres are encouraged to provide open space, such as internal atriums and courtyards, which 'replace civic space with the commodity space of the market' (Cuthbert and McKinnell, 1997: 296). These spaces 'masquerade' as social space but are actually commodified and controlled space.

Because of the continued private ownership of such urban spaces, they are open to surveillance by corporate powers, and this occurs through physical policing and technological surveillance (e.g. video cameras). The case of Jardine House

Figure 1.1   Woolworth Building, New York

in Hong Kong illustrates this. Jardine House appears to be surrounded by public space in the form of landscaped gardens, fountains and open space. Yet, this space is under Jardine's control, which was evident when it decided to stop Filipina domestic workers from using these spaces as a meeting place on their one day off work per week. The area was taped up and employees were tasked to police the area and remove 'offenders'. Video cameras were also installed for surveillance purposes (Cuthbert and McKinnell, 1997: 300–1). The right to occupy certain open spaces in the city is thus whittled away by the power of capital to control its use, which is made possible in the first place by the state which granted the domination of such space to corporate power.

Finally, as a third example, I move indoors into commercial spaces created by capital which seek to create and shape sensory experiences and coax particular behaviours, particularly consumption. The best illustration of this is the contemporary shopping mall, which is an unmistakable manifestation of the power of capital to reshape landscapes. Shopping malls developed as consumer spaces in association with the rise of the suburbs and the private car. Some of the world's major malls are fantastic places, in size and significance. For example, Hopkins (1990: 5) described the breathtaking size and scale of Canada's West Edmonton Mall, which alone takes 1 per cent of Canada's retail sales, employs the equivalent of a small town's entire population (18,000 people), and extends over a landscape of 110 acres, including an 18-hole golf course, a seven-acre water park and a replica of Miami Beach.

Drawing on the work of Jon Goss (1999), we can suggest that commercial space appropriates images and landscapes, so that malls are regarded not only as spaces for selling, though that is the primary goal. In Goss's analysis of the Mall of America, the commercial space is also an expression of the exploitation and commodification of nature. The appropriation of nature – that which is most separate(d) from humans – exemplifies the power of capital over nature. Such exploitation and commodification is undertaken to achieve various ends: to 'soothe tired shoppers, enhance the sense of a natural outdoor setting, create exotic contexts for the commodity, imply freshness and cleanliness, and promote a sense of establishment' (Goss, 1993: 36). In this presence of nature, consumption is naturalized in the hope of mitigating 'the alienation inherent in commodity production and consumption' (Goss, 1993: 36).

Several strategies of commodification are apparent at the Mall of America. For example, nature is commodified in Rainforest Café ('an enchanted place for fun far away, that's just beyond your doorstep' (Goss, 1999: 54), Camp Snoopy and the Underwater World. There are also animals for petting in shops such as Nature's Wonders, Wilderness Station and Wilderness Theater, while names of stores also evoke 'pristine and mysterious nature', such as Forever Green, Natural Wonders and Rhythms of the Earth (Goss, 1999: 60). An 'essentialist Minnesota sense of place' is recreated through a stretch of northwoods stream, 30,000 plants and trees (the largest in indoor planting in the world), and an artificial 70 foot waterfall and plaster cliffs cast from the originals along the St Croix River, complete with animatronic moose and bird noises (Goss, 1999: 50). By naturalizing consumption, particularly of goods and services that are not essential, the effect of capital is to legitimize the acts of shopping, purchasing and consuming, and to reinforce the ideology of consumerism.

New York skylines, Hong Kong's public urban spaces, and American mall interior spaces all reveal the power of capital to shape urban landscapes. Whether it is the manifestly visible skyline, a meaningful reality when viewed from a distance, or whether the gaze is turned to ground level in the access to public urban spaces, or whether it is the manufacture of interior environments, capital has the power to carve the contours of a city's physical profile, enable and limit the use of city spaces, and shape human behaviour in these spaces. Like the state, capital plays an important role in the transformation of cities.

Power and prestige in the city: intersecting powers

While I have discussed the power of the state and capital separately in the preceding sections, the intersection of state and capital in the creation of city space equally bears discussion. In this section, I will discuss the example of the Disneyfication of landscapes, a process by which the principles of the Disney theme parks are taken into urban planning and landscaping, as well as into social and economic relations. Such Disneyfication is perpetuated by the Disney Company itself, which pursues planning and design commissions, and thus intersects with the state in its planning and design functions. This example is chosen so that I might illustrate how the combined power of state and capital to transform landscapes may nevertheless invite resistance from the community.

The process labelled 'Disneyfication' demonstrates that the logic of the theme park has extended outwards beyond its confines and recast many urban landscapes as 'variations on a theme park' (Sorkin, 1992). This has been possible because state planners and designers have called on corporate experience, seeking to replicate both the look and underlying structure of Disney theme parks in cities (Blake, 1972; Boles, 1989). Thus, the 'imagineered logic' of the theme park (Relph, 1991: 104) is being extended into urban areas, with 'perceptual control, centralized provision of goods [and] services, and conglomerate organization and ownership of attractions' (Davis, 1996: 417). This has entailed planners 'seeing with "imperial eyes", that is, with little recognition of the existing human population or social history, except as potential labour and potential attraction, respectively' (Davis, 1996: 417). Principles and practices of Disney theme parks have pervaded not only the planning and design of whole cities. Specific elements of urbanscapes in the form of shopping malls, festival markets, small town main streets and residential neighbourhoods have also been 'co-opted by the mouse' (Warren, 1994: 89).

The Disney model is attractive to city planners because it appears to offer attractive solutions to urban problems, and is a 'powerful and comprehensive urban vision' (Warren, 1994: 96). This is aided by the Disney Development Company, a subsidiary dedicated to applying the lessons Disney had learnt in Disneyland to urban development in all its various forms, encouraged by its CEO of the 1980s who exhorted cities to apply Disney's principles of design, crowd management, transportation and efficient entertainment to urban spaces. However, from another perspective, Disneyfication has been described as 'sinister' (Sorkin, 1992: xiv).

Seattle Center was an ageing civic centre area originally constructed for the 1962 World Fair. Disney was engaged as urban planning consultant in order to inject the principles championed by Disney and, hopefully, to introduce its people-friendly and efficient designs. As part of its arrogance (and perceived power), Disney indicated that they wished to develop and possibly finance and operate the entire site, offering input in architecture, design, site layout, landscaping, crowd and traffic management, and security. It would 'reshape the centre, organize the chaos, and harmonize the currently inefficient use of space' (Warren, 1994: 100). For sure, it did not wish to simply finetune the amusement zones. It nevertheless assured Seattle that something unique was to be created for the area, and there would not be mere replication of Disneyland or EPCOT. Unfortunately, despite its claims, Disney seemed unable to create designs that took into account the needs and desires of Seattle residents, and would neither seek nor take advice from locals. All three plans it submitted appeared dysfunctional and unappealing, did not consider Seattle's unique character and recycled the same ideas. Indeed, Disney proposed to demolish several cherished structures, replacing them with buildings and activities deemed inappropriate for Seattle's needs. The media, citizens' groups and council chambers alike grew increasingly dissatisfied, and despite a renewed effort at a fourth plan, the damage had been done. The final straw was when Disney projected it would cost US$335 million, a far cry from the original US$60 million estimation. Disney was sacked, underscoring the fundamental contradiction that the charming fantasy and efficient infrastructure of Disney World could only be achieved with unacceptably authoritarian planning practices in real life. Local architects, planners, designers and other citizens were called in to do the job, and the newly renovated Seattle Center became symbolic of the rejection of 'an autocratic, outside force in order to retain control over their space' (Warren, 1994: 104). In this instance, the imposition of an external vision and the attempt at direct control of the cityscape, reflecting a lack of effort at persuasion and hegemony, resulted in a rejection of the imposed vision and cityscape.

The lesson to be learnt from this is that the state and capital can act as intersecting forces, each reinforcing the power of the other to effect landscape change in the city. However, as this example illustrates, where the vision and ideology that state (in this case, represented by city authorities) and capital subscribe to do not adequately dovetail, city transformation cannot proceed. Thus, the coincidence of powers is sometimes a necessary condition for urban change to occur.

Summary

Cities are the medium by which the powerful express their influence. They simultaneously represent the outcome of the impress of power. This chapter has illustrated how states, capital and individuals all have the capacity to impact urban form and the use of urban space. Where ideologies and goals converge, the power to impact urban form and use is enhanced, as in the case of Jakarta and Singapore. Where there is divergence, urban change may not occur as originally intended, as in the case of Seattle.

What this chapter has illustrated are just some of the different ways in which power is impressed on the city – through shaping its skyline, influencing its architecture, segregating its people, constructing its monuments, greening its public spaces, controlling access to urban space, and influencing urbanites' behaviour. These reflect the extensive reach states, capital and individuals can have on the landscape and its use, through direct control or ideological hegemony. What is only intimated briefly in this chapter is that power is often fractured. In the Seattle case, it is apparent that conflicting views can result in different outcomes from those originally intended, but in other situations, it is possible that resistances can emerge in everyday ways. Thus, a monument erected as an expression of the power of an individual, or a skyscraper inserted in the midst of a city because of the power of capital, may find resistance from other groups, for example through graffiti and vandalism, or through the appropriation of these spaces for unintended activities. Power – be it direct control or hegemony – is never total.

References

Anderson, K.J. (1992) *Vancouver's Chinatown: Racial Discourse in Canada, 1875–1980,* Montreal and Kingston: McGill-Queen's University Press.

Balandier, G. (1951) 'The colonial situation: a theoretical approach', in I. Wallerstein (ed.), *Social Change: The Colonial Situation,* New York: John Wiley.

Blake, P. (1972) 'Walt Disney world', *Architectural Forum* 136, 24–41.

Boles, D. (1989) 'Reordering the suburbs', *Progressive Architecture* 70, 78–91.

Cosgrove, D. (1984) *Social Formation and Symbolic Landscape,* London: Croom Helm.

Cuthbert, A.R. (1995) 'The right to the city: surveillance, private interest and the public domain in Hong Kong', *Cities* 12 (5), 293–310.

Cuthbert, A.R. and McKinnell, K.G. (1997) 'Ambiguous space, ambiguous rights – corporate power and social control in Hong Kong', *Cities* 14 (5), 295–311.

Davis, S.G. (1996) 'The theme park: global industry and cultural form', *Media, Culture and Society* 18, 399–422.

Domosh, M. (1989) 'New York's first skyscrapers: conflict in design of the American commercial landscape', *Landscape* 30 (2), 34–8.

Giddens, A. (1987) *The Nation-State and Violence,* Cambridge: Polity Press.

Goss J. (1988) 'The built environment in social theory: towards an architectural geography', *Professional Geographer* 40, 392–403.

Goss, J. (1993) 'The magic of the mall: an analysis of form, function, and meaning in the contemporary retail built environment', *Annals of the Association of American Geographers* 83 (1), 18–47.

Goss, J. (1999) 'Once-upon-a-time in the commodity world: an unofficial guide to the Mall of America', *Annals of the Association of American Geographers* 89 (1), 45–75.

Gramsci, A. (1973) *Letters from Prison*, New York: Harper & Row.

Hopkins, J.S.P. (1990) 'West Edmonton Mall: landscape of myths and elsewhereness', *The Canadian Geographer* 34 (1), 2–17.

King, A.D. (1976) *Colonial Urban Development: Culture, Social Power and Environment*, Boston: Routledge and Kegan Paul.

Kobayashi, A. (1989) 'A critique of dialectical landscape', in A. Kobayashi and S. Mackenzie (eds), *Remaking Human Geography*, London: Unwin Hyman.

Kong, L. and Yeoh, B.S.A. (1996) 'Social constructions of nature in urban Singapore', *Southeast Asian Studies* 34, 402–23.

Lee, K.Y. (2000) *From Third World to First: The Singapore Story 1965–2000, Memoirs of Lee Kuan Yew*, Singapore: Times Editions.

Macdonald, G.M. (1995) 'Indonesia's *Medan Merdeka*: national identity and the built environment', *Antipode* 27 (3), 270–93.

Relph, E. (1981) *Rational Landscapes and Humanistic Geography*, London: Croom Helm.

Relph, E. (1991) 'Post-modern geography', *Canadian Geographer* 35, 98–105.

Sorkin, M. (ed.) (1992) *Variations on a Theme Park: The New American City and the End of Public Space*, New York: Hill and Wang.

Thompson, J.B. (1981) *Critical Hermeneutics*, Cambridge: Cambridge University Press.

Warren, S. (1994) 'Disneyfication of the Metropolis: popular resistance in Seattle', *Journal of Urban Affairs* 16 (2), 89–107.

2 FAITH AND DEVOTION

Keith Lilley

This chapter

○ Provides an overview of the ways the city works as a site of religion and religiosity

○ Shows that the design of cities has been informed by particular ideas about divine order

○ Demonstrates how cities take on religious significance through particular rituals and performances

○ Suggests that in multicultural socities, the city can often become a site of conflict where different faiths struggle to chisel out spaces of religious belonging

---

Introduction

Geographers, sociologists and anthropologists have long shared an interest in the relationships between religion and the city (see Kong, 1990, 2001). Some have taken a social or demographic view, looking at where particular faith groups live, especially in cases where contemporary cities are 'divided' by religion and faith, as in modern-day Belfast and Glasgow, or Jerusalem and Beirut (e.g. Pacione, 1990; Boal, 1996; Broshi, 1996; Emmett, 1997). Others have examined the practices of the faithful and their places of worship, noting that certain groups make spaces in which their religious values are expressed (e.g. Duncan, 1991; Naylor and Ryan, 2002; Brace et al., 2006). Some, too, have used their own faith and religious experience as a starting point to reflect on issues of spirituality in certain urban settings (e.g. Graham and Murray, 1997; Slater, 2004). Then there are those whose work has examined how in the past religion shaped urban lives and landscapes (e.g. Slater, 1998; Baker and Holt, 2004), and likewise for more recent times (e.g. Scott and Simpson-Housley, 1991; Kedar and Werblowsky, 1998).

One aspect of urban geography and religion that has not been developed so much in recent years is how the city *itself* reflects and reinforces patterns and practices of faith and devotion. One notable exception to this is Wheatley's (1969, 1971, 2001) work (see Extract 2.1). He has shown how urban landscapes in some 'traditional' societies (e.g. imperial China and Japan) were sometimes physically shaped to imitate a map of the heavens and ensure that their earthly world was as ordered as the cosmos, the city being seen and understood by its inhabitants as a small version of the wider world, a *microcosm*. Nitz (1992) took a similar approach in exploring the links between the city and the cosmos in his study of Hindu 'temple cities' of southern India. However, this type of work on religion and the city is, as Kong (2001: 220) points out, rather neglected in urban geography. Instead, it is stronger in other humanities disciplines, such as anthropology, politics, archaeology and architecture (e.g. see Rykwert, 1988; Levy, 1990; Carl, 2000). Indeed, it is within these disciplines that this aspect of research on religion and the city has gained most momentum in recent years, and it is this cross-disciplinary research that forms the focus of this chapter.

Extract 2.1: From Wheatley, P. (1969) *City as Symbol*, London: H.K. Lewis, pp. 9–10.

I would like to now consider certain aspects of the pre-industrial city which have been even more than usually neglected. First, from the innumerable topics which offer themselves for discussion, I have selected one which has been ignored by virtually all students of urbanism, yet which is of fundamental importance because it pervades the whole range of activities focused in the traditional city. I am referring to the cosmo-magical symbolism which informed the ideal-type traditional city in both the Old and the New Worlds, which brought it into being, sustained it, and was imprinted on its physiognomy. This is not the place to embark on an extended discussion of the origins and nature of this symbolism, which in any case have been the subject of elaborate expositions by, among others, Mircea Eliade and René Berthelot. Suffice it to say that for the ancients the 'real' world transcended the pragmatic realm of textures and geometrical space, and was perceived schematically in terms of an extra-mundane, sacred experience. Only the 'sacred' was real, and the purely secular – if it could be said to exist at all – could never be more than trivial. For those faiths which derived the meaning of human existence from revelation no site was, apart from a possible incidental soteriological sanctity, intrinsically more holy than another; but in those religions which held that human order was brought into being at the creation of the world there was a pervasive tendency to dramatize the cosmogony by constructing on earth a reduced version of the cosmos, usually in the form of a state capital. In other words, Reality was achieved through the imitation of a celestial archetype, by giving material expression to that parallelism between macrocosmos and microcosmos without which there could be no prosperity in the world of men [*sic*].

This chapter is divided into three main parts, each taking a different perspective on how the city is formed through faith and devotion – formed that is, as an *imagined space*; materially as a *built space*; and habitually as a *performed space*. Of course it is artificial to divide up the city this way, for really it is a product of perpetual interaction between the conceived, the built and the lived. Henri Lefebvre (1991) conceptualized the 'production of space' in this way, and I have taken my lead from him, though mindful of the criticisms levelled against his triadic model (Merrifield, 1993). My aim is not to restrict myself to any one individual religion or to any single cultural context, but rather to range across time–space to show that different belief systems make similar uses of the city to connect with the divine.

The city imagined: urban mappings of the sacred

As was discussed in the Introduction, urban scholars began to take a more critical interest in texts and images in the 1980s and 1990s. Rather than simply using them to tell us about a particular place, the emphasis was more on how representations construct reality. Issues of authorship, production and circulation of 'texts' were thus scrutinized, and 'imagined geographies' examined (e.g. Duncan and Ley, 1993; Driver, 1995). In anglophone human geography, this interest in representation was particularly fostered through a collection of essays under the title *The Iconography of Landscape*, edited by Cosgrove and Daniels (1988). 'Iconography' is the study of religious imagery, and Cosgrove and Daniels were extending this approach to the study of landscape, particularly representations of landscapes in art and literature. Since then numerous studies have appeared in this genre, though curiously little work by geographers has dealt with landscapes in religious representations. Yet art historians have long been concerned with how imagery, including landscape imagery, is imbued with religious meaning and symbolism, and it is clear the same can be said for images of the city (see Frugoni, 1991).

A case in point is Jerusalem, a city important in Islamic, Jewish and Christian faith. The symbolic centrality of Jerusalem in Christianity is most obviously reflected in stylized 'world maps' (*mappaemundi*) of the middle ages, drawn showing the holy city located at the spatial centre of the world (see Woodward, 1985; Kühnel, 1998). Jerusalem itself was frequently depicted in medieval images as both a real and imagined city, as an 'earthly city' and a 'heavenly city', for not only was it a place of pilgrimage for Europeans in the middle ages, and venerated for its place in Christian doctrine, it was also significant as a symbol of salvation after the end of the world, as told in the New Testament in the Book of Revelation (Frugoni, 1991). Here Jerusalem is described as a city descending from heaven, 'four square' in shape with gates on all four sides. This powerful image was depicted in manuscript copies of the Bible, sometimes the city being shown as a circle of walls, sometimes as a square (Lilley, 2004a). Jerusalem thus became the archetypal, ideal

Christian city, and in medieval images the earthly Jerusalem itself was made to look like its heavenly archetype (see Rosenau, 1983). In images of other cities, too, the idealized shape of the divinely-ordered Jerusalem was replicated in depictions of 'real' places, such as Bristol (Figure 2.1).

Figure 2.1   The city imagined: Bristol as depicted in the later middle ages

It was not just in images that people of the middle ages drew out likenesses between the heavenly Jerusalem and the cities they knew and worshipped in. It features also in textual descriptions. For instance, from the later twelfth century comes a description of Chester, by then an important city in the north-west of England (later eclipsed in economic importance by Liverpool). The description was written as a sermon by a local monk called Lucian, and in it is revealed his symbolic interpretation of Chester's urban landscape, read in the light of his understanding of Christian doctrine and scripture (cited in Palliser, 1980: 3). The following is an extract:

> [Chester] having four gates to the four winds, looks on the east to India, on the west to Ireland, on the north to greater Normandy [Norway] and on the south to Wales. ... There are two excellent straight streets in the form of the Blessed Cross, which through their meeting and crossing themselves, then make four out of two, their heads ending in four gates ... [and] in the middle of the city, in a position equal for all, [God] willed there to be a market for the sale of goods. ... Now if anyone standing in the middle of the market turns his face to the east, according to the positions of the churches, he finds John the forerunner of the Lord to the east, Peter the Apostle to the west, Werburgh the Virgin to the north, and Archangel Michael to the south. Nothing is more true than that Scripture, 'I have set watchmen upon thy walls, O Jerusalem' [Isaiah 57: 6]. ... So behold our city, as it was predicted, entrusted to the holy guardians as it were in fourfold manner. From the east the mercy of the forerunner of the Lord supports it, from the west the power of the doorkeeper of Heaven, to the north the watchful beauty of the virgin, and to the south the wonderful splendour of the angel.

Lucian's description of Chester is allegorical and draws analogies between it and the heavenly city by pointing to their shared form. Like the celestial Jerusalem, Chester has 'four gates to the four winds'; it was the four-square city of Revelation. Chester is also 'in the form of the Blessed Cross', and thus in Lucian's mind is symbolic of the body of Christ. Through this description then, Lucian is drawing Christian meanings from the urban landscapes he sees around him, linking them to scripture, while at the same time layering these meanings in Chester's urban landscape.

In such Christian imaginings, then, the city was conceived as a sacred 'body', linking human and divine existence, the earthly and heavenly worlds. The city itself became symbolic of the human body – the microcosm – and the cosmic body – the macrocosm. The micro-macrocosm idea was imported into European Christian thinking of the middle ages (500–1500 CE) from classical natural and political philosophy, such as Plato's *Timaeus* and Aristotle's *Politics* (see Lilley, 2004b). But conceiving the city as a microcosm is not confined simply to the Latin Christian tradition. Jerusalem, for example, plays a central place in Judaic and Islamic thought as well as Christian. 'For Jews', Dan (1996: 60–1) writes, Jerusalem 'is the centre of the universe, the meeting point between chaos and creation', while 'celestial

Jerusalem represents the indestructability of perfection'. He points to the description of the cosmos in the *Sefer Yezira* (*Book of Creation*), in which the centre of the world is 'the holy Temple', uniting 'the spiritual and the worldly, the divine and the earthly, into the cosmic concept of Jerusalem as the centre of human and divine existence' (Dan, 1996: 61). Of course, Jerusalem is also 'the ideal "king's city", the city of King Solomon's palace and government', and hence there were 'two Jerusalems, the ideal and the earthly' (Dan, 1996: 62–4). The idea of the city as a microcosm also appears in Jewish thought around the Second Commonwealth period, as is made clear in an account of God's creation of the world by Philo of Alexandria:

> When a city is being founded to satisfy the soaring ambition of some king ... there comes forward now and again some trained architect who ... first sketches in his own mind well nigh all parts of the city that is to be wrought out ... and like a good craftsman he begins to build the city of stones and timber, keeping his eye upon his pattern and making the visible and tangible objects correspond in each case to the incorporeal ideas. Just such must be our thoughts about God. We must suppose that, when He was minded to found the one great city, He conceived beforehand the model of its parts. (cited in Friedman, 1974: 425)

Philo's view of an 'architect' God found subsequent resonance in Christian creationist thought. Indeed, in illuminated manuscripts of the middle ages, God is often depicted with compass in hand, an architect imparting order to the world, an *artifex principalis*.

In these shared imaginings of city and cosmos we therefore see an interweaving of Judaic and Christian thinking about the city, and about the specific place of Jerusalem in Judeo-Christian cosmology. In both traditions the city is a microcosm: through its central and axial place in cosmology and cosmogony, linking earthly and heavenly worlds, it is given symbolism through geometry, whether in depictions of Jerusalem as a quartered circle or in accounts of the world being like a city and founded by God. In Islam, too, Jerusalem is symbolic of cosmology and cosmogony, architecturally embodied and expressed in the Dome of the Rock, the Qubbat al-Sakhra, of the seventh century CE, as well as in early Islamic descriptions of the city, like that of al-Maqdis, who remarked how Jerusalem is 'the most exalted of cities as it unites in itself the advantages of both this world and the next' (Neuwirth, 1996: 111–12). Jerusalem, then, has a strong place in the imaginary of all three faiths, formed through the city's status in each of these three global religions (see Rosovsky, 1996).

But Jerusalem is not unique as a cosmic archetype, or as an idealized city, for other cities performed similar functions elsewhere. In the case of imperial China, for example, Wheatley (1971: 153), refers to the capital of the Shang dynasty, of the second millennium BCE, and its role as a seat for the emperor: 'the great mediator between heaven and earth, the son of heaven whose appropriate locale was at that axis of the universe which was also the axis of the kingdom and the only site for an imperial capital'. This divine role was immortalized in the form of an ode in the *Shih-Ching* – a record of dynastic exploits and legends:

> The capital of Siang (Shang) was a city of cosmic order,
> The pivot of the four quarters.
> Glorious was its reknown,
> Purifying its divine power,
> Manifested in longevity and tranquillity
> And the protection of us who come after
> (cited in Wheatley, 1971: 19)

Shang, like Jerusalem, was a city connecting worlds above and below, having a dual centrality in their cosmogonies, in their creation and destiny. Through its description in the *Shih-Ching*, the divinely ordained status of the city of Shang was commemorated and mythologized. This was true for later dynasties too. Around the second century BCE, as Wright (1977: 44–5) notes, a 'ritual canon' – 'ideologues in the service of the great Emperor Wu' of the Han dynasty – 'offered a coherent view of the world and the place of China and the imperial system in it', including of course the imperial capital itself, Ch'ang-an (see also Steinhardt, 1992). Representations of the city, whether in texts or in images, thus 'map' out the sacred to the faithful, influencing the beliefs of devotees in the process.

The city built: making the invisible visible

Geographies of faith and devotion also influence the building of cities in more material ways. In effect, urban landscapes provide a means to imitate in the material world the divine order of the spiritual, and so make the invisible visible. Yet urban scholars often overlook the religious symbolism inscribed within the physical layouts of cities beyond celebrated examples such as Angkor Thom in Cambodia (e.g. Kostof, 1991). In fact, cities throughout the world, both past and present, betray many signs of having been planned and built so as to reflect religious or spiritual values. Significantly, layouts of streets, blocks and plots have often been informed by religious conceits. For example, studying 'the cosmic city of the ancient east', the art historian H.P. L'Orange (1953: 13) suggested the street plan of 'residential cities' such as Firuzabad (Iran), where 'wall and fosse are traced mathematically with the compass' offered 'an image of the heavens, a projection of the upper hemisphere on earth'. He reflected too on the meaning of the plan-form of these cities: 'two axis streets, one running north–south and the other east–west divide the city into four quadrants which reflect the four quarters of the world. At the very point of [their] intersection, in the very axis of the world wheel, the [royal] palace is situated, here sits the king, "the axis and pole of the world" ' (L'Orange, 1953: 13). The founded city's shape, in an imitation of the universe, not only provided spatial order but also supported the divine order of things, the king – the city's founder – personifying the universal creator.

Bringing a new city into being has thus often involved rituals of foundation imbued with sacred meaning. This process might involve consulting oracles or divination. In India, it was the *sthapatis* who were responsible for planning a town, which began 'on a day fixed by astronomers', and involved, according to the *Manasara* (a manual of architecture and planning), selecting a site using an examination of its 'smell, colour, taste, shape, direction, sound, and touch' (Volvahsen, 1969: 49). In the foundation of Roman cities, the site for a new town was seen to be a 'direct and arbitrary gift of the gods', and it was through geomantic signs that the 'will of the gods' was revealed to the town's founder, sometimes, as in the case of founding Alba Longa, for example, with the help of animals (Rykwert, 1988). Rome itself had of course been founded, according to one story, by Romulus and Remus going to separate hilltops to watch for auspicious birds (Rykwert, 1988: 44–5) (see Extract 2.2).

Extract 2.2: From Rykwert, J. (1988) *The Idea of a Town: The Anthropology of Urban Form in Rome, Italy and the Ancient World,* Cambridge, MA: MIT Press, pp. 44–5.

… Romulus and Remus agreed to found the city near a place where they had been picked up by the she-wolf. The exact spot where this occurred was said to have been the site of the Lupercal shrine. Here the two brothers separated, and each went on a hilltop to watch for auspicious birds. This was the *inauguratio.* The *inauguratio* was a complex rite. It consisted of a prayer, a naming of signs, and a description of the augur's field of view. The augur watched for the signs and when they appeared, he determined their exact significance. The specific terms for the culminating acts were *conregio, conspicio* and *cortumio.* For the *conregio* the augur drew a diagram on the ground with his curved wand, his *lituus.* Livy gives an account of this part of the rite in his description of the inauguration of Numa as king of Rome: 'The augur, with his head veiled, took a seat on his (Numa's) left, holding in his hand a crooked and knotless staff called *lituus.* … He prayed to the gods (*deos precatus*) and fixed the regions from east to west, saying that the southern parts were to the right, and the northern to the left'. This fixing of the regions, and the naming of landmarks, such as trees, which bounded them, while he pointed to them with his staff, constituted the *conregio.* The *conspicio* seems to have been parallel to the *conregio.* The direction of the augur's eyes followed his gesture, and by taking in the whole view, town and country beyond, he contemplated it, and united the four different *templa* into one great *templum* by sight and gesture.

The foundation of imperial capitals in China also involved rituals of divination and geomancy, and involved searching for favourable settings, and 'practitioners known as *wang-ch'i-che* who surveyed the ambience or emanations of a site or situation', adjusting a site 'to the local currents of the cosmic breath (*ch'i*)'

(Wright, 1977: 46). In founding a city the Emperor was also establishing the divine basis of his authority. His role was central in the whole process. According to the *Chou li* (part of an ancient series of prescriptive texts probably originating in Han times), 'it was the sovereign alone who establishes the states of the empire, gives to their quarters their proper positions, gives to the capital its form and to the fields their proper divisions' (cited in Wright, 1977: 46).

Rituals of urban foundation, as 'simulations of cosmogony', were also ceremonial occasions (Wheatley, 1968: 11). In Etruscan and Roman Europe, for example, city founding was begun by ploughing a furrow with a bronze ploughshare ('associated with the worship of Jupiter') pulled along by white oxen, starting in the south-western corner and then proceeding in an anti-clockwise direction until a full circle was complete, the process ensuring the fertility and future strength of the city and its inhabitants (Rykwert, 1988). The circle traced by the plough defined the boundary of the future city and were the means by which the city's walls were made sacred and the new city 'fully constituted', the plough being lifted to form gaps where the city gates were to be sited (Rykwert, 1988: 65). The use of oxen for ritual ploughing to found a new city was also undertaken in India, and likewise had sacred, cosmological significance, as is revealed in the *Manasara* which states 'the wise architect should meditate on the two oxen as the sun and the moon, on the plough as the boar-god (Visnu) and on the builder as Brahma' (cited in Rykwert, 1988: 166). In the Roman ritual a hole was also dug – referred to as *mundus* ['world'] – 'the focus of the town', and filled with 'good things' by its new inhabitants with an altar set upon it complete with a fire (Rykwert, 1988: 34). The circle they had ploughed imitated the heavens, in the centre of which was their 'world' (i.e. the city).

A city shaped through cosmic rituals was thus sanctified. Its shape also reflected the order of the universe, emulating its hierarchies and boundaries. In ancient imperial China, for example, it was thought 'the earth was a perfect square', and since 'it was fitting that the ruler of all under heaven should live in a structure that was a replica and a symbol of the earth' the imperial city was itself rectangular or square shaped (Wright, 1977: 47). Again, the *Chou li* describes a city's harmony with the cosmos: 'here, where heaven and earth are in perfect accord, where the winds and the rains gather, where the forces of *yin* and *yang* are harmonised, one builds a royal capital' (cited in Wright, 1977: 47). The city's orientation, to the four corners of the world, and its four-square plan, all reflected the city's cosmic basis and assured a 'harmonious rapport between heaven and earth' for the emperor and the city's inhabitants, while within the compass of the capital itself, within the rectangular circumference of its walls, sat the imperial palace in its own enclosure, an inner sanctum (Wright, 1977: 56). The oriented, rectangular layout, with nested enclosures, is also a characteristic of Indian temple towns such as those examined by Nitz (1992) in the case of the Chola dynasty of southern India (ninth–fourteenth century CE). Here too imperial authority equated with divine authority. Supported by Brahmin priests, 'a divine superstructure of the imperial rule' legitimized the Chola kings' sovereignty over land

and people, a legitimization that literally took shape through inscribing 'the cosmic order onto the ground' by 'drawing lines upon the land' (Nitz, 1992: 108–9). The royal provincial temple towns created were in the form of the Hindu cosmos, as at Tirunannayanellur and Shrirangam; their rectangular plans consisting of a 'concentric' series of walled enclosures, the inner-most containing the temple itself, the outer ones streets and houses, forming a hierarchy of spaces imitating the universal order of things, as well as giving the city an overall symbolic order.

As well as rectilinear forms imitating the cosmos, the circle too was employed as a means of creating a city on earth in an image of the heavens. Cities of purely circular form are comparatively unusual, but not unknown. In 762 CE the caliph al-Mansar created a new city, Madanat as-Salam (Baghdad), described in literary sources as being circular in form with four opposing gateways, and divided into three concentric zones (Akbar, 1988: 90). Apparently, having consulted with various experts, al-Mansar himself then 'wished to see its actual form', and 'so he ordered the plan to be traced on the ground with lines of ashes; he then entered prospective gates and walked around. Cotton seeds were laid on the traced lines, saturated with naphtha and set on fire, enabling the caliph to see and sense the city. ... Then he ordered the foundations to be dug' (Akbar, 1988: 90). Lassner (1980), in a comprehensive study of the 'round city' of Baghdad, sets out the arguments for the cosmological basis of the city's form, though he himself remains somewhat sceptical (see also Wheatley, 2001: 270–4).

The circle and square were thus two geometrical shapes through which religious beliefs, ideas and rites of devotion were materially imprinted in the urban landscape, providing a link in the minds of the faithful between their earthly existence and the spiritual world. More complex geometries were also sometimes used to do this. For example, the Christian cosmos was reflected in the layouts of European medieval urban landscapes through subtle use of geometry to imitate divine order. In the case of new towns founded by the city-republic of Florence around 1300 CE, Friedman (1988) has shown that sine geometry was used to set harmonious proportions between street-blocks (as at Terranuova). In so doing, the town's designers imbued the town's design with a geometry that reflected the sacred proportions and compass of the Christian universe. Similarly, the layout of Grenade-sur-Garonne, a town founded in south-west France by Eustache de Beaumarchais in 1291, had a complex and 'invisible' geometry behind it (Figure 2.2). Here the street-blocks appear to have been derived using the 'rotating square' method employed by medieval architects in the construction of cathedrals (Lilley, 2005). Significantly, the proportioned layout of the town could only be fully appreciated from above, so only God (and the town's designer) could see it.

The city lived: performing the sacred

While the city thus became itself a scaled-down world, a microcosm, shaped in an image of the cosmos, it was also *lived* in ways that connected the earthly and

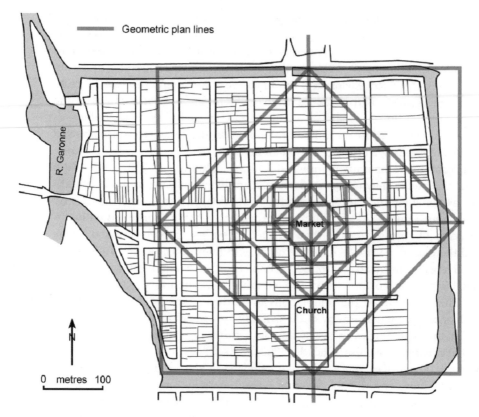

Figure 2.2   The 'hidden geometry' of Grenade-sur-Garonne (France)

heavenly worlds. Processions, pageants and pilgrimages, for example, are crucial in imbuing cities with sacred meaning, and are still very much a part of contemporary faith and devotion, as anyone who has visited a city in Spain during *santa semana* will testify (Mitchell, 1990).

In urban studies, performance has lately attracted much intellectual discussion and debate, not least by geographers interested in the 'body' and its spatialities (e.g. Pile and Thrift, 1995). Yet much of this has dealt with individual, secular experiences rather than collective practices of ritualized performance. However, religious rituals are highly significant as they 'tie together public space, individuals, social units, deities, and time into a larger assemblage' (Levy, 1990: 101), acting as an embodiment of sacred ideas (Boogaart, 2001; Lilley, 2004b). For example, in his extended ethnographic treatment of Hindu ritual and symbolism in the Newar city of Bhaktapur in Nepal, Levy demonstrates the significance of 'calendrical events' during the Hindu 'festival year' (Levy, 1990: 403). Some of these events are tied to solar and lunar cycles, and are, in Levy's view, the 'scanty echoes' of 'cosmic events' memorialized by generations of the city's inhabitants in periodic 'focal festivals' such as Biskā: – a nine-day festival marking the start of the solar New Year (Levy, 1990: 407, 463, 464). Biskā: involves the whole city in a series of linked rituals that bond civic and religious life, the basis of which is a conception of the cosmos, its order and form, which is performed in the city's streets and spaces. And not just any streets and spaces, nor any sequence, but a closely orchestrated and choreographed 'dance', as Levy (1990) calls it, involving people of the two 'halves' of the old city, and a ceremonial 'cast' comprising their deities and religious and political representatives.

Levy (1990) describes in some detail the series of events that make up the Biskā: festival. It begins four days before the solar New Year's day and involves two chariots, each containing a deity, one male (Bhairava) and one female (the 'dangerous goddess' Bhadrakālī) the former being the larger of the two. The two chariots and their deities play a central role in the festival. The Bhairava chariot is placed at the 'conceived' centre of the city, in Ta:mārī square, at the boundary between the Upper City (Datattreyasthan) and Lower City (Bansagopal), the two halves of Bhaktapur, and it is from here, a 'neutral centre between the two halves' (even though it is geographically inside the Lower City), that members of the city's halves, and their constituent neighbourhoods, begin to pull the chariot into the streets of their respective half of the city, east and west (Levy, 1990). Witzel (1997) has considered this chariot tug-of-war as signifying a cosmic struggle, as well as a tension between male and female sexualities, and the city's different social factions. The 'female' chariot is initially placed in front of Bhadrakālī's godhouse in the city's lower, western part, but then also brought into the Ta:mārī square where it is placed alongside the Bhairava chariot (Levy, 1990: 469). It is then taken to a square to the south, Gāhitī, where the Bhairava chariot should also end up 'on the evening of the first day of the festival sequence' (Levy, 1990: 474). From here the chariots with their deities are within sight of another important deity, the Yasi(n) God, which forms a related yet mythically only tangentially linked set of rituals in the Biskā: festival. The chariots with their deities ultimately return to the Ta:mārī square on the ninth and final day, and being placed there together, following the various struggles during the preceding days, mark a return of order and harmony.

Through these civic rituals and processions Bhaktapur affirms its place in the world, relating its living people with those of its past, and linking together – harmonizing – the overall social 'body' of the city as one, bringing together the two halves and ordering those within them. The two chariots with their respective deities are seen to symbolize this, as are the origin myths that underpin the two *yasi(n)*: both invoke the struggles that exist in the world between opposites at the macrocosmic level, between the gods, and those at the microcosmic level in class and gender struggles (Witzel, 1997). Levy (1990: 493) notes that the emphasis in the Biskā: festival 'is on the integration of the city as a whole and in its relation to annual time', as well as in maintaining the divine order of things, such as social and urban hierarchies. Thus, the Biskā: festival uses the city – as a social and spatial body – to unite, order and harmonize the world of its inhabitants, where the city itself, in Levy's word, becomes a link – a *mesocosm* – between macrocosmic events, such as the solar cycle, and the lives of those at the micro-cosmic scale, in and of the city itself, and their life-cycles (Levy, 1990).

The 'civic performances' practised by the inhabitants of Bhaktapur in Nepal are by no means unique, however, and parallels have been drawn between them and those of the city-states of classical Greece (Levy, 1990; Jameson, 1997), as well as elsewhere. In a detailed study of Christian rituals in the medieval city of Bruges, Boogaart (2001) finds compelling parallels between the sacred perfor-mances in Bruges and Bhaktapur, in terms of the 'symbolic parallelism and reci-procity between community and cosmos, city and hinterland, head to body, guilds to government, and parts to whole' (Boogaart, 2001: 94). The particular focus of Boogaart's study is a longstanding and still-surviving procession of the Holy Blood (*Heilig-Bloed*), an annual ritual in Bruges involving the city's veneration of a relic reputedly containing the blood of Christ. The circumstances by which Bruges came to have such a holy relic are now uncertain. By the later middle ages, how-ever, the city was one of northern Europe's wealthiest, and the procession of the Holy Blood well-established. Boogaart (2001: 70) argues that the procession of the Holy Blood at Bruges at this time fitted into a growing medieval Christian tradi-tion of Eucharistic processions celebrating the body of Christ. Circumferencing their community in a procession around the edge of the city, the procession took in the whole city, and encompassed it by circumscribing its urban space with the holy blood of Christ the redeemer.

As Boogaart (2001) makes clear, the perambulatory geography of the Holy Blood procession route was itself symbolically and cosmologically significant. The procession and its unfolding sequence began first with an evening vigil on 2 May in the church of St Basil, after which, 'at the tender hour of four o'clock' the following day, the 'holy relic was taken from its tabernacle and placed on dis-play in the Burg', the courtly and ecclesiastical heart of Bruges, where it 'remained on display until ten in the morning' (Boogaart, 2001: 87). With the relic placed 'in its carriage' and 'under its embroidered canopy', the procession led out from the Burg to the city's main market place, the Great Market, and at this point the notables and urban elite who had started off the procession were joined by the (male) citizenry at large, 'assembled by guild and parish under their banners' (Boogaart, 2001: 88). The procession then led out through the central market place in front of the community's civic offices to the city's larger market place, the Sand, where it was watched by those not processing (including pilgrims, children and women – they were spectators), and then headed in a south-westerly direc-tion to exit the outer city defences at Boeverie Gate, where it turned right and proceeded in a clockwise direction around the whole circuit of the city walls, moving through the Ezel, Spei and Catherine Gates before returning to Boeverie Gate, from which the procession returned from whence it came, returning the relic to its home in the centre of the city, in St Basil's (Boogaart, 2001: 88–91).

In being moved through and around the whole city, the holy blood of Christ had 'served an obvious sacramental function, fortifying the community against pestilence, natural disasters, demons lurking outside the city walls' (Boogaart, 2001: 89). Boogaart (2001: 89–93) thus notes that in processing the Host through urban space and reciting psalms at different urban landmarks, a parallel was drawn between Bruges and 'Heavenly Jerusalem', a significant symbolic association for the city and its people. The procession thus linked urban space and cosmic time, and connected the living urban body of Bruges with the 'living' body of Christ (his blood was said to liquefy in the phial as a sign of transubstantiation). As well as having a cosmogenic symbolism, the procession had cosmological meaning through the route it took. It moved from centre to edge, from core to periphery, and then encircled the city, spatially encompassing its body. The procession at Bruges defined the outer edges of this world, the urban sphere, the boundary marking the difference between outer 'chaos' and internal order. The living city thus embodied the cosmic body, with its hierarchies and divine order, at the inner centre of which was the 'pure', the Burg, the Church, the Holy Blood, and the outside, beyond the urban walls, the 'impure', the margins being defined by following the line of the fortifications and by moving inside and outside of the city's main gates. For the people of Bruges, the city was a 'map' of the cosmos.

The procession of the Holy Blood is instructive in giving some indication of the depth of cosmological meaning and symbolic complexity of urban performance that existed in medieval Christian Europe. It was by no means rare for towns and cities across the Latin West to perform the sacred in similar ways, not least in Eucharistic processions celebrating *Corpus Christi*. Historians and philologists have long noted the social and symbolic importance of medieval Corpus Christi processions as an urban embodiment of Christ, of how 'the procession, which was itself a central symbol of wholeness of the urban social body, gathered in unity and concord to venerate the Corpus Christi … a central symbol of social wholeness' (James, 1983: 11; see also Phythian Adams, 1972; Nelson, 1974; Travis, 1987). Certainly Christ's body was seen by medieval urban communities to be analogous with their urban body, and as a means of maintaining social order and harmony among townsfolk. In Beverley in England, for example, the municipal scribe recorded in his ledger how the annual Corpus Christi procession in the town was 'for the praise and honour of God and the Body of Christ, and for the peaceful union of worthier and lesser commons of the town' (Leach, 1900: 34). However, some medieval historians have now begun to question the assumption that the 'body' was such a unifying metaphor, and that instead of conveying a sense of 'social wholeness' the Corpus Christi procession in fact exposed urban social conflict (Rubin, 1991; Beckwith, 2001).

The civic rituals performed in medieval Europe demonstrate how city and cosmos were linked in the minds of urban inhabitants. These civic performances were themselves acts of religious faith and had as their basis a (re)enactment of cosmogony and cosmology. Through them the shared social and spatial hierarchy and embodied form of macrocosm and microcosm were brought to life. Performing the sacred thus reified the divine order of things, linking collective and individual earthly bodies of the *lived* city with the living heavenly cosmic body above.

Such urban rituals find their parallels in contemporary cities, where religion continues to be embodied in a series of rituals and performances which bring identities and cities into being. They may mark off particular urban spaces as having special significance to particular faith groups. Watson (2005) has examined the way urban Jewish communities mark off the boundaries of a symbolic space (an *eruv*) enabling them to maintain the sanctity of the Sabbath (Extract 2.3). Though the boundaries of such spaces are virtually 'invisible', their symbolic importance is clear, as they represent an attempt to claim a space for a particular religious minority, and their boundaries require constant vigilance from those who construct them. Hence, Watson (2005) suggests that their construction provokes considerable opposition from non-Jewish residents in the locale. This symbolic power of religiously ascribed space is considerable, especially in multi-cultural societies where public space is increasingly regarded as secular, and the expression of religious values often regarded with some antipathy. While for some religions faith and devotion is expressed through embodied performances of the divine, connecting city with cosmos, past and future, there are many cities through-out the world where religious practices are played out through multilayered heterogeneous urban spaces, with different groups inhabiting contemporaneous yet different urban worlds.

Extract 2.3: From Watson, S. (2005) 'Symbolic spaces of difference: contesting the eruv in Barnet, London and Tenafly, New Jersey', *Environment and Planning D. Society and Space*, 23: 597–613.

What exactly is an eruv? For traditional Jews, the Sabbath is the day which is set aside for rest and calm away from the fast pace of weekday life, which involves a cessation of labour of various kinds. Various restrictions are laid down in Jewish law that impose conditions on the Sabbath which include the carrying of objects from private to public domains and vice versa. ... The purpose of the eruv is to integrate a number of private and public properties into one large domain – or, to put it another way, to redefine the activities permitted in semi-public (or karmelite) space for the purposes of the Sabbath in order that activities normally allowed only in the private domain can be performed. This is a process of temporal spatial reordering. Once an eruv is constructed, individuals within the designated area are permitted to carry objects across what was hitherto a private–public boundary. This may include keys, bags, a walking stick, a stroller or wheelchair...

The practice of demarcating an eruv has been used by Orthodox Jews for 2000 years and is based on principles from the Torah, developed in the Talmud and codified in Jewish Law. ... For an area to be designated as a private domain it must cover a minimum of twelve square feet and be demarcated from its surroundings by a wall or a boundary of some sort or by virtue of its topography. Already existing boundaries such as fences, rivers or railways or even rows of houses can serve as the basis for the eruv, but where the boundary is not continuous ... a boundary must be constructed in order to maintain the enclosed space ... and eruv can use existing poles in the street – such as telephone, electric, cables poles – or new poles can be constructed, joined by new wire. There are eruvim in many urban areas across the globe. ... Typically, they are patrolled the day before the Sabbath to ensure that the enclosure is intact and wires are not broken, as they cease to function once a gap has emerged. The eruv is unlike other boundaries in that when it ruptures nowhere inside is safe or unaffected.

Summary

The purpose of this chapter was to explore some of the ways the city is formed through faith and devotion – formed, that is, through being conceived, built and lived as a scaled-down version of the wider cosmos, a microcosm, connecting the human and spiritual worlds. As stated at the start, dividing up the city this way is undoubtedly artificial, but by now it should be evident that practices and beliefs of faith and devotion are simultaneously mapped on and through multiple imagined, material and habitual spaces that make up the city itself. Faith and devotion thus construct the city in mutually reinforcing ways.

In sum, it is clear religion has a place in the city, while the city has a place in religion. This symbolic and symbiotic relationship between religion and the city has a long history spanning thousands of years, as well as a wide and varied geography encompassing different cultures and faith systems, some extinct, some thriving. What this chapter has shown is that there is still much potential for exploring how the city itself reflects and reinforces patterns and practices of faith and devotion. This means adopting the recent approaches of geographers and others in studying the city in its different facets – as a representation, as a built environment, as a social habitat – and bringing all of these together, into dialogue with one another. It means, too, taking a comparative approach, ranging across time and space, of linking contemporary and past cities and religions in ways that accept their differences but also look for commonalities and transversalities, some common ground that unites rather than divides. In all the cases considered here, whatever the specific religious context, comes a long-lived concern of people both past and present to connect their lives to a higher divine being. The city has always provided a means to make this spiritual connection, and in so doing has not only shaped how we picture, make and inhabit the earthly city, but also how we make sense of the celestial metaphorical 'city', the world that is our past, present and future. This transcendental approach forces us at a time of conflict and intolerance to surmount perceived religious differences and find instead a unity and harmony between city and cosmos.

References

Akbar, J. (1988) *Crisis in the Built Environment: The Case of the Muslim City*, Singapore: Concept Media.

Baker, N. and Holt, R. (2004) *Urban Growth and the Medieval Church: Worcester and Gloucester*, Aldershot: Ashgate.

Beckwith, S. (2001) *Signifying God: Social Relation and Symbolic Act in the York Corpus Christi Plays,* Chicago: Chicago University Press.

Boal, F.W. (1996) 'Integration and division: sharing and segregating in Belfast', *Planning Practice and Research*, 11: 151–8.

Boogaart, T.A. (2001) 'Our Saviour's Blood: procession and community in late medieval Bruges', in K. Ashley and W. Hüskin (eds), *Moving Subjects: Processional Performance in the Middle Ages and the Renaissance*, Amsterdam: Rodopi.

Brace, C., Bailey, A. and Harvey, D.C. (2006) 'Religion, place and space: a framework for investigating historical geographies of religious identities and communities', *Progress in Human Geography,* 30 (1): 28–43.

Broshi, M. (1996) 'The inhabitants of Jerusalem', in N. Rosovksy (ed.), *City of the Great King: Jerusalem from David to the Present*, Cambridge, MA: Harvard University Press.

Carl, P. (2000) 'City-image versus topography of Praxis', *Cambridge Archaeological Journal,* 10: 328–35.

Cosgrove, D. and Daniels, S. (eds) (1988) *The Iconography of Landscape,* Cambridge: Cambridge University Press.

Dan, J. (1996) 'Jerusalem in Jewish spirituality', in N. Rosovsky (ed.), *City of the Great King: Jerusalem from David to the Present,* Cambridge, MA: Harvard University Press.

Duncan, J. (1991) *The City as Text: The Politics of Landscape Interpretation in the Kandyan Kingdom*, Cambridge: Cambridge University Press.

Duncan, J. and Ley, D. (eds) (1993) *Place/culture/representation*, London: Routledge.

Driver, F. (1995) 'Visualising geography: a journey to the heart of the discipline', *Progress in Human Geography,* 19: 123–34.

Emmett, C.F. (1997) 'The status quo solution for Jerusalem', *Journal of Palestine Studies,* 26 (1): 16–28.

Friedman, D. (1988) *Florentine New Towns: Urban Design in the Late Middle Ages,* Cambridge, MA: MIT Press.

Friedman, J.B. (1974) 'The architect's compass in Creation miniatures of the later Middle Ages', *Traditio*, 30: 419–29.

Frugoni, C. (1991) *A Distant City: Images of Urban Experience in the Medieval World*, trans. W. McCuaig, Princeton, NJ: Princeton University Press.

Graham, B. and Murray, M. (1997) 'The spiritual and the profane: the pilgrimage to Santiago de Compostela', *Ecumene,* 4: 389–409.

James, M. (1983) 'Ritual, drama and social body in the late medieval English town', *Past and Present,* 98: 3–29.

Jameson, M.H. (1997) 'Sacred space and the city: Greece and Bhaktapur', *International Journal of Hindu Studies,* 1 (3): 487–501.

Kedar, B.Z. and Werblowsky, R.J. (eds) (1998) *Sacred Space: Shrine, City, Land*, New York: New York University Press.

Kong, L. (1990) 'Geography and religion: trends and prospects', *Progress in Human Geography,* 14: 355–71.

Kong, L. (2001) 'Mapping "new" geographies of religion: politics and poetics in modernity', *Progress in Human Geography,* 25: 211–33.

Kostof, S. (1991) *The City Shaped: Urban Patterns and Meanings through History,* London: Thames and Hudson.

Kühnel, B. (1998) 'The use and abuse of Jerusalem', in B. Kühnel (ed.), *The Real and Ideal Jerusalem in Jewish, Christian and Islamic Art,* Jerusalem: Hebrew University.

Lassner, J. (1980) *The Shaping of 'Abbasid Rule,* Princeton, NJ: Princeton University Press.

Leach, A.F. (ed.) (1900) *Beverley Town Documents,* London: Selden Society.

Lefebvre, H. (1991) *The Production of Space,* Oxford: Blackwell.

L'Orange, H.P. (1953) *Studies on the Iconography of Cosmic Kingship in the Ancient World,* Oslo: H. Aschehoug.

Levy, R.I. (1990) *Mesocosm: Hinduism and the Organisation of a Traditional Newar City in Nepal,* Berkeley: University of California Press.

Lilley, K.D. (2004a) 'Cities of God? Medieval urban forms and their Christian symbolism', *Transactions of the Institute of British Geographers,* 29: 296–313.

Lilley, K.D. (2004b) 'Mapping cosmopolis: moral topographies of the medieval city', *Environment and Planning D: Society and Space,* 22: 681–98.

Lilley, K.D. (2005) 'Urban landscapes and their design: creating town from country in the Middle Ages', in K. Giles and C. Dyer (eds), *Town and Country in the Middle Ages,* Leeds: Maney.

Merrifield, A. (1993) 'Place and space: a Lefebvrian reconciliation', *Transactions of the Institute of British Geographers,* 18: 516–31.

Mitchell, T. (1990) *Passional Culture: Emotion, Religion, and Society in Southern Spain,* Philadelphia: University of Pennsylvania Press.

Naylor, S.K. and Ryan, J.R. (2002) 'The mosque in the suburbs: negotiating ethnicity and religion in South London', *Journal of Social and Cultural Geography,* 3: 39–60.

Nelson, A.H. (1974) *The Medieval English Stage: Corpus Christi Pageants and Plays,* Chicago: University of Chicago Press.

Neuwirth, A. (1996) 'The spiritual meaning of Jerusalem in Islam', in N. Rosovsky (ed.), *City of the Great King: Jerusalem from David to the Present,* Cambridge, MA: Harvard University Press.

Nitz, H.J. (1992) 'Planned temple towns and Brahmin villages as spatial expressions of the ritual politics of medieval kingdoms in South India', in A.H.R. Baker and G. Biger (eds), *Ideology and Landscape in Historical Perspective,* Cambridge: Cambridge University Press.

Pacione, M. (1990) 'The ecclesiastical community of interest as a response to urban poverty and deprivation', *Transactions of the Institute of British Geographers,* 15: 193–204.

Palliser, D.M. (ed.) (1980) *Chester: Contemporary Descriptions by Residents and Visitors,* Chester: Council of the City of Chester.

Phythian Adams, C. (1972) 'Ceremony and the citizen: the communal year at Coventry 1450–1550', in P. Clark and P. Slack (eds), *Crisis and Order in English Towns,* Cambridge: Cambridge University Press.

Pile, S. and Thrift, N. (eds) (1995) *Mapping the Subject: Geographies of Cultural Transformation,* London: Routledge.

Rosenau, H. (1983) *The Ideal City,* London: Methuen.

Rosovsky, N. (ed.) (1996) *City of the Great King: Jerusalem from David to the Present,* Cambridge, MA: Harvard University Press.

Rubin, M. (1991) *Corpus Christi: The Eucharist in Late Medieval Culture,* Cambridge: Cambridge University Press.

Rykwert, J. (1988) *The Idea of a Town: The Anthropology of Urban Form in Rome, Italy and the Ancient World,* Cambridge, MA: MIT Press.

Scott, J.S. and Simpson-Housley, P. (eds) (1991) *Sacred Places and Profane Spaces: Essays in the Geographics of Judaism, Christianity, and Islam,* New York: Greenwood Press.

Slater, T.R. (1998) 'Benedictine town planning in medieval England: evidence from St Albans', in T.R. Slater and G. Rosser (eds), *The Church in the Medieval Town,* Aldershot: Ashgate.

Slater, T.R. (2004) 'Encountering God: personal reflections on "geographer as pilgrim" ', *Area,* 36: 245–53.

Steinhardt, N. (1992) *Chinese Imperial City Planning,* Honolulu: University of Hawaii Press.

Travis, P.W. (1987) 'The social body of the dramatic Christ in medieval England', *Early English Drama ACTA,* 13 (1): 17–36.

Volvahsen, A. (1969) *Architecture of the World: India,* Lausanne: Benedikt Taschen.

Waston, S. (2005) 'Symbolic spaces of difference: contesting the eruv in Barnet, London and Tenafly, New Jersey', *Environment and Planning D: Society and Space,* 23: 597–613.

Wheatley, P. (1969) *City as Symbol,* London: H.K. Lewis.

Wheatley, P. (1971) *The Pivot of the Four Quarters: A Preliminary Enquiry into the Origins and Character of the Ancient Chinese City,* Edinburgh: Edinburgh University Press.

Wheatley, P. (2001) *The Places Where Men Pray Together: Cities in Islamic Lands, Seventh through the Tenth Centuries,* Chicago: Chicago University Press.

Witzel, M. (1997) 'Macrocosm, mesocosm, and microcosm: the persistent nature of "Hindu" beliefs and symbolic forms', *International Journal of Hindu Studies,* 1 (3): 543–67.

Woodward, D. (1985) 'Reality, symbolism, time, and space in medieval world maps', *Annals of the Association of American Geographers,* 75: 510–21.

Wright, A.F. (1977) 'The cosmology of the Chinese city', in G.W. Skinner (ed.), *The City in Late Imperial China,* Stanford, CA: Stanford University Press.

3   SCIENCE AND TECHNOLOGY

Colin Chant

This chapter

O   Argues that the relationship between urbanization and technological innovation is both long-standing and intimate

O   Insists that this relationship is reciprocal and complex

O   Shows, through examples, that cities have been both mirror and mould of technology in pre-industrial, industrial and post-industrial times

---

Introduction

The contributions of science and technology to urban development have very often been overlooked; conversely, when they are considered they are often over-egged. A balance therefore needs to be struck. But clarity is elusive, for the 'scientific', 'technological' and 'urban' between them connote a vast range of interlocking and overlapping human activities and institutions. To consider the relations between them is potentially to raise countless questions. In a short chapter, these need to be simplified. An obvious starting question is: what part did scientific discoveries and technological innovations play in the emergence and development of urban settlements and urban ways of life? Another major question works against the causal grain of the first: how far did urbanization stimulate the development of science and technology? The fact both questions can sensibly be put supports the main thrust of this chapter: that the relations between science, technology and cities are reciprocal.

This chapter's approach to these issues will be historical. Its premise is that there is a unique (historical) relationship between every human culture and the natural environment it struggles to exploit though science and technology. In consequence, no one linear story can be told about the transition from the first agricultural villages to the vast, omnivorous modern megalopolis. Throughout history cities have waxed and waned, and displayed characteristics peculiar to the societies that built them. These societies are characterized in turn both by their embrace of technologies, and also by the specific social, economic and political pathways through which the potentials of these technologies are channelled and shaped.

Contextualization is essential to the understanding of the history of science and technology; there is nonetheless a core strand in urban history concerning the increasing power urban 'movers and shakers' have derived from science and technology – powers with increasingly far-reaching consequences, both intended and unintended. The adoption of certain major innovations – wheeled vehicles, writing, printing, steam engines, railways, electric power, steel-framed buildings, motor vehicles and computers, for example – have ensured the emergence of new urban forms and ways of life.

The importance of changing urban instruments and innovations informs the conventional, and by no means implausible, division of all urban settlements into the pre-industrial and industrial. There is now talk of 'post-industrial' cities, in which economic services rather than manufacturing shape their various structures. This economic and technological classification will sometimes be applied here as shorthand, with the qualification that such historical constructs can considerably oversimplify. Enduringly fascinating to the historian are the manifold variations in the ways scientific and technological powers have been exercised, and the diverse urban forms that have resulted, threatening the neatness of any urban taxonomy that hinges on the Industrial Revolution.

Theorizing the relationships between science, technology and the city

The problem of the relationship between scientific discovery, technological innovation and urban development is a subset of a wider collection of issues about the relations between science, technology and society. The central conundrum scholars grapple with is how to reconcile the forward-thrusting, objectifying temper of scientific research and engineering with the variegated, contingent patterns of societies and their histories. Initially, the perspective and temper of the practising scientist and engineer prevailed. Science – broadly defined as the quest to understand nature – was seen as developing logically through observation, experiment and the rigorous pruning of erroneous theories and extraneous beliefs. Technology – the theory and practice of turning nature and natural knowledge to human advantage – could likewise be regarded as a progressive sequence, gathering pace as superior, increasingly scientifically-based ways of doing things displaced less efficient methods and devices.

Faith in scientific objectivity and technological progress underpinned the 'technological-determinist' model of the relations of science, technology and society.

In fact, far fewer theorists have explicitly expounded technological determinism than have been accused of doing so (White, 1959; Smith and Marx, 1994). In their actual or presumed guises, technological determinists took the view that the main characteristics of any society flow inexorably from its adopted technologies, and by extension the science on which so much of modern technology is based. Social changes, not least urbanization, have their roots in the ultimately autonomous realms of science and technology.

For adherents of 'social constructionism', the autonomy of science and technology is an illusion. Technological determinists have overlooked the essentially social nature of science and technology. When addressing technology specifically, social constructionists insist that any technological artefact or design is always the outcome of negotiations between relevant social groups. It is the resolution of these groups' varying interests in promoting a given technology that results in its finalizing or 'closure' – not the ironing out of technological flaws, improvements in efficiency or construction, more rational production methods, and so on. Thus for social constructionists, the innovations that technological determinists privilege in their account of urbanization are themselves the product of prior social change.

A less stringent version of social constructionism is 'social shaping'. Advocates of this position focus on the effects of society on science and technology, unlike technological determinists, who explore only the effects of science and technology on society. The main difference is that room is allowed for influences other than the social, such as the natural environment and antecedent science and technologies. The most influential versions of the social shaping of technology in particular are Thomas Parke Hughes' systems approach and the 'actor-network' analysis developed by the French theorists Michel Callon and Bruno Latour. In the latter, technologies are treated as a dynamic network of social, scientific and technical actors, without privileging the social over the others (Bijker et al., 1987; Fox, 1996; Latour, 2005).

For those who find technological determinism and the varieties of social constructionism one-sided, the concept of the 'mutual shaping' of science, technology and society points to a synthesis in which science and technology are recognized as being socially shaped, but, in turn, have due weight in explanations of subsequent social change. The end-point should be a richer account of the reciprocal and mutually reinforcing historical relationships between science, technology and urban development, one that is sensitive to the inherent propensity of a given innovation to be society-shaping, and the power of a given society's cultural, economic and political structures to give special form to those potentials.

Science, technology and the city in history

The topics in the urban history of science and technology selected for this section are intended to exemplify two broad 'dialectical' relationships in which the resolution of two seemingly opposed notions deepens our understanding. One is the tension between the increasingly generalizing thrust of technological innovations and the localizing effects of particular urban actors and contexts. This is a dialectic that has rather less force in the pre-industrial period, when technologies were often more spatially limited in their applications. The other is a dialectic embracing the effects of science and technology on urban development, and the effects of urban development on science and technology – what might be called a dialectic of mutual shaping.

A further, subsidiary distinction should be made. For the past five millennia and more, there have been two main ways in which applications of science and technology have helped to shape urban settlements. First, certain technologies – above all building construction and transport, but also energy sources and military technology – have directly influenced the physical form and fabric of cities. Distinctive urban forms and landscapes have resulted, from the first compact, mud-brick settlements of the ancient Near East to today's sprawling metropolises of steel, concrete and glass. Second, there are the often indirect effects of technological applications – notably transport and communications, but also water supply and sanitation – on the social geography of towns and cities.

The varied influence of technology on the city can thus be examined by exploring the relations between urbanization and science. Here, a fundamental historical question is why settlements of urban dimensions and complexity arose during the fourth millennium BC. The answer is necessarily elusive, but holds the key to a full understanding of the unfolding process of urbanization. Of course, ancient cities did not emerge fully-formed. The capacity to build, organize, sustain and defend large, permanent settlements presupposes a repertoire of skills, honed and accumulated over several prehistoric millennia. Many of these skills were technological to be sure, but they were also social (in a broad sense that encompasses the political, ideological and economic). Technological innovation, nevertheless, has sometimes been mooted as the very wellspring of urbanization. In technological-determinist vein, the archaeologist V. Gordon Childe identified an array of Neolithic innovations that propelled the process he dubbed the 'Urban Revolution' (Childe, 1966 [1936]; Extract 3.1). However, the identification of significant technological antecedents does not make them the main historical driver. Some archaeologists and anthropologists, led by Robert McCormick Adams, stood technological determinism on its head; for them, radical social change was a necessary condition for the organization of the technological effort required for the emergence of cities (Adams, 1966; Extract 3.2). In this way Adams adumbrated the social constructionism of a later generation of science and technology theorists. Technology has therefore been seen both as the precursor of urbanization, and as instrumental in a wider process of social and economic development of which urbanization is a manifestation: a clear instance of the dialectic of mutual shaping.

Extract 3.1: From Childe, V.G. (1966 [1936]) *Man Makes Himself*, London: Fontana, pp. 8, 105.

The archaeologist's divisions of the prehistoric period into Stone, Bronze, and Iron Ages are not altogether arbitrary. They are based upon the materials used for cutting implements, especially axes, and such implements are among the most important tools of production. Realistic history insists upon their significance in moulding and determining social systems and economic organization …

The scene of the drama lies in the belt of semi-arid countries between the Nile and the Ganges. Here epoch-making inventions seem to have followed one another with breathless speed, when we recall the slow pace of progress in the millennia before the first [Neolithic] revolution or even in the four millennia between the second [Urban] and the Industrial Revolution of modern times.

Between 6000 and 3000 [BCE] man has learnt to harness the force of oxen and of winds, he invents the plough, the wheeled cart, and the sailing boat, he discovers the chemical processes involved in smelting copper ores and the physical properties of metals, and he begins to work out an accurate solar calendar. He has thereby equipped himself for urban life, and prepares the way for a civilization which shall require writing, processes of reckoning, and standards of measurement – instruments of a new way of transmitting knowledge and of exact sciences. In no period of history till the days of Galileo was progress so rapid or far-reaching discoveries so frequent.

Extract 3.2: From Adams, R. (1966) *The Evolution of Urban Society: Early Mesopotamia and Prehispanic Mexico*, London: Weidenfeld and Nicolson, pp. 11–12.

Usefully to speak of an Urban Revolution, we must describe a functionally related core of institutions as they interacted and evolved through time. From this viewpoint, the characteristics Childe adduces can be divided into a group of primary variables, on the one hand, and a larger group of secondary, dependent variables, on the other. And it clearly was Childe's view that the primary motivating forces for the transformation lay in the rise of new technological and subsistence patterns. The accumulative growth of technology and the increasing availability of food surpluses as deployable capital, he argued, were the central causative agencies underlying the Urban Revolution.

This study is somewhat differently oriented; it tends to stress 'societal' variables. … Perhaps in part, such an approach is merely an outgrowth of limitations of space; social institutions lend

(Continued)

themselves more easily to the construction of a brief paradigm than do the tool types or pottery styles with which the archaeologist traditionally works. But I also believe that the available evidence supports the conclusion that the transformation at the core of the Urban Revolution lay in the realm of social organization. And, while the onset of the transformation obviously cannot be understood apart from its cultural and ecological context, it seems to have been primarily changes in social institutions that precipitated changes in technology, subsistence, and other aspects of the wider cultural realm, such as religion, rather than vice versa.

The example of ancient Egypt adds to the uncertainty. Egyptologists have sometimes doubted whether any of the settlements in the Nile Valley really count as cities. All the broad conditions for urbanization were in place, just as they were in ancient Mesopotamia: a hierarchical society legitimated by a state religion; fertile land on a floodplain yielding an agricultural surplus; highly organized hydraulic engineering; and considerable technological expertise in building construction and crafts. Yet Egypt developed a different, seemingly less urbanized culture than Mesopotamia. There were technological and political differences between the two civilizations rooted in environmental specifics: the greater availability in Egypt of building stone; the scarcity of tin for bronze metallurgy; a benign opposition between the direction of the prevailing wind and the flow of the river Nile that favoured developments in sail boats; and a stable inundation pattern and protective geographical setting that fostered an usually unified political culture. All these considerations made redundant the embattled Mesopotamian citadels of volatile Sumer and Assyria (Moorey, 1994; Van de Mieroop, 1999; Nicholson and Shaw, 2000; Kemp, 2006).

The example of ancient Rome is equivocal about the link between technological innovation and exceptional urban development. The Romans have usually been depicted as lacking in scientific and technological creativity, though this verdict has been questioned of late (Greene, 2000; Wilson, 2002). The Romans undoubtedly learned their surveying techniques, building methods and urban designs from the Greeks, though added materially to them (Lewis, 2001). It was to a considerable extent by scaling up established technologies of transport, water management and construction that the city's leaders fed, watered and housed a population which may at its peak have numbered one million. However, one undeniably innovative contribution to the capital's built environment was the invention of Roman concrete. The novel ingredient was *pozzolana*, a locally available volcanic sand that conferred additional strength as well as hydraulic properties on the traditional mix of lime, sand and aggregate. Concrete's relative cheapness and versatility enabled Roman architects to develop remarkable structures, making unprecedented use of curved forms (arches, vaults and domes) (Ward-Perkins, 1997; R. Taylor, 2003). These could to some extent be reproduced elsewhere with more traditional materials, or

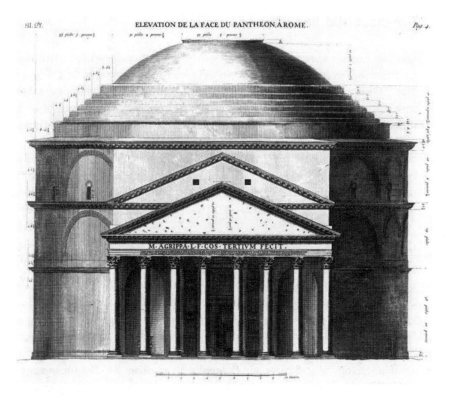

Figure 3.1 Antoine Desgodetz, 'Elevation de la face du Pantheon, a Rome', in *Les edifices antiques de Rome*. Paris: Claude-Antoine Jombert, 1779 (engraving)

with other kinds of concrete, though the use of *pozzolana* itself was of rather limited geographical scope. This says a good deal about the localized nature of many pre-industrial urban technologies (White, 1984).

The new concrete buildings were shaped not just by technology and architectural creativity, but also by the imperial capital's culture and political structures (Stambaugh, 1988; Anderson, 1997; Lomas and Cornell, 2003). The location and design of Rome's ubiquitous temples was mainly determined by religious tradition, though archaeological innovation and religious conservatism is starkly juxtaposed in the Pantheon, with its innovative concrete rotunda and traditional granite-columned portico (Figure 3.1). Like the monumental palaces, amphitheatres and the great bath-houses, the Pantheon was intended to display imperial munificence and dominion. Economic and political influences on Rome's built forms were evident in the spaces commanded by markets, concrete warehouses and granaries, the villas of the wealthy and resplendent imperial fora. These spaces squeezed the rest of the population into multi-storey concrete apartment blocks. In this way the ancient Roman built environment evinces the actualization of new technological potentials by cultural pressures: an instance of the dialectic of mutual shaping.

Pre-industrial and industrial cities

Despite Rome's special features, it clearly exemplified some of the characteristics that Gideon Sjoberg emphasized in elaborating the concept of the pre-industrial city (Sjoberg, 1960). These common features related to technology, which Sjoberg saw as the main causal variable, though he rebutted the charge of technological determinism (Extract 3.3). The technological dimension was in any case a somewhat negative consideration; it was the limitations of pre-industrial technology that explained these cities' common characteristics. They had a largely administrative, military and ideological function in an overwhelmingly agricultural economy, with craft and small-scale industrial production occupying a secondary role, often in limited quarters of the city.

Extract 3.3: From Sjoberg, G. (1960) *The Pre-industrial City*, Glencoe, IL: Free Press, p. 7.

For analytical purposes we distinguish three types of societies: the folk, or preliterate, society; the 'feudal' society (also termed the pre-industrial civilized society or literate pre-industrial society); and the industrial-urban society. …

To achieve this typology of societies, and consequently of cities, we take technology as the key independent variable – i.e., associated with varying levels of technology are distinctive types of social structure. Technology both requires and makes possible certain social forms. This viewpoint does *not* commit us to technological determinism, however, for recognized is the impact upon social structure of other variables – the city, cultural values, and social power – all of which can affect the patterning of technology itself …

Technology is not some materialistic, impersonal force outside the socio-cultural context or beyond human control; technology is a human creation *par excellence*.

The main objection to the generalized notion of the pre-industrial city, apart from the questionable privileging of technology, is that it brushes aside notable differences among settlements before the diffusion of the technologies of the Industrial Revolution. Pre-industrial cities were both fortified and non-fortified, and in some there was marked development of suburbs and some resultant social segregation. In pre-industrial China, to take one example, there were administrative settlements set out in a deliberately ideological fashion, informed by Confucianism, Daoism and *feng shui*, such as Beijing, Kaifeng and Chang'an; others were informally planned, commercial entrepôt cities such as Hangzhou and Hankou, where mercantile function and riverside location combined to create an altogether different urban morphology (Heng, 1999; Naquin, 2000). Islamic cities have been strongly identified with labyrinthine layouts, which one scholar has attributed to the replacement of wheeled vehicles by camels (Bulliet, 1975), but there were notable examples of planned Islamic cities, such as Baghdad, Cairo and Samarra (see also Chapter 2).

Technologies themselves were also far from moribund in the pre-industrial period. Lynn White, Jr. went so far as to ground modern industrial, urban society in a cluster of medieval technological innovations, notably cranks, water mills, iron ploughshares, crop rotation, horse-shoes and collar horse-harnesses (White, 1962; Roland, 2003). White's thesis has been dismissed, perhaps unfairly, as technological determinism, but the dynamism of medieval technology needs to be acknowledged. It is never more evident than in the soaring structures of the medieval Gothic cathedral, with its innovative rib vaults and flying buttresses. It is also apparent in the rapid diffusion of mechanical public clocks in fourteenth-century Europe, a phenomenon marking a fundamental transition in the regulation of urban life from ecclesiastical routines to the measured schedules of a burgeoning commercial and industrial era (Dorhn-van Rossum, 1996).

Later in the pre-industrial era, more spaces were created by mercantile and proto-industrial interests, such as breweries, silk mills, walled fairs, and mercantile exchanges (O'Brien et al., 2001; Harreld, 2003; Lanaro, 2003). Seventeenth-century Amsterdam and Paris were notable for innovations in street lighting, and the beginning of public transport (Bernard, 1970; Israel, 1995). Both sets of innovations demonstrated the dialectic of mutual shaping: they first met a latent demand for the extension of urban life in both time and space, and in so doing opened up new social and economic opportunities. Another kind of mutual shaping was evident in the relationship between military innovations and the physical structure of many an early modern European city. The development of cast-bronze, muzzle-loading cannon and gun carriages initiated a lethally escalating dialogue between the military technologies of offence and defence. An aesthetic by-product was fortified cities of striking geometrical harmony. Lofty medieval towers, crenellations and machicolations gave way to low ramparts, ditches, angled bastions and ravelins. Security as well as aesthetic considerations led military engineers to straighten street lines and widen public squares within cities. This is technology at its most forcefully influential, but nevertheless operating within a geopolitical context of continental conflict (Pepper and Adams, 1986).

There was indeed considerable technological and morphological variety in pre-industrial cities. But to give Sjoberg his due, this variety existed within broad limits imposed by the use of urban technologies that relied upon human and animal muscle power and natural forces, and organic and mineral materials requiring little more than manual manipulation. From the late eighteenth century, this was all to change as technologies wrought from Britain's plentiful coal and iron ore deposits worked their way though the British economy and its cities. These Industrial Revolution innovations were undeniably instrumental in remarkable changes to urban physical and social structures, changes that perhaps matched all those accumulated over the previous three millennia of urban history.

The crucial innovation was undoubtedly converting the energy of coal into motion. This was achieved in the eighteenth century though the expansion of steam, and from the late nineteenth century through the medium of electricity. Steam power in the form of stationary engines enabled the new cotton spinning factories to migrate from their original rural locations by fast-flowing streams into cities like Manchester and Glasgow. Second, in its mobile form on the railways, the power of steam intensified the movement of goods and people into and between towns and cities, a process already accelerated during the eighteenth century by the new system of river navigations and canals. The alliance of steam power with iron production provided the technological heart of the Industrial Revolution, and these innovations left their mark on the physical structures of cities. New industrial cityscapes were dominated by belching chimneys and blast furnaces, iron-framed factories and warehouses, and the panoply of railway engineering and architectural forms.

Investment in the railways hence transformed the British urban system. Crewe was created, and many other towns, such as Derby, Brighton and Swindon, grew rapidly; conversely, some settlements, such as Bath and Cambridge, contracted as the railways passed them by (Turnock, 1998). As railways rapidly diffused around the globe, the fortune of many an erstwhile urban backwater was made (Young, 2005). Within towns and cities, land-use patterns typically changed around stations, goods yards and railways. In particular, slums were both cleared and created by the urban swathes of the railways. Viaducts and railway land ownership also created physical and legal barriers to urban expansion. The carriage of building materials from afar, in Britain first by canal and then railway, helped homogenize the face of cities. However, technology is insufficient to account for all this transport-related change. Among the non-technological considerations were the rising costs of urban railway installations, rivalry between railway operators, legislative interventions and the often decisive – if usually only reactive – role of landowners (Kellett, 1969).

Rapid industrialization and associated population growth ran well ahead of urban amenities, and massive problems of morbidity and mortality arose from the overcrowded, polluted and unsanitary conditions. Unprecedented air pollution resulted from the increased burning of coal, both industrial and domestic (for Manchester, see Mosley, 2001; for a comparison between Manchester and Chicago, see Platt, 2005). Not only the air, but the land and water too of such locations as Glasgow, St Helens and Tyneside were contaminated by the waste products of the new heavy chemical industries. Drinking water was also threatened by the gross inadequacy of sanitation in the early decades of industrialization, and then by a later innovation, the water closet, which served at first to visit the excreta of the better off on the poor. In part, these problems were alleviated by innovative hydraulic engineering, notably Bazalgette's massive and enduring drainage scheme for mid-Victorian London and Belgrand's contemporaneous reconstruction of the sewers of Paris, part of Baron Haussmann's technocratic reworking of the French capital's fabric (Porter, 1998; Gandy, 1999). Gandy argues, however, that the technocratic impulse behind the Paris sewers was part of a much more complex process of cultural change as cities and citizens attempted to digest the scientific, technological and economic fruits of industrial capitalism (see Extract 3.4)

Extract 3.4: From Gandy, M. (1999) 'The Paris sewers and the rationalization of urban space', *Transactions of the Institute of British Geographers*, 24 (1): 23–44, pp. 24, 32.

[T]he reorganization of subterranean Paris held implications far beyond the modernization of drainage and sanitation. Metaphors of progress and the application of scientific knowledge became entangled with wider cultural and political developments surrounding the transformation of nineteenth-century Paris. ... By tracing the history of water in urban space, we can begin to develop a fuller understanding of changing relations between the body and urban form under the impetus of capitalist urbanization. ... During the last decades of the nineteenth century, more and more towns and cities across Europe became integrated into comprehensive water supply and sewerage systems, in order to accommodate the increasing demand for personal use of water. With the growing use of private washrooms, the smell of human excrement began to lose the last semblance of its rural associations with fertility: from now on it was to be indicative of disorder, decay and physical repulsion. ... The new-found bashfulness towards bodily functions in bourgeois French society emphasized the association of sewers with excrement. With the growing involvement of the state, under the guise of public health reform, the management of excrement became an increasingly rationalized activity, resulting in a steady decline in the use of cesspits, the activities of night-soil collectors and communal places for defecation. Henceforth, the 'regimes of the alimentary' were to be confined increasingly to domestic space under a new set of relationships between the body, technology and urban architecture.

It is an irony that need for and adoption of water-borne sewage disposal arose partly from the success of new water supply schemes, which threatened to over-whelm existing waste systems. As with all urban technologies, the adoption of systems of water supply or sewage and solid-waste disposal involves political and economic decision-making, and there are social groups who are better placed to turn those choices to their own advantage (Melosi, 2000). In this way, they exemplify the dialectical relationship between technologies and their contexts. They also imply a different set of relationships between science, technology and urban development. Much innovation in these areas has indeed been *reactive*, a response to urban growth partly stimulated by other technologies. Accordingly, they also present a special case of the dialectic of mutual shaping.

In the context of rapid industrialization and population growth, technology contributed to a notable inversion of the social geography of towns and cities. In Sjoberg's model of pedestrian pre-industrial cities, the rich dominated the centre

Figure 3.2   Steam-driven trams, Melbourne, 1925

of cities, with the poor and industry relegated to the periphery. However, the noise and smoke of industrial plant and railway locomotives made many city centres places to avoid. Just as the railways brought raw materials and rural migrants into the rapidly growing cities, so they and other transport innovations helped the wealthy escape to the semi-rural edges. Taking nineteenth-century London as an example, there were a variety of successive, overlapping and competing modes for travelling within the city, and commuting to it: short-stage coaches, horse-drawn omnibuses, paddle steamers, surface and underground railways, the various forms of tram (horse, cable, steam and electric) (see Figure 3.2), and right at the century's end, the motor bus. Beyond the biggest cities, steam locomotives were uneconomic and inefficient for urban transport when distances between stations had to be short; it was the quick-starting and smoothly accelerating electric motor, attached to a tram, that first brought daily travel to the working classes in most Western cities (McKay, 1976).

These transport innovations underpinned the construction of a range of urban models attempting to generalize the form of the new industrial city. In many such models, a central business district is neatly surrounded by concentric belts of residential and industrial land use. Historians inevitably object that this is too schematic to capture actual urban morphological patterns. There are in any case variations according to the dominant transport modes: the star-shaped city based on fixed-rail transport modes; infill and amorphous sprawl with rise of motor transport; and latterly the hollowing out of the centre and the emergence of 'edge cities' (Garreau, 1991; Fogelson, 2001). Transport innovations are clearly inadequate to give a full explanation of these fundamental changes. Even with regard to transport itself, policy can be as decisive as technological innovation: the choice between flat fares and graduated fares has had a significant effect on the extent of suburbanization (Divall and Bond, 2003).

The specific relationship between transport innovations and the outward growth of cities is in any case controversial. Many British historians have downplayed their significance. Thompson argued the mainspring of suburban growth was the desire for a private house and garden. He also pointed out that many suburbs preceded the railways, though it needs to be recognized that early residential suburbs such as Hillhead in Glasgow and Didsbury in Manchester were facilitated by an eighteenth-century transport innovation, turnpike roads. Thompson concluded that transport technology permitted rather than created suburban growth, and generally lagged behind the urban fringe (Thompson, 1982). There are exceptions, notably some outlying stations of the London Undergound's Metropolitan line, which were built on greenfield sites in anticipation of housing development. Tellingly, the Metropolitan was exceptional among British transport undertakings in being allowed to invest in property. This was more in line with US practice, which might help explain why American historians have usually given transport innovations (especially trams or streetcars, and latterly the automobile) more prominence in explanations of suburbanization (Jackson, 1985; Gutfreund, 2004).

Among the reasons why suburban life developed earlier in both Britain and the USA was the constraint on urban size imposed by city walls on the war-torn continent of Europe. But the age-old contest between city walls and weapons was ended by the sheer offensive power and range of nineteenth-century artillery, testimony to the awesome power of the heavy steel and chemical industries. City walls simply became obsolete. Again, military technology is heavily influential, but context-dependent too. Cities were already bursting through their fortifications from the inside. The razed walls could be converted to highways, as happened in many European cities, such as Vienna and Paris, where a fortification ring also gave way to an outer circle railway: a case of changing contexts and changing technologies.

Science, technology and the twentieth-century city

The second half of the nineteenth century saw an efflorescence of innovations with extensive urban implications: electric lighting and power, electrical communications, structural steel and reinforced concrete, and the internal combustion engine. Their embrace during the twentieth century by engineers, architects, civic leaders and the rising profession of urban planning ensured a reworking of the physical and social structures of European and American cities that surely matched that of the Industrial Revolution. The changes were both spatial and temporal: the central core reached ever upward and deeper into the night; residential and industrial suburbs sprawled further outwards.

The emblematic urban building type of this period was the tall building or skyscraper. It can be seen as a high-tech amalgam of new materials (bulk steel, reinforced concrete, plate glass), new construction methods (steel and reinforced-concrete frames, giant cranes, mechanized site processes) and other essential innovations (hydraulic and electric lifts, electric lighting, air conditioning). But there was no simple transition from these innovations to today's high-rise cityscapes: technologies were evidently necessary, but not sufficient. Economic considerations have a big part to play, especially in North America, where skyscrapers have very often functioned as speculative buildings, intended to turn a profit (Willis, 1995). This is only one side of the story: civic pride and rivalry are part of the North American mix (see Chapter 1), and in other political contexts the urge to build a towering national virility symbol can be the main driver, as with the Stalinist skyscrapers of Moscow or Kuala Lumpur's Petronas Towers (Sudjic, 2005).

There has been considerable resistance to these overbearing structures in other parts of the industrialized world, both from governments and private citizens. The strict regulation of building heights and other structural features in European cities partly explains their lower physical profile during the twentieth century. In Britain, high-rise social housing became fashionable during the 1960s, shaped not only by competing construction systems but also by conflicts among the interest groups sponsoring them: the design professions, local councillors and building contractors (Glendinning and Muthesius, 1994). Multi-storey flats fell from favour at the time of the Ronan Point collapse in London's East End in 1968, though its significance has been exaggerated (see Chapter 4). England's temporary embrace of the tower block always worked against cultural prejudices that favoured low-rise, low-density settlements, and regarded architectural modernism with suspicion (Gold, 1997; Bullock, 2002). Multi-storey living had always been more accepted in Scotland, where practices of land ownership encouraged the building of tenement blocks well before the Industrial Revolution (Rodger, 2001).

During the twentieth century, the rise to dominance of motor transport, and in particular the private car, radically affected the design of cities, housing estates and individual dwellings. This was most evident in the USA, the first motorized society (McShane, 1994). Throughout the century, the planners of most cities in the Western world sought to accommodate the car to tight street systems and housing developments predicated upon horse-drawn transport, and then fixed-rail modes. A mood of technocratic hubris in the post-war decades inspired some civic leaders to tear up existing road patterns and residential districts to make way for motor traffic: it was a period reminiscent of the remoulding of towns and cities by the railway interests. What emerged was a motorized city with its urban motorways, ring roads, underpasses and walkways, the segregation of pedestrians and traffic, multi-storey car parks, and so on. The USA again led the way in the physical and social restructuring of housing and housing estates: the substitution of the garage for the porch marked a turning of family life away from the street to the back-yard (Jennings, 1990).

Los Angeles is often cited as the classic example of the motorized city with its decentralized, sprawling city structure, and an architectural landscape adapted to the dominance of the automobile (Bottles, 1987; Hise, 1997; Longstreth, 1997 and 1999). However, it would be misleading to see the car as driving the whole process. The initial shape and suburbanizing thrust of the city in any case owes much to the earlier fixed-rail mass transport system. It also reflects the nature of its industries and their dispersal, and also owes something to the politics of its water supply. The city's subsequent embrace of the motor car says a great deal about an individualistic cultural preference for the automobile, as well as its relative cheapness compared with other parts of the world, given the proximity of Los Angeles to petrol supplies and the favouring of road travel by its climate.

These contextual arguments are reinforced by comparing the USA with Soviet Russia during the same period. The Soviet regime was keen to exploit a range of Western innovations, including electrical communications and motor vehicle production, during the industrialization drive of the Stalinist plan era. These transferred technologies were adapted to the Russian context in a variety of ways. Modernist architectural designs of the 1920s were rejected by Stalin in favour an ideologically shaped adaptation of modern materials and construction techniques. These found particular expression in Moscow's 'Stalinist Gothic' skyscrapers, but more ubiquitously in the post-Second World War accelerated programme of standardized, prefabricated housing. The Soviets' ideological commitment to public transport shaped the diffusion of transport innovations, including the restriction of the motor car to the political elite.

As well as the ideological commitment to public transport and a limited journey-to-work, Soviet planning ideals envisaged strict land-use zoning, the dispersal of industry and enforced limits on city size. The urban realities were rather different, partly through a contradiction between the planning ideals and forced production targets in the struggle to match Western military and industrial might. Great urban-industrial agglomerations sprang up, including the capital Moscow, in which housing and industry became intermingled. Nevertheless, there were marked differences between the morphologies of Soviet, West European and North American cities: a relative lack in the USSR of social segregation, and of city centres dominated by commercial interests; a flatter population density gradient; and a relative absence of the effects of motor cars for most of period (Bater, 1980).

What of the future? Innovations in information and communications technology (ICT) – from the telegraph and telephone to today's computing and telecommunications networks of satellites and fibre-optic cables – have already affected individual cities and the global urban system in a systematic, albeit often indirect and invisible fashion (Graham and Marvin, 1996 and 2001). Whereas most technologies of industrialization have acted to concentrate populations, and so increased the pace of urbanization, ICT has been heralded as an agent of dispersal, at a time when the proportion of urban dwellers continues its seemingly inexorable rise, and the carbon footprint of the great metropolises grows ever larger. Predictions of the dissolution of face-to-face contacts in the city seem premature at best. Rather than dispersal, there has been a reorganization of spatial relationships. ICT has facilitated suburban sprawl, and in particular the separation of manufacturing processes and corporate administration. Nevertheless corporate headquarters continue to cluster in high-density, high-value city centres. What needs to be weighed up is the relative contribution of ICT, in conjunction with transport developments, to the growing homogeneity of 'world cities' that is a feature of the complex phenomenon of economic and cultural globalization (Abu-Lughod, 1999; Taylor, P.J. 2003). Within that context further attention is needed on the specific role of 'technopoles': urban complexes located in an economic order increasingly based upon information technologies (Castells and Hall, 1994).

Summary

This chapter has emphasized the reciprocity of the historical relations between technological and urban change. Technological innovations were an important part of the causal mix that resulted in the emergence of urban settlements. Their subsequent spread and growth then became a spur to scientific discovery and technological innovation. This was partly because cities presented problems and opportunities increasingly seen as amenable to scientific analysis and technological remedy. It was also because the concentration of population and resources within town and cities encouraged the clustering of expertise and the exchange of information. This applied as much to groups of scientists and engineers as it did to businessmen, professionals, craftsmen, retailers, artists, criminals and so on (Inkster, 1997; Hall, 1998).

Two broad dialectical relationships have linked science, technology and urban development. The first was the realization of the general potentials of technological innovations by particular urban actors operating within specific urban contexts. This tension became more marked in the industrial and post-industrial eras. The growing power and sophistication of technologies now makes local solutions to the challenges of city-building less diverse: the cities of the early twenty-first century are more homogeneous than they ever have been. This is surely a general trend, though the universality of urban layouts under the Hellenistic and Roman empires show that ancient technologies could spread very widely and also that their diffusion owed as much to political as to technological power. The near universality of a given technology has to be put down to something more than its intrinsic superiority at meeting specific urban needs.

The second dialectic of mutual shaping embraces the effects of science and technology on urban development, and the effects of urban development on science and technology. Regarding the first thesis of the dialectic, it is clear that scientific discoveries and technological innovations have exerted a powerful influence on the physical form and social structure of cities. The form and fabric of ancient Athens and imperial Rome demonstrate the availability of a specific range of building materials, mostly but not always local, the expertise of architects and builders with certain structural forms, the prevalence of animal (including human) power sources, and the absence of any system of public urban transport. The physical and social geography of contemporary Los Angeles and Shanghai, in their turn, testify to the properties of modern steel, reinforced concrete and glass, the dissemination of electrical energy for lighting, power, communication and mass transport, and in the case of Los Angeles above all, the overriding requirements of the ubiquitous motor car.

These contrasting morphologies reflect prevailing technological instruments of city-building, but only because their adoption and implementation have suited the purposes of influential urban actors. The differences between the ancient and modern cities also testify to a massive change in the attitudes of urban elites to science and technology. Even so there are shades in between. It is surely relevant that the new and rapidly diffusing building and transport technologies that emerged in the latter part of the nineteenth century had notably different urban effects in the very divergent political and economic climates of the USA and the Soviet Union.

Technologies are essentially human means to human ends. One has therefore always to ask which ends are being met in order to account for the diffusion of technologies, and to understand their effects. The spread of a new technology may need to be explained not just for the benefits it may confer on the population at large, but for the advantage it may give to the most powerful urban groups. The global spread of a particular set of urban innovations, so much a feature of the later twentieth century, may actually be ill-suited to the needs of many citizens across the world, even though it serves the interests of transnational corporations, electronically networked planners and globe-trotting architects.

References

Abu-Lughod, J.L. (1999) *New York, Chicago, Los Angeles: America's global cities*, Minneapolis: University of Minnesota Press.

Adams, R. (1966) *The Evolution of Urban Society: early Mesopotamia and prehispanic Mexico*, London: Weidenfeld and Nicolson.

Anderson, J.C. Jr. (1997) *Roman Architecture and Society*, Baltimore, MD: Johns Hopkins University Press.

Bater, J.H. (1980) *The Soviet City: ideal and reality*, London: Edward Arnold.

Bernard, L. (1970) *The Emerging City: Paris in the age of Louis XIV*, Durham, NC: Duke University Press.

Bijker, W.E., Hughes, T.P. and Pinch, T. (eds) (1987) *The Social Construction of Technological Systems: new directions in the sociology and history of technology*, Cambridge, MA: MIT Press.

Bottles, S.L. (1987) *Los Angeles and the Automobile: the making of the modern city*, Berkeley: University of California Press.

Bulliet, R.W. (1975) *The Camel and the Wheel*, Cambridge, MA: Harvard University Press.

Bullock, N. (2002) *Building the Post-war World: modern architecture and reconstruction in Britain*, London: Routledge.

Castells, M. and Hall, P. (1994) *Technopoles of the Word: the making of twenty-first-century industrial complexes*, London: Routledge.

Childe, V.G. (1966 [1936]) *Man Makes Himself*, London: Fontana.

Divall, C. and Bond, W. (eds) (2003) *Suburbanizing the Masses: public transport and urban development in historical perspective*, Aldershot: Ashgate.

Dohrn-van Rossum, G. (1996) *History of the Hour: clocks and modern temporal orders*, trans T. Dunlap, Chicago: Chicago University Press.

Fogelson, R.W. (2001) *Downtown: its rise and fall, 1880–1950*, New Haven, CT: Yale University Press.

Fox, R. (ed.) (1996) *Technological Change: methods and themes in the history of technology*, Amsterdam: Harwood Academic.

Gandy, M. (1999) 'The Paris sewers and the rationalization of urban space', *Transactions of the Institute of British Geographers*, 24 (1): 23–44.

Garreau, J. (1991) *Edge City: life on the new frontier*, New York: Doubleday.

Glendinning, M. and Muthesius, S. (1994) *Tower Block: modern public housing in England, Scotland, Wales and Northern Ireland*, New Haven, CT: Yale University Press.

Gold, J.R. (1997) *The Experience of Modernism: modern architects and the future city, 1928-53*, London: E & FN Spon.

Graham, S. and Marvin, S. (1996) *Telecommunications and the City: electronic spaces, urban places*, London: Routledge.

Graham, S. and Marvin, S. (2001) *Splintering Urbanism: networked infrastructures, technological mobilities and the urban condition*, London: Routledge.

Greene, K. (2000) 'Technological progress and innovation in the ancient world: M I Finley reconsidered', *Economic History Review*, 53: 29–59.

Gutfreund, O.D. (2004) *Twentieth-Century Sprawl: highways and the reshaping of the American landscape*, New York: Oxford University Press.

Hall, P. (1998) *Cities in Civilization: culture, innovation and urban order*, London: Weidenfeld and Nicolson.

Harreld, D.J. (2003) 'The public and private spaces of merchants in sixteenth-century Antwerp', *Journal of Urban History*, 29 (3): 657–69.

Heng, C.K. (1999) *Cities of Aristocrats and Bureaucrats: the development of medieval Chinese cityscapes*, Honolulu: University of Hawaii Press.

Hise, G. (1997) *Magnetic Los Angeles: planning the twentieth-century metropolis*, Baltimore, MD: Johns Hopkins University Press.

Inkster, I. (1997) *Scientific Culture and Urbanisation in Industrialising Britain*, Aldershot: Ashgate.

Israel, J. (1995) 'A Golden Age: innovation in Dutch cities, 1648–1720', *History Today*, 45 (1): 14–20.

Jackson, K.T. (1985) *Crabgrass Frontier: the suburbanization of the US*, Oxford: Oxford University Press.

Jennings, J. (1990) *Roadside America: the automobile in design and culture*, Ames: Iowa State University Press.

Kellett, J.R. (1969) *The Impact of Railways on Victorian Cities*, London: Routledge and Kegan Paul.

Kemp, B.J. (2006) *Ancient Egypt: anatomy of a civilization* (2nd edition), London: Routledge.

Lanaro, P. (2003) 'Economic space and urban realities: fairs and markets in the Italy of the early modern age', *Journal of Urban History*, 30 (1): 37–49

Latour, B. (2005) *Reassembling the Social: an introduction to actor-network theory*, Oxford: Oxford University Press.

Lewis, M.J.T. (2001) *Surveying Instruments of Greece and Rome*, Cambridge: Cambridge University Press.

Lomas, K. and Cornell, T. (eds) (2003) *'Bread and Circuses': euergetism and municipal patronage in Roman Italy*, London: Routledge.

Longstreth, R. (1997) *City Center to Regional Mall: architecture, the automobile, and retailing in Los Angeles, 1920–1950*, Cambridge, MA: MIT Press.

Longstreth, R. (1999) *The Drive-In, the Supermarket and the Transformation of Urban Space in Los Angeles, 1914–1941*, Cambridge, MA: MIT Press.

McKay, J.P. (1976) *Tramways and Trolleys: the rise of urban mass transport in Europe*, Princeton, NJ: Princeton University Press.

McShane, C. (1994) *Down the Asphalt Path: the automobile and the American city*, New York: Columbia University Press.

Melosi, M.V. (2000) *The Sanitary City: urban infrastructure in America from colonial times to the present*, Baltimore, MD: Johns Hopkins University Press.

Moorey, P.R.S. (1994) *Ancient Mesopotamian Materials and Industries*, Oxford: Clarendon Press.

Mosley, S. (2001) *The Chimney of the World: a history of smoke pollution in Victorian and Edwardian Manchester*, Cambridge: White Horse Press.

Naquin, S. (2000) *Peking, Temples and City Life, 1400–1900*, Berkeley: University of California Press.

Nicholson, P.T. and Shaw, I. (eds) (2000) *Ancient Egyptian Materials and Technology*, Cambridge: Cambridge University Press.

O'Brien, P., Keene, D., Hart, M. and van der Wee, H. (eds) (2001) *Urban Achievement in Early Modern Europe: golden ages in Antwerp, Amsterdam and London*, Cambridge: Cambridge University Press.

Pepper, S. and Adams, N. (1986) *Firearms and Fortifications: military architecture and siege warfare in sixteenth-century Siena*, Chicago: Chicago University Press.

Platt, H.L. (2005) *Shock Cities: the environmental transformation and reform of Manchester and Chicago*, Chicago: Chicago University Press.

Porter, D.H. (1998) *The Thames Embankment: environment, technology and society in Victorian London*, Akron, OH: University of Akron Press.

Rodger, R. (2001) *The Transformation of Edinburgh: land, property and trust in the nineteenth century*, Cambridge: Cambridge University Press.

Roland, A. (2003) 'Once more into the stirrups: Lynn White, Jr, *Medieval Technology and Social Change*', *Technology and Culture*, 44 (4): 574–85.

Sjoberg, G. (1960) *The Pre-industrial City*, Glencoe, IL: Free Press.

Smith, M.R. and Marx, L. (eds) (1994) *Does Technology Drive History? The dilemma of technological determinism*, Cambridge, MA: MIT Press.

Stambaugh, J.E. (1988) *The Ancient Roman City*, Baltimore, MD: Johns Hopkins University Press.

Sudjic, D. (2005) *The Edifice Complex: how the rich and powerful shape the world*, London: Allen Lane.

Taylor, P.J. (2003) *World City Network: a global analysis*, London: Routledge.

Taylor, R. (2003) *Roman Builders: a study in architectural process*, Cambridge: Cambridge University Press.

Thompson, F.M.L. (ed.) (1982) *The Rise of Suburbia*, Leicester: Leicester University Press.

Turnock, D. (1998) *An Historical Geography of Railways in Great Britain and Ireland*, Aldershot: Ashgate.

Van de Mieroop, M. (1999) *The Ancient Mesopotamian City*, Oxford: Oxford University Press.

Ward-Perkins, J.B. (1997) *Roman Imperial Architecture*, New Haven, CT: Yale University Press.

White, K.D. (1984) *Greek and Roman Technology*, London, Thames and Hudson.

White, L. (1959) *The Evolution of Culture: the development of civilization to the fall of Rome*, New York: McGraw-Hill.

White, L. Jr. (1962) *Medieval Technology and Social Change*, Oxford: Clarendon Press.

Willis, C. (1995) *Form Follows Finance: skyscrapers and skylines in New York and Chicago*, Princeton, NJ: Princeton Architectural Press.

Wilson, A. (2002) 'Machines, power and the ancient economy', *Journal of Roman Studies*, 92 (1): 1–32.

Young, D.M. (2005) *The Iron Horse and the Windy City: how railroads shaped Chicago*, DeKalb, IL: Northern Illinois University Press.

4 MODERNITY AND UTOPIA

John R. Gold

This chapter

○ Suggests that the search for the ideal city has been long-standing, generating contrasting ideas of what the modern city ought to be

○ Provides in-depth consideration of two attempts to improve the city, namely the Garden City and the Modern Movement

○ Argues that the failure of utopian visions cannot be apportioned solely to the naïviety of planners and architects, but needs to be placed in the context of wider social, economic and political forces

---

Introduction

The quarter century after the Second World War saw unprecedented changes in the cities throughout many parts of the world. Largely regardless of prevailing Cold War ideological differences, authorities responsible for cities unleashed waves of change designed to redevelop urban structure and form. The results bore many similarities. These included the wholesale clearance of dilapidated districts, the comprehensive reconstruction of town centres and housing areas, the re-creation of an urban structure based on principles of single land-use zoning, and attempts to separate pedestrians and vehicles into their own spaces, with traffic-free precincts paralleled by high-capacity roadways and multi-storey car parks. Most important, however, were the changes in the appearance of the built environment. Skylines became punctuated by tall buildings. The use of modern materials and the adoption of novel building technologies contributed to a new visual vocabulary of non-traditional built forms. This affected houses, offices, factories, transport facilities, bridges and even street furniture. Concrete became a favoured building material. Official handbooks provided guidelines for landscaping, signage and even ways of trimming street trees. As Relph (1987: 141) observed, the results looked much the same whether in 'Canberra, Corby New Town in England, or Columbia in Maryland'.

The sea-change in practices that brought about these new townscapes, of course, did not occur by chance. Historians now identify the 1950s and 1960s as a key period in which the state extended its control over urban development to accommodate renovation and modernization (Cherry, 1988). Nations that had experienced war damage necessarily prioritized action to repair physical damage, renovate the urban fabric, deal with housing shortages, and repair infrastructure. Yet the shape of the resulting interventions also reflected more general social consensus about the necessity of tackling problems arising from earlier, poorly planned urban expansion, from sharp demographic increase, and from rising expectations about living standards. Certainly, few commentators actively contested the wisdom of pursuing modernity through technological progress or challenged the primacy given to planning as a prerequisite for guiding present-day development and for taking account of future needs. The formation of consensus was also assisted by the existence of several significant bodies of planning and architectural thought that offered ready-made visions of how newly created settlements or reconstructed existing cities might look and function.

This chapter explores the role of two such schools of thought: the Garden City Movement and the Modern Movement. Both, as we will see, offered sets of planning principles and supportive architectural strategies that went beyond mere prescriptions about the design of buildings, transport routes and open spaces. Indeed, both wove together ideas about design and society in the telling combination that typifies 'utopianism', the brand of social thought that sees urban transformation not just as a way of improving the built environment, but also as a vehicle for achieving the Good Life. The ensuing parts explore the conceptual models associated with the Garden City and Modern Movements, before offering some case studies that stress the *limits* to the powers of planners and architects when implementing innovative schemes. The conclusion reflects on the importance of interpreting the meaning of utopian ideas about the city in the context of the times when they were developed and implemented. Before doing so, however, it is important to clarify the nature of urban utopias, and to demonstrate how these two twentieth-century movements fit into wider histories of utopianism.

Urban utopias

Few things date as quickly as yesterday's ideas about tomorrow. In any age, there are many forms the city could have taken if, for one reason or another, it had not become what we see today (Calvino, 1974). Images of the city that might-have-been fascinate urban scholars largely because of what they convey about the ideas, hopes, aspirations, fears and forebodings current at the time when they were produced. Not all such schemes, though, had a social intent. The age-old *ideal cities* tradition, for example, saw autocrats and tyrants commission architects to produce grandiose schemes featuring an array of processional axes, distant vistas of palatial buildings, triumphal arches, and set-piece urban tableaux that might glorify their regimes (see also Chapter 5). Their appointed architects might have personally favoured improving the conditions of residents but these considerations usually took second place to creating schemes that pleased their patrons (Rosenau, 1983). *Urban utopias*, by contrast, place an emphasis on the achievement of a better society, with their creators believing social problems could be solved through properly-designed environments (Ravetz, 1980).

Urban utopian schemes may accordingly embrace grand architecture or comprehensive planning, but they always proffer some notion of social transformation. Sometimes the inspiration comes from looking backwards to a (mythical) past and recreating the city in ways that might recapture the values of a golden age. Others, like those considered here, try to capture the spirit of modernity – the feeling of living in 'modern times' – and remake the city in ways that testify how 'modern' we are.

The term 'utopia' itself stemmed from Sir Thomas More's eponymous monograph published in 1516. *Utopia* was ostensibly a travelogue about an idyllic Caribbean island kingdom written by a wandering Portuguese philosopher, Raphael Hythloday. As such, its accounts of strange lands were no more surprising than many similar reports circulating in early sixteenth-century England, but, like many later 'utopian' works, the book had a carefully coded content. More coined the word 'utopia' as a deliberate play on the two Greek words that supply the 'u' sound: *eu* (good) and *ou* (not). When taken together with *topos* (place) – the root of the second part of the word – utopia could be construed either as a 'good place' or a somewhere that does not exist. By injecting this ambiguity, More inventively posed the questions that have haunted so much subsequent discussion. Is utopia the search for the ideal or is it a fruitless quest for the impossible? Are utopias real schemes for achieving human fulfilment or are they critical commentaries, invitations 'to perceive the distance between things as they are and as they should be' (Eliav-Feldon, 1982: 1)?

Whichever the interpretation, the close relationship between cities and utopia dates back at least to Plato's *The Republic* from the fourth century BC, which offered an alternative vision of the city as a harmonious living entity governed by reason. Other works that provided a measure of continuity down to the nineteenth century included St Augustine's *The City of God* (*c.* 426), Friar Tommaso Campanella's *City of the Sun* (1602), J.H. Andreae's *Christianopolis* (1619), James Silk Buckingham's *National Evils and Practical Remedies* (1849), and Edward Bellamy's *Looking Backward* (1888). All stressed the importance of cities as crucibles for achieving social change, although the cities that some of the earlier writers had in mind were of heavenly rather than earthly creation (see also Chapter 2). Equally, all offered views about the way that the right form of city could promote the Good Life, however defined.

Utopia, of course, is incomplete in itself. Just as the dream of heaven is placed in perspective by the nightmare of hell, so the joy and ordered tranquillity of utopia gains its cutting edge by contrast with *dystopia*, its antithesis. Dystopia normally implies disturbing chaos, brutality, suffering and despair. Dystopians readily agree with utopians that the city is a crucible for social transformation, but they see that transformation as wholly negative (i.e. the city gradually erodes civility). For twentieth-century urban utopians, the prime image of dystopia was the dark, insanitary and overcrowded Victorian city, which they believed impoverished human life, crushed individualism, fostered oppression, alienated people from one another, and shut out nature. Faced with these circumstances, they campaigned for new types of city in which people might live justly, morally and healthily (Hardy, 2000; Eaton, 2002). Their imaginings followed two broad paths. One, exemplified by the work of the Garden City Movement, favoured the strategy of building new settlements as oases of sanity on greenfield sites. The second, illustrated by the projects of the Modern Movement, campaigned for the radical reconstruction of existing cities. This broad distinction – between utopias of escape and reconstruction (Mumford, 1922) – is not watertight, but provides a useful classificatory device for examining the key urban utopias.

The Garden City Movement

Extract 4.1: From Howard, E. (1898) *To-morrow: A Peaceful Path to Real Reform*, London: Swann, Sonnenschein, pp. 9–10.

But neither the Town magnet nor the Country magnet represent the full plan and purpose of nature. Human society and the beauty of nature are meant to be enjoyed together. The two magnets must be made one. As man and woman by their varied gifts and faculties supplement each other, so should town and country. The town is the symbol of society – of mutual help and friendly co-operation, of fatherhood, motherhood, brotherhood, sisterhood, of wide relations between man and man – of broad, expanding sympathies – of science, art, culture, religion. And the country! The country is the symbol of God's love and care for man. All that we are, and all that we have comes from it. Our bodies are formed of it; to it they return. We are fed by it, clothed by it, and by it are we warmed and sheltered. On its bosom we rest. Its beauty is the inspiration of art, of music, of poetry. Its forces propel all the wheels of industry. It is the source of all health, all wealth, all knowledge. But its fulness of joy and wisdom has not revealed itself to man. Nor can it ever, so long as this unholy, unnatural separation of society and nature endures. Town and country *must be married*, and out of this joyous union will spring a new hope, a new life, a new civilisation. It is the purpose of this work to show how a first step can be taken in this direction by the construction of a Town-country magnet...

These words, from Ebenezer Howard's book *To-morrow* (1898), set the agenda for a British planning movement that quickly developed an international presence (Ward, 1992). Its success stemmed largely from the scale and vision of the underlying concept. In terms of scale, the course of urbanization prompted large-scale schemes to cope with the numbers of people to be housed, but these mostly followed standardized forms governed by the by-laws passed in the late nineteenth century. Howard's achievement, along with architects like Raymond Unwin (who helped translate his ideas into practice), was to produce a radically different notion of urban form that also embraced the requisite scale needed in development. In terms of vision, Howard's book appeared when Britain was experiencing 'a wave of self-doubt and introspection', with 'international tensions, a growing arms race, severe agrarian recession, urban overcrowding and unemployment, evidence of considerable poverty and fear of class warfare' (Simpson, 1985: 9–10). Against this background, Howard's proposals for medium-sized, decentralized and freestanding garden cities provided an original solution to such problems.

Howard's starting point was to consider 'town' and 'country' as magnets, with each having the power to attract and repel. What he deemed necessary was the creation of a third, 'town–country' magnet that might attract by marrying together the best qualities of town and country and shedding their worst characteristics. The underlying conception saw Garden Cities as an expression of this spirit, as 'true utopia(s)' in which it was possible to realize 'all material, moral and spiritual values' (Hall et al., 1973: 103). Howard proposed limiting each Garden City to 32,000 people. It was suggested that they would be built as comprehensive units on greenfield sites, at a distance from any parent city, and separated from it by an inviolable green belt. Each Garden City would have its own industry and a full range of services. Development would take place in single land-use zones, using a unified system of land ownership and long leaseholds in an attempt to rationalize public interest and personal choice.

Howard had no commitment to any specific urban form, but provided diagrams of a circular city with a radius of 1,240 yards. Features of particular interest included a central park, around which were placed civic buildings and a 'crystal palace' shopping arcade. Six wide avenues divided the town into equal-sized wards. The residential neighbourhoods would have roughly 20 dwelling units to the acre, but overall urban densities would be around eight to the acre. Howard intended the Garden City to maintain close links with the surrounding countryside, which would provide leisure opportunities and a source of produce. Further, it was suggested that Garden Cities should be constructed at suitable distances from each other – closely connected by 'easy, rapid, and cheap communication' that would allow people to enjoy the 'higher forms of corporate life' while dwelling in 'a region of pure air' close to the countryside. The result would be a 'carefully-planned', polynucleated cluster of towns known as the 'Social City' (Howard, 1898: 131).

This has perhaps proved the most seriously neglected aspect of *To-morrow*. Howard outlined the possibility of Social Cities themselves aggregating into larger entities. Indeed, he argued that they could proliferate almost without limit, becoming the basic settlement form. The fact that editors replaced this diagram with a truncated version in subsequent publications of Howard's book meant that most readers failed to grasp that the Social City, rather than the isolated Garden

City, was the physical realization of the 'third magnet' (Hall and Ward, 1998: 25). The piecemeal development of Garden City schemes before 1945 reinforced this misinterpretation. Relying on limited supplies of private funding and enterprise, Garden City schemes were primarily isolated schemes, as with the two prototype English Garden Cities at Letchworth and Welwyn and the Garden Suburb schemes such as Brentham and Hampstead (England), Margarethenhöhe and Hellerau in Germany, or Forest Hills Gardens (New York). They indicated possibilities by providing contrast, but these abbreviated versions of the original conceptions were essentially places set apart. They did little to dispel their founders' sense of disdain for the cities of their time or to convey the impression that they might act as progenitors of a new urbanism.

The Modern Movement

Extract 4.2: From Le Corbusier (1967) *The Radiant City*, London: Faber and Faber, p. 94.

The problem is to create the Radiant City. … I shall explain the plan for this city, and the explanation will be neither literary nor an approximation. It will be technical and rigorously precise.

The general characteristics of the plan are as follows: the city (a large city, a capital) is much less spread out than the present one; the distances within it are therefore shorter, which means more rest and more energy available for work every day. There are no suburbs or dormitory towns; this means an immediate solution to the transportation crisis that has been forced upon us by the paradox of the city and garden cities. The garden city is a pre-machine-age utopia.

The population density of the new city will be from three to six times greater than the idealistic, ruinous and inoperative figures recommended by urban authorities still imbued with romantic ideology. This new intensification of population density thus becomes the financial justification for our enterprise: *it increases the value of the ground*. The pedestrian never meets a vehicle inside the city. The mechanical transportation network is an entirely new organ, a separate entity. The ground level (the *earth*) belongs entirely to the pedestrian. The 'street' as we know it now has disappeared. All the various sporting activities take place directly outside people's homes, in the midst of parks – trees, lawns, lakes. The city is entirely green; *it is a Green City*. Not one inhabitant occupies a room without sunlight; everyone looks out on trees and sky. The keystone of the theory behind this city is the *liberty of the individual*. Its aim is to create respect for that liberty, to bring it to an authentic fruition, to destroy our present slavery. The restitution of every individual's personal liberty. Waste will also have its throat cut. The cost of living will come down. The new city will break the shackles of poverty in which the old city has been keeping us chained.

Its growth is assured. It is the Radiant City. A gift to all of us from modern technology. Those are the outlines of this new city. And I intend to fill in those outlines later, down to the smallest detail.

Unlike the Garden City Movement, the Modern Movement had no single point of origin to parallel Howard's *To-morrow*. The work of science-fiction illustrators and filmmakers foreshadowed some of its ideas and its component imagery came from a widely scattered group of architects. German architects like Bruno Taut and Ludwig Hilberseimer, the Swiss architect Le Corbusier, the Russian urban decentralists and the Italian Futurists Antonio Sant'Elia and Mario Chiattone all contributed images of urban futures that embraced new technologies of building and traffic management, but varied profoundly in design and aesthetics. Some were compact schemes that capitalized on the dynamics of metropolitan growth; others were decentralist schemes that exploited the potential of mass transportation. The 'city of towers' (*la ville-tour*), a geometrically arranged and high-density agglomeration with a pronounced vertical dimension, exemplified the former, with 'linear cities' arranged in corridors along high-speed transport lines illustrating the latter. Almost all these schemes were experimental statements; few regarded them as blueprints intended for immediate replication.

The *Ville Radieuse* (Radiant City), often cited as a representative example of a modernist utopia (e.g. Jacobs, 1958; Hall, 2002), was Le Corbusier's most complete inter-war city plan. Its rationale stemmed partly from the challenge of the Garden City as a 'pre-machine-age utopia' (see Extract 4.2), and partly from further development of schemes that Le Corbusier had pursued since the early 1920s. His *Ville Contemporaine* ('Contemporary City for Three Million', 1922), for instance, offered an exuberant sense of scale and degree of centralization, espousing space, speed, mass production and efficient organization. The use of the vertical dimension, especially with its 60-storey office buildings in the centre, freed the ground, allowing parkland to penetrate close to the heart of the city. Although Le Corbusier worked on other schemes that developed the 'city of towers' principle further, he also worked on linear city schemes that reconfigured the functions of the city in corridors centred on high-speed transport routes.

The *Ville Radieuse* (1929–30) was a prime example of the latter. Le Corbusier (1967: 170) envisaged a settlement arranged along a spinal transport corridor, roughly 8 miles long and 3.75 miles wide. Like earlier projects, it set the city in a park and featured strict functional differentiation of land-uses and a three-dimensional approach, segregating pedestrians and vehicles vertically as well as in different movement systems. Le Corbusier responded to criticisms that previous schemes lacked the possibility for organic growth, particularly for the residential sector, by placing housing districts in the central part of the city. Changes in the arrangement of land-use, with the business and industrial districts placed either side of the residential area, were also intended to reduce internal travel times. Unlike earlier schemes, which accepted the existing class-based conception of life by housing different socio-economic classes separately, Le Corbusier now added a sociological dimension that placed his work firmly in the utopian camp. He abandoned the idea of segregation of housing by social class by advocating undifferentiated high-rise dwelling units (*unités*) for 2,700 people. Each would have services that included crèches, communal kitchens, shops and gymnasia.

The guiding rule for housing allocation was family size rather than the worker's place in the industrial hierarchy. These housing units were envisaged as an essential ingredient in constructing a 'classless society'.

In characteristic fashion, this scheme contained imagery derived from the preferences of the creator. Although Le Corbusier intended this imagery to show concern for living conditions in the future city and to soften previous suggestions that the quality of life was secondary to the needs of production (Frampton, 1980: 155), he faced a dilemma. He realized the envisaged benefits would come about only if individuals willingly embraced the living patterns of the new urban utopia, but saw residents as having little say in the shape of a future that only awaited 'a "yes" from a government with the will and the determination to see it through!' (Le Corbusier, 1967: 94). He therefore set about building the case that architecture itself would play a socially transformative role. His arguments replayed 1920s German discussion (e.g. Pehnt, 1995; Ward, 2001) that new enlightened beings would emerge from the radically changed environment. These people, for example, would unswervingly favour leisure pastimes that worked to improve social cohesion and generated land-use demands that, as if by chance, did not conflict with the city plan. They inherently preferred communal, disinterested projects 'achieved by the harmonious grouping of creative impulses directed toward the public good' or team sports designed to enhance individual health and social harmony (Le Corbusier, 1967: 65). Whether this had any known basis in real-world user preferences was immaterial. Le Corbusier was simply availing himself of the *tabula rasa* – that favoured device of utopians – remaking the world in the manner of his choosing and freely deciding which activities were socially productive and which pursuits were not.

Constellations

Using Howard's writings and Le Corbusier's Radiant City to identify these two utopian movements and introduce their thinking is inevitably only a starting point. Over time, the Garden City and Modern Movements both developed *constellations* of design ideas and sociological precepts. The word 'constellations' is significant here. It suggests identifying groups of utopian ideas 'with reasonably well marked time–space parameters' that collectively form a recognizable pattern (Manuel and Manuel, 1979: 13). Garden City thinking developed steadily through the experience of the early private Garden Cities and suburbs (see above). American practice contributed the key ideas of using cellular neighbourhood units as the building blocks for new settlements and of arranging housing layouts to separate pedestrian and vehicular circulation systems (Buder, 1990; Fishman, 1992). However, developments such as Radburn (New Jersey) – the 'garden city for the motor age' started in 1928 – made the underlying ideas palatable to Americans by stripping out the collectivist principles underpinning earlier English schemes (Macy and Bonnemaison, 2003: 172). The movement also absorbed a preference for picturesque neo-vernacular style for buildings, derived from links with the English Arts and Crafts Movement.

For their part, the Modern Movement voraciously accumulated new ideas from practice in Europe, the USA and Latin America. Perhaps the most important forum for development of ideas about planning came in the form of the Congrés Internationaux d'Architecture Moderne (CIAM). Founded in 1928 as an international body for the discussion and dissemination of modern architecture, the Congresses and working committees of the CIAM consistently asserted the architects' right of involvement in city planning (Gold, 1998). Although far from being harmonious or single-minded, it promoted a style of functional land-use analysis that tried to translate the 'revolutionary premises of work, housing, transportation and recreation' into planning principles (Holston, 1998: 43). Although far less successful in dealing with the vexed question of town-centre design (see below), the CIAM was closely associated with advocating design principles that rejected the traditional street (with its supposed tensions) in favour of functionally rational segregation of pedestrians and vehicles (Talen, 2005: 51).

This constellatory character was a vital part of the utopian vision. In the first place, it made the imagery malleable. It allowed the accumulation of design and sociological principles in support of a broad vision of the future city without necessarily drawing attention to inconsistencies in the positions of different theorists. Second, it produced visions of the future that were self-insulated from criticism. Those who developed such visions claimed their proposals derived from the application of rational and moral principles to the needs of society. Criticisms of the design of the future city could always be answered by reference to the needs of the future urban society; needs the designers themselves specified. Seen in this way, the sociological precepts offered by these movements appear as justifications used to legitimize cherished ideas about design. Only the lessons of experience would undermine these initial premises and open up the fault lines of previously unrecognized divergences and incongruities (Gold, 1997: 231–3).

Building the modern world

At first glance, the challenges facing urban policy-makers in 1945 should have contributed to making this a supreme utopian moment. Taking the UK as an example, policy-makers found themselves faced with the immediate tasks of initiating reconstruction to rectify bomb damage, dealing with the housing shortage, coping with demographic increase, restarting slum clearance, rebuilding the shattered economy, and addressing the deep-seated problems left by inadequate pre-war city planning. Resource availability, however, effectively dictated that comprehensive approaches had to await the return of more affluent times. Instead, the immediate future held necessity-driven programmes for housing, schools and factories. The main exception was the New Towns programme, initiated by the Labour government to the surprise of many given that it was not an election promise. The legislation raced through parliamentary procedures since it enjoyed all-party support but also benefited from the 'careful groundwork that had been laid in the war years' (Hardy, 1991: 282). The Town and Country Planning Association (the successors to the Garden City Association) and its key supporters had lobbied tirelessly for the implementation of Garden City ideals. In some respects, the New Towns' connection with Howard's ideals was more symbolic than substantive, with careful publicity tending to depict them as places set apart even though most were receptacles for planned overspill from the major cities (Gold and Ward, 1997). Despite this, they took enough from the Garden City Movement in the terms of structure, layout, density and architecture to substantiate their claims to be the embodiment of the utopian ideal.

Urban reconstruction presented a different picture. The housing drive by the local authorities reached its peak in 1953 with approval for construction of over 195,000 new dwellings in England and Wales, but conventional houses still represented more than three-quarters of the total (Layton, 1961: 16). While use of flats grew steadily from this point, only a short-lived wave in the 1960s brought any extensive use of the type of system-built, high-rise, flatted estates often taken as quintessentially linked to modernist town planning. Town-centre renewal followed a similar chronology. With a few exceptions, of which Coventry was the most notable (Tiratsoo et al., 2002), redevelopment and town-centre renewal scarcely started before the official end of the 15-year period of building controls in November 1954. This spurred building activity but primarily by creating a 'developer-friendly system' that primarily benefited the private sector, giving developers sufficient confidence in the longer-term outlook to commence speculative redevelopment of office blocks and commercial premises (Ward, 2004: 131). Full-scale local authority involvement awaited the early 1960s when the combination of increased funding and the return of modernizing vigour after the disappointments of the last 15 years saw city after city unleash ambitious renewal programmes that indelibly changed the appearance, skyline and texture of the urban environment. Given that modernism supplied the visual language for urban renewal, it seemed the wider utopian predispositions of planners and architects had also triumphed (Gold, 2007).

Yet it is dangerous to depict the rebuilding of modern British cities as simply the product of imposing utopian visions. Part of the reason lay in the passage of time before giving serious attention to town-centre renewal, which had foreclosed options in many towns. The property boom had seen the piecemeal accumulation of office blocks, often in superficially modernistic styles which paid little attention to the wider townscape. Only in rare instances was there any serious resistance. Road authorities, for their part, created highways without close coordination with plans for the built environment. In Birmingham, for instance, the city's engineers pushed forward the long-awaited Inner Ring Road in 1956–7, deliberately choosing not to adopt a plan for the city centre in case its stipulations put off the commercial sector. Instead, they sought to capitalize by offering potential developers sites along the line of the Ring Road. This led, for example, to the Smallbrook

Ringway, one of Britain's first urban motorways, becoming lined with continuous ribbons of offices and shops – an extraordinary negation of the principles of pedestrian–vehicle segregation so cherished within modernist circles. Moreover, the new road was a fait accompli that bore little attention to changing needs. Hence, at the Bull Ring shopping centre and later around Paradise Circus, the new Inner Ring Road neatly bisected important civic spaces, showing how, in the absence of a coordinated view, it was often not a ring road at all (Higgott, 2000: 158).

This saga was not atypical. A prevailing view held that, with suitable surgery, the older urban cores could accommodate modern traffic and communications flows on an indefinite basis (Hamer, 2000). This view enjoyed powerful support from the road lobby, a powerful coalition of government, industry and the civil engineering and surveying professions, whose members were firmly ensconced in local authorities and who already enjoyed considerable influence over the planning process. Their preferences for accommodating demand through large-scale road construction helped to make advanced road building, often with pedestrian–vehicle segregation, the sine qua non of modernity.

The built and unbuildable

Extract 4.3: From LCC (London County Council) (1961) *The Planning of a New Town: Data and Design for a New Town of 100,000 at Hook, Hampshire*, London: London County Council, p. 53.

The central area at Hook must inevitably be a complex mechanism. It is planned to provide the main focus of the town's social life and be the centre of specialised amenities and services for the local and surrounding population. Difficult problems of servicing, of the delivery and dispatch of goods, of people arriving and departing by bus, car or on foot, and moving about from one part of the central area to another must be resolved. The needs of these various functions could best be met if pedestrian and vehicular traffic were completely segregated. This could only be achieved in the central area by keeping all pedestrian movement on a platform, raised to a sufficient height to enable a network of roads to service the area from beneath and to cover the necessary car parks.

From this network underneath, shops, offices and public buildings could be serviced by means of hoists and lifts adjacent to the service roads; cars could be parked, and their occupants, together with passengers arriving by bus, could reach the pedestrian platform above by means of staircases and escalators. The platform itself would thus cover the network of roads (except for openings for access, light and ventilation), and provide a traffic-free 'deck' to serve as the new 'ground floor' for the inner core of the central area. A series of intermediate levels, gradually decreasing in height, could link this platform, via the higher density housing surrounding it, with the pedestrian ways in the housing layouts outside the central area.

(Continued)

Thus the conception of the pedestrian platform or deck would enable pedestrian movement to be planned without the need to equate it with design-speeds and turning-circles of vehicles – movements which have long proved incompatible and which are responsible for so much of the chaos in our towns and cities today. It would furthermore enable the bulk of the town's population to live within a half-mile walking distance of the central area without an intervening ring of car parking and servicing, thus making possible a close degree of integration between the central area and the remainder of the town.

The publication of the report (LCC, 1961) on the London County Council's proposed New Town at Hook (Hampshire) aroused great interest. Here, it seemed, was a project that combined Garden City–New Town tradition with sensitivity to modern design. Hook took the spirit of town planning from the Garden City without the low-density residential neighbourhoods, cottage-like housing and semi-public open spaces found in the Mark I New Towns. From the Modern Movement, it derived building design, greater clustering to make it effectively a walking-scale city, and a functionally organized road system. At its heart lay a multilevel town centre intended to serve as a genuine focus for the gathering community, featuring shops and amenities placed on a pedestrian deck with cars and servicing beneath (Figure 4.1). These ideas of combining motor car usage and town-centre functions within the same space by vertical segregation aroused enormous interest. Not surprisingly, for some years to come, coaches brought visitors from continental Europe and North America to inspect the progress at Hook.

Their journeys, however, were fruitless. Had their organizers read the report's Introduction more carefully, they would have found that local authority politics had intervened. Faced by resistance from Hampshire County Council and the displeasure of Government, the LCC had abruptly cancelled the New Town project in 1960 in favour of 'Expanded Town' (Town Development) schemes at Basingstoke, Tadley and Andover. No construction work ever materialized at Hook, with the LCC only publishing the report to prevent the complete wastage of the development work. Despite becoming the LCC's most successful publication with three reprintings, the Hook study testified to the flimsiness of vision rendered obsolete overnight by changes in political realities.

Other touring parties learned rather more from visiting the two major influences on Hook, namely, Vällingby (Sweden) and Cumbernauld (central Scotland). Unlike Hook, Vällingby and Cumbernauld were no mirages. Situated in a suburban district nine miles west of central Stockholm, the city authorities initiated the New Town project at Vällingby in 1954. Developed on four square miles of farmland purchased in 1930, the project proceeded on the so-called ABC (*Arbete–Bostad–Centrum* or Work–Dwelling–Centre) principle. As such, the plans

Figure 4.1   Hook New Town

not only envisaged accommodating 23,000 new residents, but they also provided work, shopping and service facilities for them and a surrounding population of 60,000 people (Guerin, 1958: 444). A railway provided rapid connection with the cultural and retailing sectors of central Stockholm. The fundamental principle employed impressed several of the leaders of the Hook project as creating a settlement 'big and dense enough' to be 'an effective and attractive alternative and counter-magnet to the central area' (Chamberlin et al., 1958: 347). In contrast, Vällingby's town centre, built on a deck over the railway station, received less praise. While incorporating the favoured deck principle, *inter alia* they criticized the incomplete pedestrian–vehicle segregation, with car parking occupying a large part of the 'useful' central area (Chamberlain et al., 1958: 331).

Vällingby, of course, was a suburb. Cumbernauld, in contrast, was a full town. Designed virtually from scratch in a sparsely populated area around 15 miles north-east of Glasgow in a region pockmarked by the remains of mineral workings, Cumbernauld had set the trajectory of British New Towns away from neighbourhood units and cottage architecture of the first generation. Designated in 1955, its plan envisaging a road pattern capable of handling large throughputs of traffic to create a town *engineered* – a verb commonly applied at the time – for the motor age. The New Town developed on a compact and nucleated basis around the hill capped by the town centre, which, as at Hook, was its most notable feature. A conscious break from the high streets and one-level pedestrian precincts of the Mark I New Towns, its designers chose the grandiloquent gesture of a megastructural edifice running for a half-mile along the ridge of the central hill. Rising seven-storeys at its highest point, with a spinal high-density road passing beneath, it combined a ready embrace of modernity with the symbolic appearance of a medieval Italian hilltop citadel (Gold, 2006).

At the outset, all seemed well. Cumbernauld's innovative planning and architecture won awards, including the town being the first recipient of the Institute of American Architects' R.H. Reynolds Memorial Award for Community Architecture – presented at the opening of the first phase of the town centre in 1967. Initially, at least, its innovative designs also earned widespread praise from professional observers (Gold, 2007). Yet problems soon arose that illustrate wider themes in twentieth-century planning and utopianism. With regard to the town generally, constructional and economic problems blighted both the housing design and the provision of infrastructure. Political decisions altered the growth targets of the town and drastically constricted the flow of resources to construction projects. The limitations of the site, especially the steep terrain, prevented the scheme fulfilling its potential as a walking-scale city.

Yet perhaps the greatest problems occurred with the Central Area, the town's showpiece. The decision to build the town centre as a megastructure – a multistorey structure that brought together a desired mix of complementary land-uses and activities within a single integrated framework – placed Cumbernauld at the forefront of architectural innovation (Banham, 1976). Yet only the first two of its five phases of construction proceeded according to the original design, as the project ran into a tangled web of design, structural and commercial problems. Some of its problems lay at the door of the architects and planners, because there were many alternatives to building a town centre in this manner. They certainly failed to appreciate the problems associated with its experimental construction. These included poor understanding of unfaced concrete as a building material, naïve faith in the quality of work the construction industry could offer, failure to come to terms with microclimate, inadequate regard for the needs of the users and the inflexibility of the design. The centre's Brutalist concrete carapace turned out to be inimical to change, the opposite of the modular flexibility and extensibility then claimed for megastructures. Nevertheless, there were other problems not directly attributable to the designers. Political pressures led to cost-cutting that reduced the quality of materials and finishes. The completion of the centre in piecemeal and uncoordinated portions made nonsense of the original conception. The contractor suffered bankruptcy and left no records of major elements of design, such as with regard to the steel reinforcement in the main concrete structures. Vandalism thrived, given the lack of security patrols in the pedestrianized centre, especially at night. With changing aesthetics and commercial trends, potential developers found the centre unappealing, internally and externally (Gold, 2006: 121–3).

Critique and backlash

What will the projects look like? They will be spacious, parklike, and uncrowded. They will feature long green vistas. They will be stable and symmetrical and orderly. They will be clean, impressive, and monumental. They will have all the attributes of a well-kept, dignified cemetery. (Jacobs, 1958: 157)

Jane Jacobs' words indicate that the mounting critique that emerged in the late 1960s and 1970s had antecedents. For her part, Jacobs, an associate editor of the *Architectural Record*, was one of the first to take serious issue with prevailing approaches to urban renewal. Influenced by her experience of living in New York's West Greenwich Village, then threatened by street widening and renewal proposals, Jacobs warned about the implications of a predominantly visual-artistic approach to urban redevelopment that failed to take account of the city's vitality, complexity and intensity (Punter and Carmona, 1997). Her subsequent book, *The Death and Life of Great American Cities* (Jacobs, 1961), celebrated the virtues of the older mixed land-use neighbourhoods of New York and similar American cities, championing them against an orthodoxy that favoured clearance. In particular, she attacked the *utopianism* of renewal strategies. Jacobs argued the City Beautiful, Garden City and Modern (or Radiant City) Movements – the three key movements that had influenced American city planning – had roots in idealis-tic thinking that made little contact with the realities of the 'workaday city'. Although disagreeing about many aspects of city design, they collectively saw little worth in the traditional city as a suitable home for modern society: 'Unstudied, unrespected', cities served as 'sacrificial victims' in the search for something better (Jacobs, 1961: 25). These movements also broadly agreed about the need for a comprehensive approach to development, functional differentiation of land-use, urban hierarchy, and the link between progress and modernity. The result effec-tively fused the principles and imageries of these movements into an amalgam she called the 'Radiant Garden City' (Jacobs, 1961: 59).

Jacobs' arguments initially drew scornful rebukes from those who were their targets (e.g. Hughes, 1971), but her analysis foreshadowed a wave of criticism of post-war urban renewal policies that effectively amounted to a 'moral revolution' (Esher, 1981: 72). Much of it blamed planners, architects, and the utopian move-ments to which they supposedly subscribed for the failings of those policies. This sustained and multifaceted attack had a substantial impact that deeply wounded the reputation of town planning and architecture. However, the preoccupation with identifying culprits and with constructing simplistic chains of cause-and-effect distorted understanding of processes of urban change. Visionary architec-ture and uncompromising plan-making became decontextualized, somehow endowed with mesmeric qualities capable of leading naïve decision-makers astray. The planners and architects who were authors of those schemes appeared as autonomous actors divorced from the social, political and cultural environ-ments of which they were part.

Inevitably, the question of responsibility arose. It is easy to place planners and architects firmly in the dock, but the situation was frequently more complex than appears from first glance. Planners and architects were undoubtedly complicit in encouraging aspects of urban transformation that they considered positive, which frequently served to advance their professional interests. Yet while they could use their 'professional' knowledge to telling effect, their freedom of action and their input into the decision-making process, even as expert advisers, were often limited by the financial regimes, purchasing policies and organizational structures that surrounded the UK plan-making process. Economies introduced by local authorities or by governmental stringencies, especially in the field of housing, often stripped away elements that were important to the social functioning of estates, such as the provision of landscaping or caretakers. The result was modernist aesthetics without proper attention to the social context in which the architecture operated.

Inter-professional rivalries also played their part. Architects, in particular, struggled to make an impact against better-placed professional rivals. Despite presumptions that they were the utopian authors of the vision shaping urban reconstruction, many major towns and cities lacked a Chief Architect until well into the 1960s. Architects employed by UK local authorities tended to find that they lacked powers outside the field of housing and, even there, frequently found their remit confined to building design rather than having influence over wider issues of site selection and road planning. Such was their marginalization in town-centre renewal that Colin Buchanan, whose Committee's Report provided the apogee of the view that cities needed to adapt to the motor car (Buchanan Report, 1963), felt the need to offer a voice of support. A qualified architect and town planner, Buchanan suggested that existing approaches in the early 1960s were too narrow:

> What we are really up against is a problem of architecture, and, that unless we realise this, we run the risk of doing serious damage to our towns and cities. I hasten to make it clear that I am in no way trying to slight civil engineers, to run down traffic engineers, or to boost architects … but I think it absolutely essential to grasp the true nature of the problem and then to distinguish the various contributions that the professions can offer. (Buchanan, 1962: 2)

Arguing for a more pluralistic approach to the relationship between vision and implementation than commonly adopted, of course, does not mean architects or their counterparts in planning were blameless for the problems that emerged from ill-judged town-centre redevelopments, somehow passive victims of a process beyond their control. Rather, it emphasizes that their culpability varied over time, between sectors of constructional activity, and from town to town. It also recognizes that design professionals operated within an unprecedented social and political environment that willingly supported the idea that radical measures were necessary to solve pervasive urban problems. In such circumstances, it is not surprising that recourse to readily available and fervently advocated utopian approaches should flourish; it would have been more surprising if they had not.

Summary

This chapter has considered a number of plans for utopian settlements and revealed the gap between vision and reality, which is painfully evident in a litany of failed dreams. In the process, we may discern important lessons about the anatomy of urban utopian thought. The Garden City and Modern Movements were essentially broad churches that attracted people with very different views but with a similar propensity to appropriate the pristine imageries of shining new cities as an expression of the modernity and social progress of their age. It was only subsequently that the lessons of practice revealed the raw assumptions, inexperience and inconsistencies that were encapsulated in the plans, models and diagrams.

The strategy for town-centre development illustrates this point admirably. The Garden City Movement produced no specific model for town centres. Howard originally placed a central park at their core and his successors had not arrived at prototypes that met universal support. The Modern Movement had also blatantly failed to resolve the question of how to devise attractive town centres that gave expression to the constitutive ideas of a community or to the social structure (Norberg-Schulz, 1961: 17). 'Classical' functionalism had produced too narrow a definition of the building task to discern the necessary features of liveable town centres and there was unease about the type of 'new monumentality' that such areas might require (Giedion, 1944; Various Authors, 1984). Failure to reach any agreement about the problems of the 'heart of the city' at the eighth Congress of CIAM in 1951 marked the start of the dissolution of that organization. Strategies for renewal, therefore, proceeded on thin conceptual foundations and an even thinner body of practice. Most radical attempts to carve apart the existing town centres in search of visionary futures, especially when dealing with long-established towns and cities, created more problems than they cured. Efforts to turn to largely untried prototypes to fill the gap fared little better, as the experience of Cumbernauld's Central Area suggests.

To revisit the utopian plans of the early post-war period, then, is to invoke the spirit of a time in which planners, architects, engineers and politicians believed passionately in a future achievable through progress and technological advance, even if we now know enthusiasms were often misplaced. Read with hindsight, the plans for Hook (see Figure 4.1) suggest the inadvertent replacement of utopia by dystopia. Despite Hook's location, the artist provides an image that contrived to reassemble an impressive range of urban problems in the rolling fields of Hampshire – a portrait of a subterranean, fume-laden and noisy environment polluted by idling buses. The café tucked away in the darkness of a maze of concrete ramps and escalators seems a uniquely unpromising attraction. To describe the centre in this way, however, robs the illustration of its original visual power. In 1961, this was a simple evocation of the principle of vertical pedestrian–vehicular segregation, a much-heralded and widely approved way of returning city space to its residents safe from the hazards of cars. It now takes an effort of will to make that conceptual leap, but it is essential if urban scholars are to understand the dynamics of a period that brought unprecedented change to the contemporary urban environment – and left a legacy with which planners are struggling to come to terms.

References

Banham, P.R. (1976) *Megastructure: urban futures of the recent past*, London: Thames and Hudson.

Buchanan, C.D. (1962) 'Comprehensive redevelopment: the opportunity for traffic', in T.E.H. Williams (ed.), *Urban Survival and Traffic*, London: E. and F.N. Spon.

Buchanan Report (1963) *Traffic in Towns: a study of the long-term problems of traffic in towns*, London: HMSO.

Buder, S. (1990) *Visionaries and Planners: the Garden City Movement and the modern community*, Oxford: Oxford University Press.

Calvino, I. (1974) *Invisible Cities*, London: Martin Secker and Warburg.

Chamberlin, P., Powell, G., Bon, C., Shankland, G., Gregory Jones, D. and Millett, F. (1958) 'The living suburb', *Architecture and Building*, 33: 323–62.

Cherry, G.E. (1988) *Cities and Plans: the shaping of urban Britain in the nineteenth and twentieth centuries*, London: Edward Arnold.

Eaton, R. (2002) *Ideal Cities: utopianism and the (un)built environment*, London: Thames and Hudson.

Eliav-Feldon, M. (1982) *Realistic Utopias: the ideal imaginary societies of the Renaissance, 1516–1630*, Oxford: Oxford University Press.

Esher, L. (1981) *A Broken Wave: the rebuilding of England, 1940–1980*, London: Allen Lane.

Fishman, R. (1992) 'The American Garden City: still relevant?', in S.V. Ward (ed.), *The Garden City: past, present and future*, London: E. and F.N. Spon.

Frampton, K. (1980) *Modern Architecture: a critical history*, London: Thames and Hudson.

Giedion, S. (1944) 'The need for a new monumentality', in P. Zucher (ed.), *New Architecture and City Planning*, New York: Philosophical Library.

Gold, J.R. (1997) *The Experience of Modernism: modern architects and the future city, 1928–1953*, London: E. and F.N. Spon.

Gold, J.R. (1998) 'Creating the Charter of Athens: CIAM and the functional city, 1933–43', *Town Planning Review*, 69: 221–43.

Gold, J.R. (2006) 'The making of a megastructure: architectural modernism, town planning and Cumbernauld's Central Area, 1955–75', *Planning Perspectives*, 21 (1): 109–31.

Gold, J.R. (2007) *The Practice of Modernism: modern architects and urban transformation, 1954–72*, London: Routledge.

Gold, J.R. and Ward, S.V. (1997) 'Of plans and planners: documentary film and the challenge of the urban future, 1935–52', in D.B. Clarke (ed.), *The Cinematic City*, London: Routledge.

Guerin, J.J. (1958) 'Vällingby', *Architecture and Building*, 33: 444–64.

Hall, P. (2002) *Cities of Tomorrow: an intellectual history of urban planning and design in the twentieth century* (3rd edition), Oxford: Blackwell.

Hall, P., Gracey, H., Drewett, R. and Thomas, R. (1973) *The Containment of Urban England*, Vol. 1: *Urban and Metropolitan Growth Processes, or Megalopolis Defined*, London: George Allen and Unwin.

Hall, P. and Ward, C. (1998) *Sociable Cities: the legacy of Ebenezer Howard*, Chichester: Wiley.

Hamer, D. (2000) 'Planning and heritage: towards integration', in R. Freestone (ed.), *Urban Planning in a Changing World*, London: E. and F.N. Spon.

Hardy, D. (1991) *From Garden Cities to New Towns: campaigning for town and country planning, 1899–1946*, London: E. and F.N. Spon.

Hardy, D. (2000) *Utopian England: community experiments, 1900–1945*, London: E. and F.N. Spon.

Higgott, A. (2000) 'Birmingham: building the modern city', in T. Deckker (ed.), *The Modern City Revisited*, London: Routledge.

Holston, J. (1998) 'Spaces of insurgent citizenship', in L. Sandercock (ed.), *Making the Invisible Visible: a multicultural planning history*, Berkeley: University of California Press.

Howard, E. (1898) *To-morrow: a peaceful path to real reform*, London: Swann, Sonnenschein.

Hughes, M.R. (ed.) (1971) *The Letters of Lewis Mumford and Frederic J. Osborn: a transatlantic dialogue, 1938–70*, Bath: Adams and Dart.

Jacobs, J. (1958) 'Downtown is for people', in W.H. Whyte, Jr. (ed.), *The Exploding Metropolis*, Garden City, NY: Doubleday and Co.

Jacobs, J. (1961) *The Death and Life of Great American Cities*, New York: Random House.

Layton, E. (1961) *Building by Local Authorities: the report of an inquiry by the Royal Institute of Public Administration into the organization of building and maintenance by local authorities in England and Wales*, London: George Allen and Unwin.

LCC (London County Council) (1961) *The Planning of a New Town: data and design for a New Town of 100,000 at Hook, Hampshire*, London: London County Council.

Le Corbusier (1967) *The Radiant City*, London: Faber & Faber.

Macy, C. and Bonnemaison, S. (2003) *Architecture and Nature: creating the American landscape*, London: Routledge.

Manuel, F.E. and Manuel, F.P. (1979) *Utopian Thought in the Western World*, Oxford: Basil Blackwell.

More, T. (1516) *Utopia*. Version used here was E. Surtz and J.H. Hexter (eds) (1965) *The Complete Works of Sir Thomas More*, vol. 4, New Haven, CN: Yale University Press.

Mumford, E. (2000) *The CIAM Discourse on Urbanism, 1928–1960*, Cambridge, MA: MIT Press.

Mumford, L. (1922) *The Story of Utopias, Ideal Commonwealths and Social Myths*, New York: Harrap.

Norberg-Schulz, C. (1961) *Intentions in Architecture*, London: George Allen and Unwin.

Pehnt, J. (1995) 'The "New Man" and the Architecture of the Twenties', in J. Fiedler (ed.), *Social Utopias of the Twenties: Bauhaus, Kibbutz and the dream of the New Man*, Wupperthal: Müller and Busmann Press.

Punter, J. and Carmona, M. (1997) *The Design Dimension of Planning: theory, content and best practice for design policies*, London: E. and F.N. Spon.

Ravetz, A. (1980) *Remaking Cities: contradictions of the recent urban environment*, London: Croom Helm.

Relph, E.C. (1987) *The Modern Urban Landscape*, London: Croom Helm.

Rosenau, H. (1983) *The Ideal City: its architectural evolution in Europe*, (3rd edition), London: Methuen.

Simpson, M.A. (1985) *Thomas Adams and the Modern Planning Movement: Britain, Canada and the US, 1990–1940*, London: Mansell.

Talen, E. (2005) *New Urbanism and American Planning: the conflict of cultures*, New York: Routledge.

Tiratsoo, N., Hasegawa, J., Mason, T. and Matsumura, T. (2002) *Urban Reconstruction in Britain and Japan, 1945–1955: dreams, plans and realities*, Luton: University of Luton Press.

Various Authors (1984) 'Monumentality and the city', *Harvard Architecture Review*, special issue, 4 (Spring): 6–208.

Ward, J. (2001) *Weimar Surfaces: urban visual culture in 1920s Germany*, Berkeley: University of California Press.

Ward, S.V. (ed.) (1992) *The Garden City: past, present and future*, London: E. and F.N. Spon.

Ward, S.V. (2004) *Planning and Urban Change* (2nd edition), London: Sage.

5 MONUMENTS AND MEMORIES

Lisa Benton-Short

This chapter

○ Suggests that monuments and memorials play an important role in the life of cities as sites of civic identification, gathering and celebration

○ Demonstrates that the power of monuments and memorials is related to their location in accessible public spaces

○ Shows that the meaning of monuments and memorials is contigent and contested

---

Introduction

Almost from the time that cities emerged, inhabitants have filled them with statues and monuments. Each generation leaves a repository of civic and/or national memory, something tangible that defines virtues, celebrates heroes and speaks to the great events that built that society. Today, most cities have sites where monuments, memorials and public space mark important places of gathering, celebration, protest and reflection. These sites and places inevitably reflect certain memories and not others and thus are important symbolic messages in the urban landscape.

This chapter considers monuments and memorials in public space as part of the wider urban landscape of commemoration. I draw heavily on my own research on the National Mall in Washington, DC, where, on 21 October 2002, four dozen citizens 'marched' to protest against the siting of the Mall's most recently approved memorial, the World War II Memorial. Because the National Mall is one of America's most symbolically-charged spaces, the location of the new memorial was perceived to dramatically change the use and symbolic meaning of this urban public space. Over the next two years the memorial was hotly debated, lawsuits were filed, but in the end the memorial was built in the planned location, and officially dedicated on Memorial Day in 2004. Although critics lamented its enormity (more than seven acres in size), and criticized both its design and its location, the memorial has become the most visited memorial on the National Mall, with more than five million visitors annually. While the debate about the World War II memorial is over, the legacy of the controversy has generated ongoing public discussion about how to balance open space and memorial space and challenged us to reflect critically on the process of commemoration.

Monuments, memorials and public space

The terms *monuments* and *memorials* are often used interchangeably, although memorials tend to be a response to loss and death, while monuments are often more celebratory in tone. Traditional monuments are created through a formal process that seeks permanence. The creation of traditional monuments involves raising of funds, forming of committees, and selecting of designs, sculptors or architects. In contrast, spontaneous memorials are informal, unplanned offerings that tend to be ephemeral and fragile. Examples include the flowers, candles, messages, posters and teddy bears left at crash sites. Monuments and memorials can be built on private space (an estate, for example), quasi-public space (such as a cemetery or church grounds) or public space (city squares or parks). In cities, *public space* is a place where anyone has a right to come without being excluded because of economic or social conditions. Public space does not typically have an entrance fee or a time limitation. Such spaces could include civic plazas and town squares, city streets, parks and playgrounds and other areas found in the heart of the city used for community gathering (Jacobs, 1961; Johnston et al., 1994; Knox and McCarthy, 2005). While urban public space is defined as space open to all forms of use and expression, in this chapter we are concerned with public space that is also considered commemorative space – either because of the presence of monuments and memorials, or because of events that have shaped the role of a particular public space in collective life (and hence historical memory).

Public space is often understood to play an important role in society as a site for the expression of civil rights and assembly, recreation, open and contemplative space, gathering spots and as sites of memory and memoralization. Public space contributes to and reflects social, economic or political changes. As cultural geographer Don Mitchell notes, public space is crucial to social change because it provides a space for representation (D. Mitchell, 2003). Many consider public space a vital conduit for the formation of citizenship and broader discussion on civil society, national identity and memory. Because so many of the most important and visited monuments and memorials are located in urban public space, recent challenges and restrictions to public space are worrisome to those who have analyzed the historically important role of public space, particularly for those marginalized. Scholars have also described the decline of urban public space at the expense of the rise of more privatized spaces that don the carnival mask or provide a theme park environment (Sorkin, 1992) (see also Chapter 1). Not all public space is equally important, however. Some public space serves as an active site from which to challenge and confront issues such as national identity, citizenship, freedom, democracy and justice. These types of public space allow 'active citizenship' and are invaluable. Active citizenship includes political protests, national celebrations and public gatherings. Active participation seeks to include rather than exclude 'undesirables' and it often attempts to extend or reinforce notions of citizenship and democracy.

Some of the most important active public spaces are also memorial landscapes, which because of their large expanse can accommodate large groups. These are public spaces that play dual roles: memorialization and civic engagement. Such examples include the Mall in Washington, DC, Tiananmen Square in Beijing, The Mall leading to Buckingham Palace in London, the Grande Axis of Paris and Red Square in Moscow. Certain public spaces, like the National Mall or Red Square in Moscow also play key roles in representing nations. These examples are simultaneously public space and landscapes of commemoration. These multi-layered spaces usually have more than one monument or memorial and encompass vast acres of space for both the location of memorials and contemplation. As a result, we can classify some urban spaces as urban memorial landscapes because they also serve as sites of civic and national pageantry, political celebration and sometimes political protest. The multiple functions of these important urban memorial landscapes add to their symbolic power within the national imagination. A visit to the National Mall in Washington, DC, for example, is considered a civic rite of passage for many school children and their families; exploring the museums, monuments and memorials is one way they become aware of the history and political ideology of the USA.

Defining memorials and monuments

While public art is defined by its presence in public space, its definition is complex. However, public art is often understood (perhaps erroneously) to be synonymous with monuments and memorials. One type focuses on war or battle and implies a belief in the rightness of the cause, in the perceived glory of dying bravely, and in the honour of self-sacrifice. These monuments and memorials often carry messages of nationalism and national identity. A second type of memorial focuses on the individual. These monuments suggest that the individual can make a difference, and that one's contribution to society is worthy of *memorialization*. A third type of monument celebrates technological progress or particular values and beliefs. In general, scholars have tended to focus their studies on traditional monuments and memorials, although the subject of spontaneous memorials has begun to receive some attention.

Debates surrounding monuments and memorials located in urban public space are informed and situated at the intersection of such concepts as iconography,

landscape representation, public memory and the politics of memory and identity. In recent years, scholars have increasingly studied the social production of
national and cultural identities through memorials. Monuments and memorials
have been considered by political geographers examining national identity, or
political identity or by cultural geographers interested in the landscape representation and cultural identity (see Crampton, 2001; Leib, 2002; K Mitchell, 2003).
What many of these studies share, however, is that the monument or memorial
under examination, more often than not, is located in urban space.

Exploring memorials and monuments is accordingly a way to uncover the
dynamics at work in shaping the historical and contemporary landscape. Cultural
geographers Cosgrove and Daniels (1988: 1) describe a landscape as a 'cultural
image, a pictorial way of representing, structuring or symbolizing surroundings'.
Cultural geographers have referred to places in the landscape that symbolize
particular memories and meanings, including messages about power and politics
as landscapes of power (Zukin, 1991). This definition of landscape would certainly include memorials and monuments and sites of commemoration.
Monuments and memorials are imbued with social, cultural and political meaning that speak of the past and present. They are not merely ornamental features
in the urban landscape, but are highly visible signifiers that confer meaning and
thus concretize the politics of power (Whelan, 2002).

For such reasons, monuments and memorials often become the focal point of
conflict over urban space. Memorials are intended, if not explicitly then implicitly, to stimulate debate. The debate often revolves around the interpretation of
history, the meaning of the event or person, and how this meaning should be con-
veyed in the built form. Conflicts over monuments and memorials revolve not
only around the subject and design, but also its location, an inherent geographic
issue. Geographers have shown the location of memorials is equally politicized.
While memory and identity are abstract concepts, there is always a geography to
memory and memorials. Often, the location for a memorial is tied to where the
event occurred or the person was born (or died). In other cases, however, there is
no site-specific connection to events and people commemorated, thus the location
of a memorial is dependent upon how it is interpreted within the broader symbolism of city space. This, too, can be highly politicized and contentious (Leib,
2002). The location of memorials thus provides a glimpse of competing interpretations of memory as well as the power relationships that can ultimately determine its realization.

The role of monuments and memorials is intricately connected to more abstract
concepts of heritage and memory. Geographers and historians who conduct
research in historic preservation and on heritage have convincingly argued that
these concepts are socially constructed (Glassberg, 2001; Graham et al., 2000). The
French historian Pierre Nora, for example, has observed that memory is distinguished from history in that history is more rooted in evidence, while memory is
imagined. The process of constructing a memorial, he notes, activates national self-
consciousnesses; yet memory is also shaped among competing spheres of political

or ideological influence (Nora, 1996). Memory is constantly shifting, and hence the meaning of symbolic sites – such as memorials – takes on new meaning with the passage of time. He notes that the Eiffel Tower originally expressed the symbol of revolutionary modernism, but has since given way in the contemporary realm of memory to become part of the nostalgic architectural 'poetry of Paris' (Nora, 1996: xiii). Nora defines the sites of memory, 'lieux de me moiré', as 'any significant entity, whether material or non-material, which by dint of human will or the work of time has become a symbolic element of the memorial heritage of any community' (Nora, 1996: xvii).

Similar to the literature on memorials and monuments, much of the recent work on heritage and memory suggests that it is 'invented', or socially constructed. Heritage and memory exist because a particular society says an object, event, person or place is valuable enough to ensure it is passed to the next generation. Since heritage, national identity and memory are socially constructed they are also inherently contested (K. Mitchell, 2003). We can refer to this contest as the politics of memory. Those with the power to prevail get to 'write the message' of the monument or memorial; hence we can consider this an expression of power. For example, the use of sculpture, statues, murals and inscriptions and other symbols on and in monuments and public buildings can explicitly convey political messages. Many consider Trafalgar Square in London an exhibition of British imperialism and military might that serves as a focal point of patriotism. The square was laid out between 1829 and 1841 to commemorate Admiral Nelson's victory at the Battle of Trafalgar in 1805. Dominating the square, on a column that is 185 feet high, is the 17-foot statue of Nelson himself. Four giant bronze lions surround the base of the columns. Since its completion, Trafalgar Square has been a favourite meeting place for tourists as well as demonstrators and marchers trying to gain attention to their cause. It is also a site for the contestation of hegemonic values. In September 2005 a 12-foot marble statue of a pregnant woman with no arms was unveiled. The artist, Marc Quinn, said he sculpted his friend, Alison Lapper, because disabled people were under-represented in art, noting that 'Alison's statue could represent a new model of female heroism'.

Some monuments express power through their omnipresent visibility, such as the Statue of Liberty, the towering St Louis Arch, or the Eiffel Tower. In Baghdad, Iraq, Saddam Hussein's regime erected The Victory Arch in 'celebration' of his purported victory over Iran (in reality, the conflict was a stalemate). The Victory Arch consists of two gigantic forearms holding inclined sabres (the forearms were said to be Saddam's own). For the Iraqis, the arch was to serve as the equivalent to the Arc de Triomphe in Paris. Each monument and memorial, therefore, symbolizes the expression of a particular ideology, economic interest or political position. Public statues celebrating and venerating particular political or military leaders, or the naming of public buildings after local or national figures confer an importance to their economic, political or military legacy. In North Korea, the extreme personality cult of Kim-Il-Sung, and more recently of his son and successor, Kim-Jong-Il, pay homage to their leadership and the destiny of the nation through public statuary. In Pyongyang and cities around the country, there are thousands of statues of both men, often of colossal dimensions. Conversely, Atkinson and Cosgrove (1998) have demonstrated the rhetorical meaning of monuments can be subverted over time. They note that the Vittorio Emanuele II monument in Rome, originally designed to express Italy's cultural and political revival, has become an object of irreverent amusement and scornful ridicule (Atkinson and Cosgrove, 1998).

Extract 5.1: From Atkinson, D. and Cosgrove, D. (1998) 'Urban rhetoric and embodied identities: city, nation, and empire at the Vittorio Emanuele II Monument in Rome, 1870–1945', *Annals of the Association of American Geographers*, 88 (1), 28–49, pp. 45–6.

From the perspective of the late twentieth century, the Vittorio Emanuele II monument in Rome may seem initially to offer little more by way of urban aura and meaning than a rather embarrassing reminder of late-nineteenth-century bourgeois bombast. By over-elaborating its iconographic and symbolic context, those who commissioned and designed this memorial to the founding monarch of modern Italy succeeded in producing the very opposite to their rhetorical intention. Rather than challenging the dome of St. Peter's on the urban skyline as a lasting and dignified expression of Italy's cultural and political revival, of the nation's claim to a worthy place among the European powers, and of Rome's claim to recognition as a modern European metropolis, in the postwar era, the Vittoriano became an object of irreverent amusement among Romans and scornful ridicule among architects and urbanists.

... Our close examination of the Vittoriano's evolution, and specifically a willingness to 'read' it within a geographical context of imperial, national, urban, and corporeal rhetoric, yield a complex example of how official rhetoric is concretized and performed in urban space... the case of the Vittoriano in Rome cast light upon the continuing renegotiation of meanings and identities, of histories and memories, that marked the evolution of the modern European capital.

Monuments, memorials and other forms of heritage are created in a social/political context where culture, location, class, power, religion, gender and even sexual orientation will influence what is considered to be worthy of preserving and celebrating (Graham et al., 2000). The literature has revealed that heritage (and by extension a monument or memorial) is potentially subject to controversy, contest and continued negotiation among interested parties. There are always multiple voices and interpretations to the politics of heritage, memory and memorials. Those with power can shape the creation of heritage or the location of a memorial; shifts in power or ideology can introduce new criteria. Each memorial represents a moment in time when a particular vision has captured hegemonic status, albeit briefly (Johnson, 1994). Monuments and memorials are constructed by those with power, those who have won. However, because one side's victories are another side's defeats, one side's heroes are another side's villains, rarely are memorials or monuments uncontroversial.

Accordingly, monuments and memorials, while often permanent structures in space, are actually dynamic over time. Initially, a memorial reflects the power of those who control the design and site-selection process. Rarely is this without some level of controversy, particularly in democratic and open societies. In these early years nearly all memorials generate controversy in part because memorials and monuments will remember some things and forget others; they are partial and selective. Most monuments and memorials have a controversial past. For example, before the Jefferson Memorial was built on the Mall in Washington, DC, people tied themselves to the trees to stop bulldozers from starting work, complaining that it was another example of Greek architecture. Once in place, however, monuments and memorials do not always reflect the ideas, history or memory initially intended. Monuments and memorials can change through time in several ways. First, memorials may be forgotten, as time moves on the event or person recedes in history and current generations fail to know its significance. When this occurs, monuments and memorials may become merely decorative elements of the streetscape – meaningless distractions or disconnected sculpture rather than the powerful signifiers they are meant to be. This is true to many of the hundreds of memorials to generals and other military leaders from the Revolution for Independence or the War of 1812 that sit in street intersections in Washington, DC. Most residents and tourists alike fail to recognize the names of these statuary heroes (this may point more to the limited historical knowledge of many Americans, of course). Second, a monument or memorial may make the transition from controversial to beloved. The 1982 Vietnam Veterans Memorial was initially derided by critics as a bleak tombstone and an inappropriate comment on the war, yet it has become one of the most popular memorials in the USA. Lastly, monuments and memorials may be neglected, or abused, or even removed – a sign of conflict over interpretation or outright rejection. This represents a transition in economic, cultural or political power that seeks to re-interpret or even destroy old symbols. The dynamic character of a monument implies its meaning can shift and hence its relationship to the urban landscape will also transform.

Monuments, memorials and identity

Urban researchers have thus suggested that monuments and memorials in urban space can be 'read' to reveal the expression of social, political and economic power. Dolores Hayden has noted that 'identity is intimately tied to memory ... urban landscapes are storehouses for these memories' (Hayden, 1995: 9). While scholars have observed that the vernacular landscape is encoded with meaning, monuments and memorials are explicitly designed to structure and shape collective memories and historical narratives (Hayden, 1995). Memorials, in particular, can inscribe political power in space, conferring an economic, ideological and political message at many scales. For example, while we may commonly think of national identity in the abstract, it is often in cities where we find the most visible articulation of national history and national ideology as expressed in monuments and memorials. These urban sites sustain the political ideology of certain nations.

Hence, while the discussion of nationalism has tended to focus on the national scale, in recent years political geographers have begun to expose how localities play an important role in the articulation of national identity. Monuments allow a focus on the use of public space in the production of national images. They can serve as sites at which national traditions are invented and situated in the popular imagination and they may embody the ideals of citizenship that are connected to the nation's historical development. Nuala Johnson, for example, has focused on the monumental landscape in an Irish context as it reflected the emergence of a particular articulation of nationalism (Johnson, 1994, 1995). She notes that before the 1850s only two types of statue existed in Dublin: royal monuments and memorials to British military heroes. Both reinforced British rule in Ireland. However, as Irish nationalism grew, so did demands for the representation of Irish heroes. Johnson argues that the debate over who was worthy of a memorial is important for revealing that 'statuary offers a way of understanding nation-building which moves beyond top-down structural analysis to more dialectical conceptualizations' (Johnson, 1995: 57).

Another type of memorial space serves as a sombre reminder of the obligations of citizenship to the nation. In cities around the world, there are often memorials to war and to the dead. Memorials to the dead of the nation document the ultimate sacrifice for loyalty to their nation (Heffernan, 1995). In Paris, the Arc de Triomphe and its 'Eternal Flame' is such an example. Similarly, in the USA, Arlington Cemetery, situated across the Potomac River from Washington, DC, serves as a burial place for veterans of the country's armed forces. It is a 'National Cemetery'. There are more than 250,000 graves at Arlington, and, most are marked by simple, government-issued headstones. The rows resemble soldiers standing at attention, and while most of the gravestones are identical, each represents the life of an individual and hence consequently tells a unique story. Buried there are presidents, including John F. Kennedy, whose grave is marked by an eloquent and modest Eternal Flame, and prominent citizens, such as former Supreme Court Justice Oliver Wendell Holmes, the orator William Jennings Bryan, the boxer Joe Louis and even former slaves. Interestingly, despite a long history of racial discrimination and segregation in US society, there is no racial segregation at Arlington.

There are perhaps no more arresting emblems of nationalism than cenotaphs and tombs of Unknown Soldiers (Anderson, 1983: 9). This is because the Unknown Solider, being no one, could be anyone. Therefore, all families who have lost a loved one whose remains could not be identified could worship at the grave of this unknown. It also represents the more general sacrifices expected of loyal and dutiful citizens. At the heart of Arlington Cemetery sits the Tomb of the Unknown Soldier, where a sentry is kept on duty at all times. During the day, an Old Guard sentinel marches 21 steps in front of the tomb, then stops and stares at the grave site for 21 seconds. This solemn ritual, watched by hundreds of silent tourists, is the equivalent of a 21-gun military salute.

Yet, if monuments and memorials provide clues as to what is important to remember or know about the past, they can often serve as a way to *contest* or challenge national or cultural identity. For example, political geographer Jonathan Leib has examined the debate over a proposed statue to Arthur Ashe on Monument Avenue in Richmond, Virginia (Leib, 2002). By the mid-twentieth century, Monument Avenue had become the focal point of Confederate monuments, parades and celebrations. The monuments and memorials that lined the Avenue celebrated heroes of the 'Lost Cause' of the Civil War. Yet Richmond, home to a significant African-American population, had few monuments to black soldiers or other prominent citizens. The proposed statue to Richmond native Arthur Ashe, tennis star and civil rights activist, generated heated debate over whether Ashe should share the same Avenue with the Confederates. The debate that ensued was partly about redefining this public space to be more inclusive of African Americans and less exclusive as a symbol of Southern white history. His analysis shows that while on the surface the debate was over the location of the statue, the underlying debate was centred on issues of race relations, identity and power in Richmond at the close of the twentieth century (Leib, 2002). Leib concluded that increasingly racial conflict is often played out thorough cultural politics, such as the process of commemoration. As this example illustrates, monuments and memorials can be focal points for contesting identity, history and the use of public space.

The Voortreker Monument in Pretoria, South Africa, has also become a contested symbol. Built in the 1930s, the Voortreker Monument commemorated the Great Trek of the 1830s, the movement of primarily Dutch and German farmers from the coast and into the interior. The migration was triggered by growing discontent with English colonial authority, and particularly by the effort to end slavery. The farmers who became Voortrekers used black slave labour on their farms and saw the domination of one race by another as God's will. The movement inland generated conflict between the Voortrekers and the Zulus who occupied the land. Violent conflict resulted, and in the Battle of Blood River, the Voortrekers, armed with guns, killed some 3,000 Zulus, armed with wooden spears. The Voortrekers interpreted the lopsided victory as an act of divine intervention and confirmation of God's sanction of white dominance over blacks. The Monument, which portrayed these ideas in its design and artwork, became the symbol of apartheid when commissioned by the Afrikaan-dominated National Party. In a post-apartheid South Africa that now seeks reconciliation, it is interesting that the Voortreker Monument has not been torn down or closed. Since a monument's meaning is never fixed, contestations around a monument and the temporal shift in meaning are always possible. Once prevented from visiting the Monument, African blacks may now interpret the Monument as a marker of political and social change, a celebration of a new political and social order in South Africa.

Public space itself can also be a site of contest and civic discourse. The National Mall in Washington, DC, for example, is not only home to various memorials, it is also a national rallying point (see Figure 5.1). The historian Lucy Barber has explored the evolution of the National Mall as a site of public protest (Barber, 2002: 28). She notes that the Mall, particularly the area around the Washington Monument and the Lincoln Memorial, has become an important public space for people to congregate and express ideas. She has commented that 'marching on Washington' has contributed to the development of a broader and more inclusive view of American citizenship and has transformed the capital (and the Mall) from the exclusive domain of politicians into a national stage for Americans. Critical public protests, beginning with the 1894 protest by unemployed civil worker Jacob Coxey, to the Women's suffrage procession in 1913, to the Veterans' March of 1932, to the Civil Rights March in 1963, to various anti-war marches during the 1970s, have made a powerful claim to the public spaces of the capital, and have linked democratic ideals with public space. Some protests succeed in challenging and contesting the economic, social or political status quo. The iconic image of Martin Luther King delivering his 1963 'I Have a Dream' speech from the Lincoln Memorial steps remains powerfully associated with public protest to extend citizenship to African Americans. More recent protests and marches, such as the March for Gay Rights (1993), Million Man March (1995), the Million Mom March (2000), and annual marches such as the March for Life and annual celebrations such as Earth Day, have reaffirmed the importance of the Mall as a place to express visions of national politics and identity (Barber, 2002: 228).

It is also important to remember that the *absence* of memorials speaks volumes about who is marginalized and silenced in a society. Those monuments and memorials that have not been built tell us about what or who is not celebrated, commemorated or acknowledged. In the late twentieth century, we experienced the emergence of a memorial culture (K. Mitchell, 2003). Numerous individuals and groups challenged existing interpretations of history and national memory. In many cases, the impulse to commemorate is part of a broader social reconstruction of national or cultural identity. Seen in a positive light, proposals for memorials represent an expansion of history and identity by including the previously marginalized or ignored groups. In the USA, for example, the demand for new memorials is part of a larger politics of identity that seeks to make memorial spaces more reflective of a multicultural America both in history and today. The addition of more memorials is a process of expanding and acknowledging different contributions and identities that have helped to form history and memory. Because of the high profile of the National Mall as a public space of national significance, many of the demands for memorials involve requests for locations on the Mall. The proposed Martin Luther King, Jr. Memorial (approved but not yet

Figure 5.1 Aerial view of the Washington Monument with the White House in the background

constructed) acknowledges the struggle for civil rights among African Americans, and has been long absent on the Mall; similarly, the newly completed National Museum of the American Indian (also on the National Mall) reflects Native American experiences and history. Women, who still lack a presence on the Mall in the form of statuary, have made their presence known by marching in protest on several occasions, such as the women's suffragist movement and more recently the Million Mom March. Homosexuals have also been moving from the margins to the centre with protest marches and the AIDS Quilt display. The Mall is thus being 're-written' by those who have traditionally been excluded (Native Americans and African Americans, for example). Such important public spaces act as a stage where social processes contest the ideology of nationalism, or seek to include those once silenced. Just as parades and marches can be used by minority groups to contest dominant discourses of community, national identity and national history, demands for memorials or monuments – whether built or not – signal an attempt to represent the previously marginalized.

Public art and memorials can thus serve as a powerful locus for the contestation of identities, and may attract significant controversy and opposition. In this respect, while monuments and memorials are intended to be permanent parts of the urban landscape, they do not always remain. This has been particularly true in many cities in Eastern Europe and the former Soviet Union, where memorials served as powerful propaganda tools for 'remembering', particularly under

Stalin's tyranny. The plethora of statues of Stalin that appeared during his lifetime across the Soviet Union depicted him as a benevolent leader. In the years immediately following the Second World War, there were few monuments commemorating the victims of Nazism or praising the final victory in Western Europe, but cities in Eastern Europe were coerced into 'voluntarily' erecting countless monuments expressing their gratitude to their Soviet liberators. In public plazas, city centres and in other prominent locations, memorials to the 'liberating' force of the Red Army were erected. These monuments generally consisted of a medium-sized obelisk with a red star and a statue of a soldier holding a flag in front of it. In Berlin and Vienna, two cities that the Soviets shared with their Western allies, the complexes built were meant to glorify the wartime sacrifices and the victories of the Soviet army, but also to signal Russia's claim to a large chunk of Europe (Michalski, 1998). By the 1950s, the communist parties of Eastern Europe were firmly entrenched and thus there came the task of paying homage to Stalin. Cities such as Prague, Budapest and Warsaw dedicated great palaces or avenues to Stalin; others erected statues of him. Many of these monuments were not only a panegyric to the great leader, but also served as a tribune from which communist leaders could address the masses or view May Day parades (Michalski, 1998). After Stalin's death, Khrushchev initiated the 'de-Stalinzation' process by gradually dismantling many of the Soviet and East European Stalin statues. Most were gone within a few years; still there remained statues to Stalin, Lenin and Marx in many city centres.

But controlling commemoration and the depiction of history does not necessarily control public memory. Many of the monuments and memorials erected during the communist era were intended to interpret and remind society about the ideals of socialism, nationalism and national identity. Ultimately, these monuments proved an equally powerful symbol of major political and social changes. With the momentous events in Eastern Europe in 1989, followed by the collapse of the Soviet Union in 1991, the process of destruction, displacement and dismantling of symbols became a highly visual way to signal change. City street names were changed, and monuments were sprayed with graffiti or demolished, as revolutionary ideology aimed to destroy symbols of the Soviet past.

The sight of seemingly 'timeless' monuments entering new semantic contexts – the decapitated head of Lenin, for example, the scrawling of graffiti, or the disposal of broken parts – were powerful and provocative. Many of the statues of Stalin, Lenin and Marx were removed. Still others were left standing, but were left to deteriorate. In either case, this was a powerful statement of the new political order. Today, the level of sympathy with or ongoing communist structures can be gauged accurately by the extent to which statues of Lenin or Marx have been preserved (Michalski, 1998).

Most discarded monuments have been destroyed, melted down, or put in storage. Some however, have had more unanticipated fates. The toppled and headless statue of Lenin that once resided in St Petersburg, now sits in Freedom Park in Arlington, Virginia, an outdoor museum celebrating the spirit of freedom and the struggle to protect it. Today, the headless statue of Lenin stands alongside pieces of the Berlin Wall and stones from the Warsaw Ghetto: now they are celebrated as symbols with a different meaning – the effort to overcome tyranny.

Recent trends in commemoration

Currently, urban researchers are not merely casting their gaze back to landscapes and memories of past events, but are also exploring ongoing trends in commemoration and public space, including the rush to commemorate, the commodification of memorial space, and the discarding of memorials. One specific focus has been an interest in spontaneous memorials. Unlike most of the memorials and monuments mentioned thus far, spontaneous memorials are not commissioned, planned by a committee or orchestrated by a government. They appear spontaneously as a first reaction to the unexpected and violent loss of life. They are often an expression of love and loss, grief and anger. They can be a modest expression of grief, such as family and friends who place flowers, candles, stuffed animals and notes on a tree by the roadside, or, as in the case of September 11, a spontaneous memorial that blossoms in thousands of places at the same time, often gathering in civic plazas, public spaces and churches (see Low, 2002; Sturken, 2005). In the case of 9/11, for example, American embassies around the world became the sites of spontaneous memorials in a massive display of emotion. Spontaneous memorials are often ephemeral and fragile; some are short-lived, others might eventually become part of a permanent memorial or provide the template for its location. Consider the following brief examples.

The death of Princess Diana in August 1997 prompted an outpouring of grief expressed in public spaces around the world, and particularly in London. Her death and the emotion it prompted proved to be a turning point in the relationship between the British people and their Royal Family. Thousands of bouquets, as well as toys, candles and photographs, were left at the gates of both Buckingham Palace and Kensington Palace, creating a vast sea of tributes in the days after her death. In Paris, adjacent to the tunnel where Diana was killed, is a hundred-year-old monument with a gold Liberty Flame. This monument was also appropriated as a monument to Diana, in part because of its location near the accident site. The spontaneous monument consists of notes, letters, photos and flowers, which often wash away or deteriorate (see Figure 5.2). Yet new ones continually appear in their place: it has become more than just a fleeting expression, but time will tell if it remains a monument permanently transformed into a memorial for Diana. It will also be interesting to see how the existing memorial is transformed by the spontaneous one to Diana.

Spontaneous memorials articulate a 'community in bereavement' and may be short-lived. But they can also transform public areas into 'sacred spaces', as was the case with Ground Zero in New York City after 9/11. The capacity of spontaneous memorials to render public space sacred makes them an important element

Figure 5.2   Spontaneous memorial to Diana, Princess of Wales, in Paris

in the geography of memorials. In the hours and days after the horrific events on
11 September, families and friends of missing people papered sections of New
York with posters and pictures of their loved ones. This has been referred to as a

'Grief Ritual'. In a *New York Times Magazine* article, writer Marshall Sella traced the evolution to these posters. Initially, they were a frantic effort to gain information; then the posters began to include increasingly detailed physical descriptions, apparently to make identification of the bodies possible. Most of these were hung in areas around Ground Zero, on chain-link fences. In a final evolution, posters began to address the missing people directly as 'good-bye letters'. Ground Zero has become a sacred space not only because it was the site of tragedy and death, but also because it became a place of spontaneous commemoration, which many of the survivors and victims' families want to see transformed into a permanent memorial (Sorkin and Zukin, 2002).

Extract 5.2: From Goldberger, P. (2004) *Up from Zero: Politics, Architecture and the Rebuilding of New York*, New York: Random House, pp. xi–xiv.

There is no instruction manual to tell a city what to do when its tallest buildings are suddenly gone, and there is a void in its heart. There is no road map to lead its officials and its citizens along the route of renewal, no guidebook to help them figure out whether renewal, in fact, is even what they want. When the twin towers of the World Trade Center – the two tallest skyscrapers in New York and each the second tallest in the US – were destroyed on September 11, 2001, there was not only no precedent for dealing with the enormity of the loss, there was no system for figuring out what should happen next. Should the towers be put back? Should they be replaced by a triumphant substitute or by a sober memorial? Or both? Or would it be better if nothing at all were built where these buildings had been? Who should be empowered to decide all of this, and by what means? Should it be done right away, with the wounds raw, or in a few years, when people would feel different about this piece of land? And who should pay for it?

   Although it took less than two hours for the towers for the World Trade Center to collapse, it has taken years to figure out these questions, and some of them still have no answers. It has seemed at various times to have been an architectural debate, a political one, an economic one, and a cultural one. … It was clear that whatever was built on the sixteen acres of Ground Zero was to carry a symbolic weight far greater than that of any other building project of our time. And thus amid complex politics and a widespread sense of passion began the most challenging urban-design problem of the twenty-first century.

Some commentators have thus concluded that contemporary society rushes to commemorate (and that we live in more emotive times). Historically, it was not uncommon for decades to pass before a monument or memorial was erected. The Washington Monument, for example, was completed just prior to the 100th anniversary of Washington's death. Today, weeks after a tragic event, communities may be discussing how best to commemorate or memorialize loss. The rush to commemorate, however, may prove not so much a process of memorialization as of healing, not so much remembrance as honouring. It often occurs when those family and friends who remain try to make meaningless death meaningful by transforming victims into heroes. Some argue that letting time elapse is important because it allows historical perspective as well as a sense of whether the event or individual made a lasting contribution. The seemingly urgent need to plan and construct memorials also raises the more difficult question of whether it is appropriate for survivors and the victims' families to be intimately involved in the commemoration process, or whether this is best left for another generation to do. In the USA, for example, Congress acknowledged the pressure for new monuments and memorials and passed the Commemorative Works Act of 1986. This stipulated that a monument may not be erected on federal public land until at least 25 years had passed since the event took place or the individual had died. This was in recognition that there were increasing demands for new monuments and memorials, often within weeks or months of a death or event.

Despite the new law, the 'rush to commemorate' has not abated. Instead, it has become a 'right' of those affected by the event. Consider the following example: on 19 April, 1995 at 9:02 am, Timothy McVeigh detonated a two-ton fertilizer bomb packed in a rental truck in front of the Alfred P. Murrah Federal Building in Oklahoma City. The explosion, which destroyed the entire front of the building, killed 168 people, including 15 children who were in the daycare centre. It was the largest domestic terrorist attack in US history (until 11 September 2001). Within days of the bombing, the Mayor's office, the Governor's office, non-profit agencies and citizens of Oklahoma City began to receive suggestions, ideas and offers of donations related to the creation of a memorial. Within a few months, Oklahoma City Mayor Ron Norick appointed a 350-member Memorial Task Force charged with developing an appropriate memorial to honour those touched by the event. Members of the Task Force included family members of those killed in the bombing, survivors of the blast and volunteers with expertise in areas ranging from mental health, law and the arts, to fund-raising. From the summer of 1995 until the spring of 1996, members of the Memorial Task Force gathered input from families, survivors and the general public about what visitors to the Memorial should feel and experience. According to the architects, the design process was heavily influenced by family members and survivors, who insisted that the names of those who died were incorporated into the memorial. They also noted that the process involved a great deal of compromise on their part with family members.

Within five years the Memorial had been designed and dedicated. Today, the site of the Murrah building is occupied by a large memorial. This memorial, designed by Oklahoma City architects Hans and Torrey Butzer and Sven Berg, includes a reflecting pool bookended by two large 'doorways', one inscribed with the time 9:01, and one with 9:03, the pool between representing the moment of the blast. On the south end of the memorial is a field full of symbolic bronze and stone chairs – one for each person lost – arranged according to what floor they were on. The seats of the children killed are smaller than those of the adults lost.

On the opposite side is the 'survivor tree', part of the building's original landscaping that somehow survived the blast and the fires that followed it. The memorial left part of the foundation of the building intact, so that visitors can see the scale of the destruction. Around the western edge of the memorial is a portion of the chain-link fence erected after the blast on which thousands of people spontaneously left flowers, ribbons, teddy bears, and other mementos in the weeks following the bombing. The memorial contends that 'Few events in the past quarter-century have rocked Americans' perception of themselves and their institutions' as did the bombing of the Murrah Building. This may prove true. But such a quick process of commemoration challenges urban scholars to reflect critically – on the event and well as the process.

Alongside many of the aforementioned examples, the Oklahoma Memorial also raises issues around commodification of grief. After all, monuments and memorials are often accompanied by visitor centres, gift shops or kiosks, selling an assortment of 'souvenirs': books, videos, refrigerator magnets and t-shirts, none of which interpret or enhance the memorial experience. The result, some worry, is that visits to monuments and memorials may become a passive, rather than active, experience and engagement with the past. A visit to a memorial need not necessarily be passive – particularly in those cases where monuments and memorials act as radical political statements. But late twentieth-century trends in memorial design and visitor services have tended to create memorials as spaces to be passively consumed. Some have referred to this trend as 'edutainment' and note that there has been a recent impulse to 'explain' through our monuments rather than commemorate, assuming visitors are both ignorant and impatient. John Urry has commented upon the emergence of a postmodern 'tourist gaze' which seeks to consume 'visually' – to take pictures, buy postcards and souvenirs – without necessarily thinking through the complexities of the experience (Urry, 2002). In other words, society prefers to see, not think. This promotes a more passive consumption of memory. Consider again the case of the National Mall: in the last ten years, gift shops have been built in or next to the Jefferson, Lincoln and FDR Memorials and the Washington Monument. Similar stores exist, even at memorials focusing on tragic loss. Off to the side of the Oklahoma City Bombing Memorial, which more than many reflects sadness and loss, the Memorial Store sells jewellery, t-shirts and stuffed animals. However, when visitors spend more time shopping for souvenirs of the memorial than they do thinking about the memorial, and contemplating its significance, then the memorial has become commodified – something to be consumed. This may ultimately trivialize the message of the memorial.

Summary

Since David Harvey (1979) analyzed the ideological contests that surrounded the building of the Sacre Cour Basilica at Montmatre in Paris, work on memory and landscape has developed significantly. Urban scholars have begun to incorporate theories of political economy, cultural studies or semiotics in their analyses of urban space. Subsequent work has produced well-established critiques of memorial landscapes, a sensitivity to the polyvocality of memorialization, and an awareness that the meaning of these spaces is frequently contested. This body of work has contributed to a broader sense of the connection between urban studies and cultural studies. The robust literature that now exists on monuments, memorials and public space has moved beyond the traditional and simplistic definitions of public/private space to see cities as active, lived experiences where culture is varied and contested in urban space.

Monuments and memorials can be sites of celebration, gathering or sometimes protest and conflict, but they are never neutral. Ultimately, their presence enriches the urban experience in vital ways. In cities around the world, monuments and memorials in public space are among the most important and most visible expressions of power, ideologies and history in the urban landscape, yet they also serve to thicken the urban plot, adding texture and meaning to our urban existence.

References

Anderson, B. (1983) *Imagined Communities*. London: Verso.

Atkinson, D. and Cosgrove, D. (1998) 'Urban rhetoric and embodied identities: city, nation, and empire at the Vittorio Emanuele II Monument in Rome, 1870–1945', *Annals of the Association of American Geographers*, 88 (1): 28–49.

Barber, L. (2002) *Marching on Washington: the forging of an American political tradition*. Berkeley: University of California Press.

Cosgrove, D. and Daniels, S. (1988) *The Iconography of Landscape: essays on the symbolic representation, design and use of past environments*. Cambridge: Cambridge University Press.

Crampton, A. (2001) 'The Voortrekker Monument, the birth of apartheid, and beyond', *Political Geography*, 20 (2): 221–46.

Glassberg, D. (2001) *Sense of History: the place of the past in American life*. Amherst: University of Massachusetts Press.

Goldberger, P. (2004) *Up from Zero: politics, architecture and the rebuilding of New York*. New York: Random House.

Graham, S., Ashworth, G.J. and Tunbridge, J.E. (2000) *A Geography of Heritage*. London: Belhaven.

Harvey, D. (1979) 'Monuments and myth', *Annals of the Association of American Geographers*, 69(3): 362–381.

Hayden, D. (1995) *The Power of Place: urban landscapes as public history*. Cambridge, MA: MIT Press.

Heffernan, M. (1995) 'For ever England: the Western Front and the politics of remembrance in Britain', *Ecumene*, 3 (2): 293–324.

Jacobs, J. (1961) *The Death and Life of Great American Cities*. New York: Random House.

Johnson, N. (1994) 'Sculpting heroic histories: celebrating the centenary of the 1798 rebellion in Ireland', *Transactions of the Institute of British Geographers*, 19 (1): 78–93.

Johnson, N. (1995) 'Cast in stone, monuments, geography and nationalism', *Environment and Planning D: Space and Society*, 13 (1): 51–65.

Johnston, R., Gregory, D. and Smith, D. (eds) (1994) *The Dictionary of Human Geography* (3rd edition). Cambridge, MA: Blackwell.

Knox, P. and McCarthy, L. (2005) *Urbanization: an introduction to urban geography* (2nd edition). Upper Saddle River, NJ: Prentice-Hall.

Leib, J. (2002) 'Separate times, shared spaces: Arthur Ashe, Monument Avenue and the politics of Richmond, Virginia's symbolic landscape', *Cultural Geographies*, 9 (2): 286–312.

Low, S. (2002) 'Lessons from imagining the World Trade Center site: an examination of public space and culture', *Anthropology and Education Quarterly*, 33 (3): 395–405.

Michalski, S. (1998) *Public Monuments: art in political bondage 1870–1997*. London: Reaktion Books.

Mitchell, D. (2003) *The Right to the City: social justice and the fight for public space*. New York: Guilford Press.

Mitchell, K. (2003) 'Monuments, memorials, and the politics of memory', *Urban Geography*, 24 (5): 442–59.

Nora, P. (1996) *Realms of Memory: the construction of the French past*. New York: Columbia University Press.

Sorkin, M. (1992) *Variations on a Theme Park: the new American city and the end of public space*. New York: Hill and Wang.

Sorkin, M. and Zukin, S. (eds) (2002) *After the World Trade Center: rethinking New York City*. New York: Routledge.

Sturken, M. (2005) 'Memorializing absence', Social Science Research Council. Accessed at www.ssrc.org/sept11/essays.

Urry, J. (2002) *The Tourist Gaze: leisure and travel in contemporary society* (2nd edition). London: Sage.

Whelan, Y. (2002) 'The construction and destruction of a colonial landscape: commemorating British monarchs in Dublin before and after independence', *Journal of Historical Geography*, 28 (4): 508–33.

Zukin, S. (1991) *Landscapes of Power: from Detroit to Disney World*. Berkeley: University of California Press.

SECTION TWO
ECONOMIES AND INEQUALITIES

The relationship between cities and economies is at the heart of urban studies. The economic development of cities is tied to theories of urban change and development while the functioning of economies is associated with urban inequalities and social identities. All sorts of urban phenomenon can be seen through the lens of the econ-'omy, past and present.

In this section we encourage a range of contributions to consider this city–economy nexus. Kevin Ward outlines a Marxist approach. He outlines the basic Marxist premises of historical materialism. His story in Chapter 6 begins in the city of Manchester, one of the 'shock' cities of the nineteenth century and the setting for Friedrich Engels' formulations about the inhumanity of cities in which capital accumulation came to determine the rhythms of life. Drawing on Marxist and neo-Marxist theory, Ward develops the theme of the urbanization of injustice and draws upon case studies of gentrification and business improvement districts to higlight the continuing relevance of Marxist and neo-Marxist approaches in contemporary urban studies.

Urban economic relations are enmeshed in both national and global economic relations. In Chapter 7, Yeong-Hyun Kim explores the theorizations of the city as a site and platform for globalization. She situates urban growth within the context of urban hierarchies that extend across time and national space. She outlines the reasons behind the growth of global and globalizing cities, highlights the contributions as well as the limitations of the 'globalization and the city' literature, showing how most emphasis has been placed on economic globalization. The role of cities in cultural and political globalization is, however, less theorized.

In the context of economic globalization, one of the reasons that investment is attracted to major world cities is that these are regarded as sites of immense creativity and innovation. In Chapter 8, Andy Pratt unpacks some of the confusion over the terms 'innovation' and 'creativity' and shows the role of both in understanding the dynamism of the contemporary city, especially in relation the creative industries. He untangles the populist consideration from the more formal debates about creative economies and cultural industries.

We have also recently heard much of cities breaking away from national systems of regulation. The 'end of the nation-state' thesis has often been interpreted as related to the rise of the city as a locale of independent power and authority. Chapter 9, written by Angus Cameron, shows how and why the city is still situated in and enveloped by nation-states. With particular reference to taxation, money and law, he shows how the historical legacy of nation-state evolution limits the relative power of individual cities.

Yet urban economies do not only revolve around the world of narrowly-defined work. In the current epoch, consumption too is serious business. David Bell and colleagues thus highlight the role of leisure and recreation in the city. Drawing on the economic history of the modernist city and economic developments in the postmodern city, in Chapter 10 they show, with particular reference to eating and drinking, how urban spaces are reconfigured and how social power is both implemented and resisted in such moments and spaces of consumption. In so doing, they remind us that everyday activities (eating, shopping, homemaking) should not be dismissed as somehow inconsequential in understanding the city: as Lefebvre's work on everydayness insisted, sometimes reflecting on the banality of urban life produces moments of profound illumination.

6 CAPITAL AND CLASS

Kevin Ward

This chapter

○ Considers the changing role of the city as a site of capital accumulation

○ Reviews the impact of Marxist ideas in urban studies from the nineteenth century onwards

○ Shows how recent urban restructurings can be understood as the outcome of particular conflicts between capital and class

---

Introduction

I write this chapter from the same city that was the home to Friedrich Engels for over twenty years, and in which he spent almost two years conducting the research that would form the basis of *The Conditions of the Working Class*. Written over one hundred and sixty years ago, between 1844 and 1845, this 'consideration of cities in the early development of the English working class' (Katznelson, 1992: 30) has proved a landmark text in Marxist analyses of the city and urbanization. Not far from my office (in a university that was itself established five years after Engels finished his masterpiece) is the area of Manchester where the consequences of urbanization for workers became clear to him. Walking around the central city, in and out of areas such as Gibraltar and Little Ireland, as they were then known, he was clear that what he was witnessing was of immense importance. He saw evidence of the emergence of what he believed was a 'modern working class' with a shared sense of identity and self, forged in the squalor that characterized their everyday home and work experiences. For capitalists – those who owned the factories and who rented the houses to the workers – this new way of organizing space offered the potential of increased profits. New

Figure 6.1   Engels' plan of Manchester, 1844

technologies, in the form of steam engines and the machinery for working cotton, overhauled the city, as, according to Merrifield (2002a: 35), the 'two terrible beauties were born': the industrial and the urban revolution.

Manchester was becoming the world's first industrial city. The result was a city in which inequality was pervasive. Workers lived in and around the centre, in 'cattle sheds', rented for high prices as capitalists found ways of clawing back as much of the wage as they could. On the outskirts of the city, in areas such as Chorlton upon Medlock and Whalley Range to the south, lived the wealthy (see Figure 6.1). Engels noted that Manchester had developed to allow the wealthy to move in and out of the city and not to have to see the consequence of their actions. He termed this a 'hypocritical plan', and proclaims never to have seen 'so systematic a shutting of the working class from the thoroughfares, so tender a concealment of everything which might affront the eye and the nerves of the bourgeoisie' (Engels, 1844: 87). What made Engels so fascinated and so appalled was that he saw Manchester as emblematic of the conditions under which workers in the subsequent decades and centuries would have to live and labour. A future of terrible suffering, yes, but also of immense possibilities, as the capitalist metropolis created the circumstances ripe for revolution, for this was the only way out for workers: a rejection of the whole capitalist system.

Extract 6.1: From Engels, F. (1892) *The Condition of the Working Class in England in 1844.* London: George Allen and Unwin, pp. 45–7.

Manchester lies at the foot of the southern slope of a range of hills … and contains about four hundred thousand inhabitants, rather more than less. The town itself is peculiarly built, so that a person may live in it for years, and go in and out daily without coming into contact with a working-people's quarter or even workers, that is, so long as he confines himself to his business or pleasure walks. This arises chiefly from the fact, that by unconscious tacit agreement, as well as with outspoken conscious determination, the working-people's quarters are sharply separated from the sections of the city reserved for the middle-class; or, if this does not succeed, they are concealed with the cloak of charity. Manchester contains, at its heart, a rather extended commercial district, perhaps half a mile long and about as broad, and consisting almost wholly of offices and warehouses. … With the exception of this commercial district, all Manchester proper, all Salford and Hulme, a great part of Pendleton and Chorlton, two-thirds of Ardwick, and single stretches of Cheetham Hill and Broughton are all unmixed working-people's quarters, stretching like a girdle, averaging a mile and a half in breadth, around the commercial district. Outside, beyond this girdle, lives the upper and middle bourgeoise, the middle bourgeiose in regularly laid out streets in the vicinity of the working-man's quarters, especially in Chorlton and the lower lying parts of Cheetham Hill; the upper bourgeoise in remoter villas with gardens in Chorlton and Ardwick, or on the breezy heights of Cheetham Hill, Broughton and Pendleton, in free, wholesome country air, in fine, comfortable homes, passed once every half or quarter hour by omnibuses going in to the city. And the finest part of this arrangement is this, that the members of this money aristrocracy can take the shortest road through the middle of all labouring districts to their places of business, without ever seeing that they are in the midst of the grimy misery that lurks to the right and to the left. For the thoroughfares leading from the Exchanges in all directions out of the city are lined, on both sides, with an unbroken series of shops, and are so kept in the hand of the middle and lower bourgeoise, which, out of self-interest, cares for a decent and cleanly external appearance and *can* care for it. True, these shops bear some relation to the districts which lie behind them, and are more elegant in the commercial and residential quarters that when they hide grimy working-men's dwellings: but they suffice to conceal from the eyes of the wealthy men and women of strong stomachs and weak nerves the misery and grime which form the complement of their wealth.

And yet, in the twenty-first century, capitalism remains in place, and continues to expand geographically. The urbanization of the global south – what we used to call 'the third world' – continues apace. The 'shock cities' of the late twentieth century, such as São Paulo, Pusan and, today, Ciudad Juárez, Bangalore and Guangzhou, have roughly approximated this classical trajectory, as urbanization and industrialization have occurred in tandem, like Manchester in the mid-nineteenth century. As urbanization continues in this part of the world, however, this tight coupling is showing signs of coming undone. Urbanization is continuing to transform many nations in the global south, but is occurring on its own, without industrialization. As a result, we are witnessing the 'urbanization of poverty' in many countries in this region (Davis, 2004). Yet despite these differences, from his analysis of Manchester, Engels would surely have recognized the plight of many now living in the rapidly urbanizing slum areas of countries such as Brazil, China and India. While the commonalities between mid-nineteenth century Manchester and early twentieth-century Bangalore should not be overblown, neither should they be understated. The exploitation of humankind in the name of capital continues to be a feature of today's urban societies as it was in Engels' Manchester, even if the scale of the poverty is now much deeper.

Perhaps surprisingly, Engels' book, which was not translated into English until 1892, failed to herald a rash of work by Marxists on urbanization and the city. It was not until the late 1960s that social scientists began to think about the city in relation to the work of Engels, and also, with reference to the writing of his close friend, Karl Marx. And this is the focus of my chapter: to outline briefly how Marxists of different varieties understand the city in relation to the wider processes of urbanization. In the next section of this chapter I turn to outline some of the fundamentals of Marxism, which underpinned Engel's own analysis of Manchester in the middle of the nineteenth century, and which have provided the basis of such seminal studies of the city and of urbanization as *The Urban Revolution* by Henri Lefebvre (2003), *The Urban Question* by Manuel Castells (1977) and *Social Justice and the City* by David Harvey (1973). These three foundational texts are simply the tip of a Marxist iceberg. Since the 1970s a large body of work has emerged giving urban studies a sense of definition and theoretical thrust (Katznelson, 1992), and despite its decentring in recent years, as other alternative explanations have been developed, Marxist urban research still retains a strong position in disciplines such as geography, planning and sociology. In the third section of this chapter I take the basic building blocks and apply them to the city, using two real world examples to bring to life what some understand as an outdated and rather abstract theoretical approach. My first example is that of the renewal of downtown or city-centre housing, a process known as gentrification. Using the work of Neil Smith I present a Marxist analysis of this phenomenon, of this new way of *living in the city*. My second example is the international trend in entrepreneurial urbanism, of the construction and management of a good 'business climate' and what this means for *loafing in the city*, using the example of Business Improvement Districts. In the final section I return to some of the basics in the hope of convincing readers of the continuing fruitfulness of a Marxist approach to the city.

The conceptual building blocks of Marxism

Marx did not write about the city explicitly. It was his friend, Friedrich Engels, who first wrote about the city from what today we might think of as a Marxist perspective when he produced *The Condition of the Working Class in England* (1892). Rather,

Marx was interested in capitalism as a mode of production, a system with particular economic and social characteristics. What sets one mode of production apart from another, capitalism from feudalism, for example, is whether one class works for another, how labour is coordinated and how the ruling class extracts the surplus produced by the workers.

By 'class' Marx means a group of individuals who have common interests, who acknowledge these, and who act in unison. For Marx, capitalism consists of two classes: the capitalists or the 'bourgeoisie' and the workers or the proletariat. After investing in the upkeep of buildings and the maintenance of technology, and paying the workers their wages, capitalists retain some money in the form of profit. This is what Marx termed surplus value. So, under capitalism, capitalists are able to command labour, and to accumulate more and more money though the ownership of wealth. On the other hand, the proletariat have to sell their labour power. Workers thus rely on capitalists for a wage, and for their continued existence. The relationship between bourgeoisie and proletariat, between capitalists and workers, need not be anything more than an economic one, an exploitative one resting on the production and extraction of surplus value. And, of course, the goods or services produced by workers (from which capitalists generate surplus value) are then sold.

Capitalists have to realize their profits or surplus value by exchanging them in return for money. Thus we have an exchange sphere – a market – in which capitalists pay workers with money, who in turn use that money to buy products to survive, while capitalists plough back some of their surplus value into new equipment. So we have a vicious or virtuous circle, depending on where you lie in the arrangements, in which goods are produced, sold and bought.

To understand the city, and urbanization, from a Marxist perspective, it is important that you get to grips with the notion of contradiction. Put simply, all relationships are inherently contradictory. Nowhere is this more apparent than tension between the possibilities opened up by urbanization for capitalists and for workers. For workers, their concentration offered the possibility that they could better organize, meet more regularly, launch campaigns, and, in the longer term, form a new mass, a self-conscious class that would be in a position to revolt. On the other hand, however, urbanization involved the creation of cities in which workers live and work under poor conditions, in which their social worlds remained separate from those of their employees, as we saw in the case of Manchester in the middle of the nineteenth century. Capitalist urbanization produces profoundly unequal cities, which make it more *and* less likely that workers might mobilize to reject capitalism. As Marx (1977: 815) himself put it: 'The more rapidly capital accumulates in an industrial or commercial town, the more rapidly flows the stream of exploitable human material, the more miserable are the impoverished dwellings of the workers', and out of these conditions Marx argues workers are more likely to realize their common enemy.

In his work on the USA, the late David Gordon (1978: 40) argued that 'large cities became increasingly dominant as sites for capitalist factories because they provided an environment which more effectively reinforced capitalist control over the production process'. Cities worked for capitalists because workers were close by, and their everyday activities inside and outside work could be closely monitored, giving the capitalists a hitherto unprecedented amount of control over the lives of workers, allowing them to inculcate a particular type of 'work ethic' among the working class. We can think of the industrial city as the context in which the contradictory relationship between capitalists and workers was first observable. First, the largest factories were located in the centre of cities, in the US downtowns, near transport infrastructure, such as the rail and water outlets. In some places, such as Manchester, it meant locating next to the canals that punctured much of the city. In others, such as Pittsburgh, it meant locating near the ports, to be close to where goods would be shipped in and out. Second, working-class areas emerged in the form of tenements, houses in which large numbers of families lived in cramped and often unhygienic conditions. Capitalist owners of these properties could maximize the total rent by dividing the tenements into multiple dwellings. These were close to the centre of cities, and hence to factories, to minimize travel to work times, and to reinforce the importance of capitalists in the lives of workers outside the workplace. Third, the middle and upper classes settled on the outskirts of cities. Capitalists wished to escape the conditions experienced by workers, and so relocated as far as they could afford, with the result that physical distance from the centre became a proxy for wealth and for status. Fourth, and seemingly running against the last two characteristics, shopping districts emerged in the centres or downtowns, outlets in which only the middle and upper classes could afford to shop. Despite being often quite close to the tenements lived in by the working classes, however, these districts were quiet discrete, consisting of expensive cafés and shops. Lines were clearly drawn. From this outline it should be clear that the spatial organization of the industrial city results from the contradictions inherent in the relationship between capitalist and worker.

The focus for Marx is on the products, or, as he put it, commodities, that go into production. Of course, there is a big difference between labour and the other commodities that go into the production process. Labour is what is known as a pseudo-commodity; it is a living, thinking being, a person like you or me. This matters in three ways. First, workers are only ever temporary commodities. Workers assume the form of a commodity during their working day, but despite this, they were not born and raised to be commodities. Second, because workers are physiologically and mentally complex beings, capable of independent thought and action, in contrast to most commodities, they have *agency*, the ability to act, to resist, and to make a difference to the conditions under which they live and work. Third, this means that workers, unlike other commodities – necessarily enter into a social relationship with their employers. They interact, talk, laugh, and argue with those who pay their wages. In Marxian language, this is a class relationship, since workers' pseudo-commodity status is what distinguishes them as a social group from the relative minority of capitalists who purchase their labour power. It is what distinguishes the proletariat from the bourgeoisie.

All commodities have two types of value, an exchange value and a use value. Marx argues that when commodities are exchanged the use and the exchange values are totally independent. Exchange value is based on the amount of labour power embodied in the commodity. It is labour power and the amount expended that determines the exchange value of a commodity. Under the capitalist system, exchange values dominate over use values, Marx writes not of labour but labour power. The distinction is an important one. Let us think about a standard working day, say eight hours. According to Marx, the capitalist does not buy eight hours of labour, because he would have to give the equivalent of eight hours of labour in return. In this situation the capitalist would make no profit. Rather, what the capitalist buys is the labour power, the right to call on a person's labour for the eight hours. During this time the worker does eight hours labour, but is paid less than the equivalent of eight hours labour (say five hours). Marx conceives the working day as consisting of two parts, what he calls the 'necessary working time' and the 'surplus'. The first of these – the necessary time – would be, in our example, the five hours it takes the worker to produce the equivalent of the value of being employed and to earn enough to keep working. The second part, the surplus, according to Marx, is the other three hours in our example, the rest of the working day. This is the source of the capitalist's profits, the surplus value.

The inherently antagonistic relationship – between capitalists and workers – underscores all forms of social relations in a capitalist society. As Harvey (1978a: 100) stressed, '[t]he essential Marxian insight ... is that profit arises out of the domination of labour by capital and that the capitalists as a class must, if they are to reproduce themselves, continuously expand the basis for profit'. The accumulation of more and more profits by capitalists – what Marxists term capital accumulation – is the logic that organizes and structures all aspects of our lives, including the production of the urban built environment.

Extract 6.2: From Harvey, D. (1978a) 'The urban process under capitalism: a framework for analysis', *International Journal of Urban and Regional Research* 2 (1), pp. 100–1, 106.

Within the framework of capitalism, I hang my interpretation of the urban process on the twin themes of *accumulation* and *class struggle*. The two sides are integral to each other and have to be regarded as different sides of the same coin – different windows from which they view the totality of capitalist activity. The class character of capitalist society means the domination of labour by capital. Put more concretely, a class of capitalists is in command of the work process and organizes that process for the purposes of producing profit. The labourer, on the other hand, has command only over his or her labour power, which must be sold as a commodity on the market. The domination arises because the labourer must yield the capitalist a profit (surplus value) in return for a living wage. All of this is extremely simplistic, of course, and actual class relations (and relations between factions of classes) within an actual system of production

(Continued)

(comprising production, services, necessary costs of circulation, distribution, exchange, etc.) are highly complex. The essential Marxian insight, however, is that profit arises out of the domination of labour by capital and that the capitalists as a class must, if they are to reproduce themselves, continuously expand the basis for profit. We thus arrive at a conception of society founded on the principle of 'accumulation for accumulation's sake, production for production's sake'. The theory of accumulation which Marx constructs in *Capital* amounts to a careful enquiry into the dynamics of accumulation and an exploration of its contradictory nature. This may sound rather 'economistic' as a framework for analysis, but we have to recall that accumulation is the means whereby the capitalist class reproduces both itself and its domination over labour. Accumulation cannot, therefore, be isolated from class struggle.

... The capitalist form of accumulation therefore rests upon a certain violence which the capitalist class inflicts upon labour. ... The individual labourer is powerless to resist this onslaught. The only solution is for the labourers to constitute themselves as a class and find collective means to resist the depredations of capital. The capitalist form of accumulation consequently calls into being overt and explicit class struggle between labour and capital. This contradiction between the classes explains much of the dynamic of capitalist history and is in many respects quite fundamental to understanding the accumulation process.

Much of what constitutes Marxian theory is written at a fairly abstract level that attempts to get at the essential elements of a phenomenon before examining the specific manifestations or outcomes that we are able to observe (what we might call the *concrete*). Of course the relationship between abstract and concrete is not a straightforward one – Marx was certainly concerned with moving between the two. He was fully aware that struggles between capitalists and workers take place in a context, and how this class struggle plays out varies depending on these contextual conditions. However, these conditions are not pre-given but are made under a capitalist society. Marx, then, attempts to capture the sense that capitalists and workers come together in struggles in which there is scope, or wiggle-room, for workers to be able to extract concessions from capitalists, such as better pay, shorter working hours, free transport to and from the place of work, etc., when he argued that '[people] make their own history, but they do not make it as they please; they do not make it under self-selected circumstances, but under circumstances existing already, given and transmitted from the past'. In the final instance, however, regardless of the victories that labour might be able to secure, capitalism prevails, and, as such, so does the subordination of workers by capitalists in the name of profit.

Marxism and the city: some basics

While Marx might not have written much about cities, his perspective, of seeing all things as being in a state of continuous change and evolution, does allow us to understand the ebb and flow of urbanization, the daily struggle of life in the metropolis. 'Cities, after all, help expand and socialize the productive forces, are the foundation of the division of labour [between workers and capitalists], reign as seats of government and power, exhibit class distinctions and residential ghettoizations, and bear the imprint of geographical uneven development', as Merrifield (2002b: 155) puts it. According to Marxism, the class struggle, between those who own the means of production and those who have to sell their labour power to get by, permeates all sections of society, and so, unfolding in a complicated and a contradictory fashion, it shapes the nature of urban space.

According to Harvey (1978b: 9), 'capitalist society must of necessity create a physical landscape – a mass of humanly constructed physical resources – in its own image, broadly appropriate to the purposes of production and reproduction. ... [T]his process of creating space is full of contradictions and tensions and the class relations in capitalist society inevitably spawn strong crosscurrents of conflict.' Harvey stresses here both the ways in which the logic of capitalism is behind the organization of cities, and how in and through the production and reproduction of this built environment – roads, offices, shops, cinemas, etc. – capital and labour come into conflict. While the conflict is always an uneven one – for capitalists hold the upper hand – in practice it is possible for workers, or those acting in their interests, to secure concessions. In some cases this might be better pay or longer holidays; in others, though, it might be gains not related directly to the workplace, such as better and cheaper public transport or improved childcare facilities. This distinction between the working place and living place is, of course, an artificial one (literally in some cases, with an increased amount of work now taking place outside the workplace, at airports, on buses and at the home). Both places are part of the same struggle, that of labour to control the terms and condition of its own existence. And this struggle is over space, for labour needs space to live, while capitalists need space to accumulate capital. It is this conflict over land, and the way it plays out day-in-day-out, which produces the built environment we see around us. As Harvey (1978b: 11) puts it: 'Labour, in seeking to protect and enhance its standard of living, engages, in the living place, in a series of running battles over a variety of issues which relate to the creation, management, and use of the built environment.'

We can develop our understanding of this general thesis in relation to particular examples of urban conflicts between capital and labour. The first example is the recent gentrification of the inner cities of North American and European cities in which the competitive drive for profits has been behind the substantial reinvestment in the built environment, the revalorization of downtowns or city-centres, and the accompanying growth in low-wage jobs. Coined by sociologist Ruth Glass in 1964, 'gentrification' involves 'the invasion by middle-class or higher-income groups into previously working-class neighbourhoods or multi-occupied 'twilight areas' and the replacement or displacement of many of the original occupants' (Hamnett, 1984: 284). Despite the process first being noted by Glass in her work on 1960s London, the bulk of the initial empirical research on gentrification was performed in New York in the context of the city's economic crisis of the 1970s. Significant de-population and under-investment in the built environment had led to substantial drops in the rental values of residential and retail properties. For Marxist urban geographers, the emphasis of traditional studies on individual consumer choice, exercised under economic constraints, did not acknowledge the importance of the producers as well as the consumers of gentrification. Smith's (1979, 1982) seminal Marxist analysis, on the other hand, rested on the belief that gentrification could be explained by, first, the earlier withdrawal of capital from the centres of cities, and then its return, at the expense of the working class who had remained in residence in the inner city. His 'rent gap' thesis understood gentrification as stemming from changes in the returns on capital investment in other areas of the economy. As these declined, and a gap emerged between the potential ground rent level and the rent that was being realized through the current land use, due to the early withdrawal of capital and the subsequent under-investment in the built environment, so it became more profitable for capital to return to the cities and to invest in renovating the housing stock.

Extract 6.3: From Smith, N. (1996) *The New Urban Frontier: gentrification and the revanchist city.* London: Routledge, pp. 70–1.

Gentrification is a structural product of the land and housing markets. Capital flows where the rate of return is highest, and the movement of capital to the suburbs, along with the continual devalorization of inner-city capital, eventually produces the rent gap. When this gap grows sufficiently large, rehabilitation (or, for that matter, redevelopment) can begin to challenge the rates of return available elsewhere, and capital flows back in. Gentrification is a back-to-the-city movement all right, but a back-to-the-city movement by capital rather than people.

The advent of gentrification in the latter part of the twentieth century had demonstrated that contrary to the conventional neoclassical wisdom, middle- and upper-middle class housing can be intensively developed in the inner city. Gentrification itself now has significantly altered the urban ground rent gradient. The land value valley may be being displaced outward and in part upward as gentrification revalues central city land … and as disinvestment is displaced outward to the closer, older suburbs leading in turn to a new flurry of complaints that middle-class suburbs now face 'city problems' …

Gentrification has been the leading residential and recreational edge (but no way the cause) of a large restructuring of space. At one level, restructuring is accomplished according to the needs of capital, accompanied by a restructuring of middle-class culture. But in a second scenario, the needs of capital might be systematically dismantled, and a more social, economic and cultural agenda addressing the direct needs of people might be substituted as a guiding vision of urban restructuring.

While this Marxist analysis of gentrification has not gone uncontested (Hamnett, 1991; Ley, 1986), it provides a powerful example of how the built environment is temporary home for value, and is the source of capital accumulation, tied as it is to wider circuits of capital. Since this groundbreaking research in Manhattan, a range of work has emerged on cities in the industrialized nations. Amsterdam, Bristol, Chicago, Leeds, London, Manchester, Nottingham, Sydney, and Vancouver have all had their built environments restructured through the inflow of capital investment. The remnants of past investments have been refashioned, torn down or made up, as rents have risen and a fraction of the upper and middle classes have returned to live in city centres. While this capital reinvestment and human displacement has been sharpest in New York, in other cities, such as Manchester, England, the process has, in its own way, been as pronounced. Large portions of the built environment have been brought back to life; new capital has flowed into the city, followed closely by the middle classes. New loft apartments have been built, standing shoulder to shoulder with the revalorized warehouses that remain as reminders of Manchester's past, the one that Engels wrote about so many years ago. Accompanying them has been an explosion in the number of bars, cafés, gyms, and restaurants, capital invested in the built environment to meet the consumption needs of the new urban residents, on which the bourgeoisie can spend their surplus value. Of course, as we have witnessed a rise in the urban service economy, associated with work in cafés, hotels and restaurants, so we have seen the expansion of low-paid jobs. Workers from Manchester's less affluent neighbourhoods enter the city centre to perform these jobs.

A necessary element to cities competing against each other is securing the conditions favourable to the continued investment by capital in the built environment. In a growing number of industrialized nations this has involved capitalists and local government officials working in coalitions or partnerships. The purpose of these initiatives is, according to the Marxist framework, to ensure that nothing detracts from the city's role as, first and foremost, a site for profit-making by capitalists. Let us consider the example of Business Improvement Districts.

A Business Improvement District is an alliance of government and business owners, formed to protect exchange values. The idea first emerged in Toronto, Canada, in the 1970s, but it was in the 1980s and in US downtowns that the model, which has found itself imported into Japan, New Zealand, South Africa, among other countries, was founded. Now, with over 400 Business Improvement Districts in the USA, and a few thousand more worldwide, this way of managing the urban built environment has gained global acclaim as a means of improving the 'quality of life'. Of course, as Marxists would explain, the notion 'quality of life' is more often than not defined by capitalists, and actually reflects those commodities that can be produced for profit in particular places at particular times. Quality of life is not a value-free term. Someone, or some group, is doing the defining, and under the capitalist system it is normally capitalists. In the case of Business Improvement Districts, local governments collect a levy from local businesses within the BID jurisdiction, and then hand this money over to the BID board. This grouping of local business representatives, landowners and local politicians decide how it spends this money, although there is a shortlist of activities that most Business Improvement Districts tend to undertake. These include advertising and marketing, security and street cleaning.

Underpinning what Business Improvement Districts do is to secure the conditions for capital accumulation, with any recognition of the consequences for social justice restricted to ensuring an able-to-work workforce and the maintenance of what Marx calls the consumption fund for labour, in order that commodities are consumed and capital accumulated. The recent UK legislation on Business Improvement Districts is the latest example of a country's policy-makers embracing this US model of city-centre or downtown management. Announced in 2001 by the Prime Minister, Tony Blair, the national legislation was put in place at the end of 2004. This created the legal conditions for the owners of the built environment and local politicians, in cities and towns around the country, to come together to establish a Business Improvement District. Already, twenty-four months after the legislation took effect, England has fifty. However, the securing of capital's new political role in the management of UK cities has not been without its hiccups. There have been a number of occasions where local retail and residential property owners have voted against the establishment of a Business Improvement District. Local politics took hold in each case, with different fractions of capitalists voting in different ways, rather than voting as a class in and for itself. Nevertheless, the onward march of this particular model of urban management continues.

According to Marxist thinking, Business Improvement Districts represent an attempt by capital to intervene in the governance of the city centre, or the downtown, to secure the conditions for capital accumulation. As such, they do not really constitute a break from past efforts by governments to support the needs of capital, but rather represent the latest version of a complex and at times contradictory relationship. This working together, of government and the capitalist class, has characterized the histories of most advanced capitalists countries. What though, perhaps, has changed in the last three decades is a transformation in the nature of this working, with governments becoming more and more entrepreneurial in terms of pro-actively organizing, and working with others, particularly those who seek to make profit through the use or rental of urban space. At a very general level we can say that since the mid-1980s we have witnessed some convergence in the logic underpinning the ways in which governments have exerted their power in urban policy (Harvey, 1989). Of course this does not mean that all governments act in the same way at the same time. They never have, nor is it likely that they ever will. Within any one country it is possible to discern differences in how one urban government acts to produce a good business climate. Nevertheless, there is 'a certain pattern and underlying rationales for these interventions' (Harvey, 1978b: 14). The general shift in the types of policies pursued by governments of all shapes, sizes and political stripes, is, according to Marxist analysis, done in the name of improving the 'business climate', defined as the factors that affect the ease and profitability of doing business in a particular city or region. It is this role with which Business Improvement Districts have often been charged.

Summary

The Marxist analysis of the city, and of the wider process of urbanization, provides a powerful way of revealing the forces at work in the production of the urban built environment. Sometimes these forces, and the impact they have, are not always obvious. As Harvey (1978b: 30) puts it, 'the surface appearance of conflicts around the built environment – the struggles against the landlord or against urban renewal – conceals a hidden essence, which is nothing more than the struggles between capital and labor'. A Marxist analysis means accepting that something called capitalism exists, and that this sets the conditions under which capitalists and workers occupy different positions in society. These two different positions reflect each group's relationship to the means of production. Despite this rather abstract framework, in which all of the population can be divided into just two groups, there is scope for difference: difference among capitalists and among workers; difference from one place to another; difference from one time to another. Surplus value – the profit – always belongs to the capitalist, and the labourer never accumulates enough to own her or his own means of production. Hence, the capitalist always reproduces her or his capital and maintains her or his ownership of the means of production.

As such, the relationship between the two groups is continually reproduced; the labourer's separation from the means of production is reproduced. However, certain labourers acting alone or with others can gain concessions, secure ownership over one part of the total means of reproduction. They have 'wiggle-room'. Sometimes workers and a particular faction of capital might work in unison. Staving off the relocation of a retailer from the downtown to a suburban mall might, out of necessity, involve both groups. The point is, from a Marxian position, to hold on to the insights and the understandings of the general framework, in which capitalism acts as a system to advantage one group at the expense of another, while allowing for agency, the capacity of individuals to act in ways that do not always adhere to their class position.

References

Castells, M. (1972) *La question urbaine*, Paris: François Maspero [published in English as Castells, M. (1977) *The Urban Question: A Marxist Approach*, London: Edward Arnold].

Davis, M. (2004) Planet of slums, *New Left Review*, March–April: 5–34.

Engels, F. (1892) *The Condition of the Working Class in England in 1844*, London: George Allen and Unwin Ltd.

Gordon, D. (1978) Capitalist development and the history of American cities, in W. Tabb and L. Sawers (eds), *Marxism and the Metropolis: New Perspectives in Urban Political Economy*, Oxford: Oxford University Press.

Hamnett, C. (1984) Gentrification and residential location theory: a review and assessment in D. Herbert and R.J. Johnston (eds), *Geography and the Urban Environment: Progress in Research and Applications*, New York: John Wiley and Sons.

Hamnett, C. (1991) The blind men and the elephant: the explanation of gentrification, *Transactions of the Institute of British Geographers*, 16: 173–189.

Harvey, D. (1973) *Social Justice and the City*, Baltimore, MD: Johns Hopkins University Press.

Harvey, D. (1978a) The urban process under capitalism: a framework for analysis, *International Journal of Urban and Regional Research*, 2: 100–31.

Harvey, D. (1978b) Labor, capital and class struggle around the built environment in advanced capitalist societies, in K. Cox (ed.), *Urbanization and Conflict in Market Societies*, London: Methuen.

Harvey, D. (1989) From managerialism to entrepreneurialism: the transformation in urban governance in late capitalism, *Geografiska Annaler* B, 71: 3–17.

Katznelson, I. (1992) *Marxism and the City*, Oxford: Clarendon Press.

Lefebvre, H. (1970) *La Révolution urbaine*, Paris: Gallimard [published in English as Lefebvre, H. (2003) *The Urban Question*, Minneapolis: University of Minnesota Press.]

Ley, D. (1986) Alternative explanations for inner-city gentrification, *Annals of the Association of American Geographers*, 70: 521–35.

Marx, K. (1976) *Capital, Volume: A Critique of Political Economy Volume 1: The Process of Capitalist Production*, New York: International Publishers.

Merrifield, A. (2002a) *Metro-Marxism: A Marxist Tale of the City*, London: Routledge.

Merrifield, A. (2002b) *Dialectical Marxism: Social Struggles in the Capitalist City*, New York: Monthly Review Press.

Smith, N. (1979) Towards a theory of gentrification: a back to the city movement by capital not by people, *Journal of the American Planning Association*, 45: 538–48.

Smith, N. (1982) Gentrification and uneven development, *Economic Geography*, 58: 139–55.

Smith, N. (1996) *The New Urban Frontier: Gentrification and the Revanchist City*, London: Routledge.

7 GLOBAL AND LOCAL

Yeong-Hyun Kim

This chapter

○ Demonstrates the importance of considering the contemporary city in the context of debates about globalization

○ Identifies the importance of particular 'world cities' as focal points for global trade and business

○ Suggests that globalization's impacts are felt at all levels of the urban hierarchy, with 'local' and 'global' forces evident in both 'world' and 'ordinary' cities

---

Introduction

The interface between globalization and urbanization is currently a major research topic in urban studies, in much the same way as globalization is now a central analytic position in the social science literature. External factors have long affected cities, yet there is a general consensus that the geographical scale of causes, processes and outcomes of urban changes has become more transnational, if not global, in the past couple of decades. A rather simple acknowledgement that many large cities, all with their distinctive history and socio-political system, are undergoing somewhat similar economic, cultural and spatial changes lies at the heart of the so-called globalization–urbanization nexus literature. As such, globalization is interpreted not only as a major source of urban changes but also a process that has been facilitated, or even enabled, by these changes.

The urban impact of globalization has been observed in various urban sites around the world, yet a limited number of large cities, so-called 'global cities' or 'world cities', namely London, New York and Tokyo, and, to a lesser degree,

Los Angeles and Paris, have received a large portion of the academic scrutiny. A conceptual distinction could be made between global cities and world cities, yet as noted in Short (2004: 2), the distinction is loose. In this chapter, the two terms are used interchangeably.

Global cities have been conceptualized as sites for the ever more complex, intensified global networks of businesses, markets, (non)governmental organizations and migrants to develop – the 'central places where the work of globalization gets done' (Sassen, 2002: 8). The argument that globalization causes and, at the same time, reflects various changes in the urban economy, culture, politics and space of global cities has heightened the significance of global cities, not nation-states, as a more appropriate analysis unit in the research on the process of globalization. Taylor (2004) even promotes a shift from the state-centric view of the world towards a city-centric view in our efforts for a better understanding of globalization and the unfolding world system. Consequently, some key questions on the relationship between globalization and urbanization include: What roles do cities, and large cities in particular, play in globalization? What exactly constitutes a global city? Whether and how can a city achieve global city status? How does globalization affect different cities differently? How are cities related to others in a globalizing world? Attempts to answer these questions and more have generated a great deal of both theoretical interrogation and empirical development in the global cities literature.

Despite its immense contribution to a better understanding of globalization as well as contemporary urbanization, the global cities literature has received some mixed reviews so far. There has been a concern that its overwhelming focus on emergent global cities could mislead our understanding of the urban impact of globalization on other cities, including both smaller cities in the developed world and almost all cities in the developing world (McCann, 2004; Robinson, 2002; Short, 2004). Another criticism of the global cities literature has pointed to the economic aspects of city life being excessively emphasized, while the various other aspects of urban changes neglected (Hannerz, 1996; King, 1996 and 2004).

This chapter critically reviews the global cities literature and criticisms of its growing dominance in current urban studies. Here I divide the literature into four broad groups for the sake of comparison, although there is always a great danger of over-simplification and omission in this categorizing endeavour. First, many have focused on the unique nature of emergent global cities as the command and control centre of the globalizing economy. Drawing heavily on Saskia Sassen's works (1991 and 1994), this group of researchers has paid particular attention to the financial institutions and business services that have separated London, New York and Tokyo from all other cities in the contemporary world. Second, a large number of researchers have attempted to draw a comprehensive, if not exhaustive, hierarchical mapping of major cities across the world. There has been a general consensus that John Friedmann's works (1986, 1995) have inspired many urban scholars to look at inter-city relations beyond the national scale and more towards the transnational/global scales. A third group of researchers has been concerned less about what global cities have, but more about who defines what a global city should have and what non-global cities should do to achieve such status. This line of thinking pays a great deal of attention to the political and public accounts of global cities which are highly subject to the manoeuvre of political elites, business coalitions and/or local media (Paul, 2004). Finally, a growing body of literature has pointed out the lack of scholarship on non-global cities which are not as economically powerful as global cities but, nonetheless, are undergoing various forms of restructuring with relation to the forces of globalization (Robinson, 2002). This emerging trend is often called 'a globalizing city approach' compared to the existing 'global cities approach'. Each of the research trends is reviewed in the next four sections.

The interface of globalization and urbanization has been a core theme for global cities studies, yet it is certainly true different research groups have focused on different cities and different aspects of urban change in their case studies. For example, the studies that have identified global cities with the powerhouse of the world economy have focused exclusively on top-tier cities, notably London, New York and Tokyo, while those on politics of global cities have drawn examples from large cities where city elites considered the city to be one mega-urban project short of global city status. A highly positive sign of recent trends in global cities research and, in a broader context, urban studies is that more and more urban scholars are looking into the relationship between urbanization and globalization in under-researched and 'ordinary' cities across the world.

Command and control centres of the global economy

The conceptualization of global cities as the command and control centres of a globalizing economy draws greatly on Sassen's works (1991, 1994, 2002) (see Extract 7.1). In her explanation of the emergence of global cities, Sassen points to the opposing geographical trends of dispersal and centralization that globalization has involved. The global dispersal of economic activities, helped by both space-shrinking technologies and deregulation measures, has created a huge demand for expanded central control and management functions, such as corporate headquarters and advanced business services, including accounting, advertising, consulting, financial and legal services. These services, according to Sassen, tend to be disproportionately concentrated in large global cities, such as London, New York and Tokyo, where their operation can benefit from 'territorialized business networks' – in other terms, institutional thickness (Thrift, 1994) and territorial embeddedness (Budd, 1999).

Extract 7.1: From Sassen, S. (1991) *The Global City: New York, London, Tokyo*, Princeton, NJ: Princeton University Press, pp. 3–4.

The point of departure for the present study is that the combination of spatial dispersal and global integration has created a new strategic role for major cities. Beyond their long history as centers for international trade and banking, these cities now function in four new ways: first, as highly concentrated command points in the organization of the world economy; second, as key locations for finance and for specialized service firms, which have replaced manufacturing as the leading economic sectors; third, as sites of production, including the production of innovations, in these leading industries; and fourth, as markets for the products and innovations produced. These changes in the functioning of cities have had a massive impact upon both international economic activity and urban form: Cities concentrate control over vast resources, while finance and specialized service industries have restructured the urban social and economic order. Thus a new type of city has appeared. It is the global city. Leading examples now are New York, London, and Tokyo. …

    As I shall show, these three cities have undergone massive and *parallel* changes in their economic base, spatial organization, and social structure. But this parallel development is a puzzle. How could cities with as diverse a history, culture, politics, and economy as New York, London, and Tokyo experience similar transformations concentrated in so brief a period of time? Not examined at length in my study, but important to its theoretical framework, is how transformations in cities ranging from Paris to Frankfurt to Hong Kong and São Paulo have responded to the same dynamic. To understand the puzzle of parallel change in diverse cities requires not simply a point-by-point comparison of New York, London, and Tokyo, but a situating of these cities in a set of global processes. In order to understand why major cities with different histories and cultures have undergone parallel economic and social changes, we need to examine transformations in the world economy. Yet the term global city may be reductive and misleading if it suggests that cities are mere outcomes of a global economic machine. They are specific places whose spaces, internal dynamics, and social structure matter; indeed, we may be able to understand the global order only by analyzing why key structures of the world economy are *necessarily* situated in cities.

It is still open to debate whether global cities represent a new type of city in the age of globalization (Sassen, 1991: 4) or whether their emergence marks 'a qualitatively new phase in urban development' (Taylor, 2004: 27). However, there has been a general consensus that global cities are the key locations of most aspects of globalization, with the continuing globalization of finance and its demand for centralized management a major cause of economic, cultural and spatial restructuring in global cities. An understanding among urban scholars that globalization and urban changes in global cities are mutually constitutive, instead of the latter being a mere outcome of the former, has indeed spurred the formation of so-called 'global (world) city school' in urban studies in the past decade (Knox and Taylor, 1995).

Emergent global cities' control and command functions have been affirmed by their housing of large numbers of corporate headquarters (Short and Kim, 1999), business service firms (Martin, 1999; Taylor et al., 2002) and high-paid professional jobs (Florida, 2005; Hartley, 2005). In an effort to explain the spatial clustering of business services, many have also argued that the place-specific determinants of global cities are less economic and more social and cultural, such as the business and personal networks of leading financial experts that could not be easily replicated elsewhere (Budd, 1999; Thrift, 1994). Alongside managing and servicing the global economy, cherishing cosmopolitan cultures is considered another very important characteristic of global cities (Hannerz, 1996; Yeoh and Chang, 2001). According to Hannerz (1996), global cities, including New York, London, Paris, Los Angeles and Miami, have built the status of global cultural market-place on the presence, transitory and permanent, of transnational business elites, immigrants from developing countries, expressive specialists and international tourists.

While the commonalities of a handful of global cities have been emphasized, ranging from clustered financial firms through to high rents in central business districts to growing social polarization among urban residents, there are also pronounced differences among them. For example, Tokyo's global city status has often been attributed to the economic prowess of Japanese multinational corporations, while both London and New York have been able to complement their economic power with a long history of cosmopolitanism. In the meantime, New York is situated in a relatively decentralized national urban hierarchy, yet London and Tokyo each in a highly centralized one. More comparative studies could look into issues of ethnic communities and urban politics of globalization in these global cities.

Global cities at the apex of the global urban hierarchy

A major trend in the global cities literature is its focus on connections and hierarchical relations between major cities around the world, compared to Sassen's emphasis on the attributes held by emergent global cities. Heralded by John Friedmann's seminal works on the world city hypothesis (Friedmann and Wolff, 1982; Friedmann, 1986, 1995), many have attempted to draw a global hierarchy of cities indicative, or at least suggestive, of individual cities' influence in the current world economy (Knox and Taylor, 1995; Short et al., 1996; Smith and Timberlake, 2002) (see Figure 7.1). Arranging world cities hierarchically, 'in accord with the economic power they command' (Friedmann, 1995: 25), would naturally beg the question of indicators and data. However, the lack of data availability has hampered efforts in measuring individual cities' economic command across the globe, which has been a major deficiency of world city research. Indeed, the nature and extent of relations between cities has long been researched

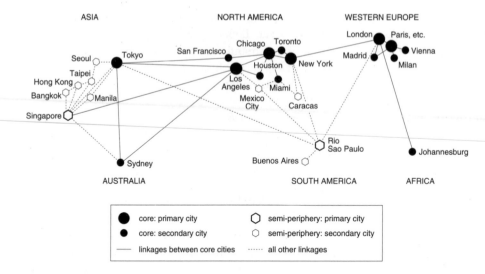

Figure 7.1   Friedmann's hierarchy of world cities

in urban system studies, which included the hierarchical classification of cities along population size, spatial diffusion analysis and air/transportation network analysis. Prior to Friedmann's (1986: 69) attempt to 'link urbanization processes to global economic forces', however, most of the studies on inter-city relations had aimed at finding the urban centres at the regional/national scale. There is little debate that Friedmann's world city research has contributed immensely to a better understanding of large metropolises and connections between them by setting its analysis unit at a global scale.

Despite much appreciation of Friedmann's contribution to the theoretical development in urban studies, there has been a growing criticism of his classification of world cities, called a 'world city hierarchy' (Friedmann, 1986) and 'spatial articulation of 30 world cities' (Friedmann, 1995). Much of the criticism has centred on the empirical data, or the lack of them, used in establishing this urban hierarchy (Godfrey and Zhou, 1999; Short et al., 1996; Smith and Timberlake, 1995, 2002; Taylor, 2004), while some have questioned the relevance of the hierarchical approach to our understanding of those 'ordinary cities' which do not feature in these hierarchies (Robinson, 2002). Friedmann's (1986) list of selection criteria for world cities includes a major financial centre, headquarters for multinational firms, international institutions, business services, an important manufacturing centre, a major transportation node and population size, but he fails to explain the relationships between these criteria and his urban hierarchy. This absence of specifications on evidence is described as 'an empirical cop-out' in Taylor (2004: 34).

In recent years, a great deal of scholarship has paid attention to data sources that might offer a more empirically grounded, if not sound, hierarchy of world cities. Many have focused on one or two indicators such as business services, multinational firms and/or international airline networks, instead of exhaustive aspects of cities' economic power, to complete their hierarchy. Some have explored new indicators that would be more telling about individual cities' influence in a culturally globalizing world, including the host of international sports competitions (Short, 2004; Short et al., 1996), the concentration of global entertainment industries (Abrahamson, 2004) and the presence of the creative class that creates 'meaningful new forms' in arts, writing, designing, architecture, science and business management (Florida, 2005: 34). Some have attempted to devise a composite index that could quantify various qualities of a global city (for example, Cai and Sit, 2003), yet with little success in producing any convincing results.

The endeavour to tackle the issue of 'empirical poverty' has been led by the Globalization and World Cities Study Group and Network (GaWC), which was created in 1998 by Peter Taylor and Jon Beaverstock at Loughborough University. GaWC has carried out unprecedentedly extensive data collection and quantitative network analysis to establish inter-city relations at the global scale (Taylor, 2004). In addition to creating a large data matrix, GaWC has gained a high reputation for its core members' productivity in research. Indeed, its website has emerged as the essential place for global cities research, listing latest research outcomes and, consequently, controlling major issues and topics related to global cities in the past few years. Although GaWC's website covers almost all aspects of global cities and their network, its core members have paid particular attention to the role of advanced business services in both globalizing the current world economy and connecting major cities around the world. One emerging data matrix, named the GaWC 100, is comprised of headquarter and subsidiary locations of 100 leading accountancy, advertising, banking and finance, insurance, law and management consultancy firms across the globe. Based on this matrix, Taylor (2004) establishes the 'world city network' in which global cities function as the nodes in transnational economic flows. Independent of GaWC, Alderson and Beckfield (2004) also seek to grasp the world city system through a network analysis of the location of the world's 500 largest multinational firms' headquarters and subsidiaries.

It is still too early to evaluate thoroughly and fairly GaWC's contribution to our understanding of global cities and, in broader terms, the globalization–urbanization nexus, since a large number of its projects are still underway. While his data collection and network analysis is innovative and impressive indeed, Taylor's (2004) maps and configurations demonstrating global network connectivities of cities do not look as sophisticated as his theoretical argument for inter-city relations at the global level (Figure 7.2). Another criticism that has been raised is that GaWC pays little attention to numerous large cities which do not represent a market for leading service firms (Short, 2004: 50) calls them 'black holes' of advanced global capitalism. However, it should be noted that GaWC has heightened the importance of empirical, quantitative bases in the discussion on the global urban

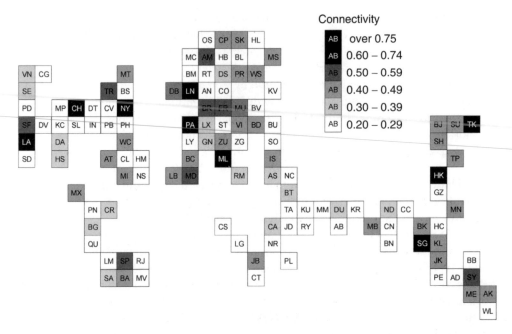

Figure 7.2 GaWC's global network connectivity (for city codes, see Taylor, 2004: 72)

hierarchy where lists of global cities were often presented without minimum specifications.

Global city status as a political project

With the growing use of the term global cities in public and political accounts and the situations it applies to, many have grown interested in the politics of global cities. An increasing number of urban studies have been conducted to probe urban and/or national governments' efforts and, more precisely, political interests in boosting their city's global citiness (Kim, 2004; Rutheiser, 1996; Todd, 1995). Needless to say, it is still very arguable what exactly constitutes a global city and, furthermore, how individual cities can achieve global city status. Indeed, this vagueness in conception has allowed politicians and related interest groups to pursue freely the coveted status and to promote a set of strategies that might benefit themselves more than their city's pursuit of that status. It is not uncommon to see that selected images of a global city are crafted, articulated and subsequently promoted to serve highly politically motivated agendas, with politicians making somewhat arbitrary claims of their city possessing global qualities and potential connections (Paul, 2004; Yeoh, 2004).

The politics of local economic development has been examined extensively in the new urban politics literature that highlights political and business interests behind prevalent urban boosterism and entrepreneurial governance in the USA and beyond (Hall and Hubbard, 1998; Logan and Molotch, 1987). It is argued that the notion of being or becoming a global city has added one more dimension to the existing debates on successful, desirable urban governance (Jessop, 1998; Keil, 1998; McCann, 2004). Of the growing scholarship on the politics of global city status, the following three aspects stand out. First, local politicians aggressively promote the vision of their city finally becoming a global city, when they solicit public support for such controversial plans as urban redevelopment projects (Todd, 1995), large-scale infrastructural upgrading (Jenks, 2003; Shachar, 1994; Wu and Yusuf, 2004) and hosting/bidding for high-profile international events (Hiller, 2000; Kim, 2004). Although it is very debatable that such projects or events would bring global city status to the city in question, the global city rhetoric often proves to be worth exploiting when the public react rather positively to the political manoeuvre of their city being recognized globally.

Second, the global city rhetoric is also employed widely by politicians and business associations who argue for large-scale tax incentives and business-friendly economic policies (Hall and Hubbard, 1998). They often associate the city's global-cityness with its homegrown companies' global reputation and/or its ability to attract internationally famous businesses while attempting to fulfil their ambitions, including dominating public debates on good governance.

Third, the global city rhetoric is often capitalized when there needs to be heightened multiculturalism and, in a broader sense, cosmopolitanism among city residents (Keil and Ronneberger, 2000; Machimura, 1998; Yeoh, 2004; Yeoh and Chang, 2001). The imagineering of a global city that nurtures cosmopolitan cultures is a very effective tool for city governments to solicit public support for costly cultural projects, such as the construction, or renovation, of opera houses and the staging of ethnic festivals.

While highlighting the absurdity, rather than legitimacy, of global city status, studies on the politics of global cities have also paid attention to the role of the media in global city projects (Schuster, 2001; Wilson, 1996). Local/national news media are primary agents in not only transmitting selected images of a global city, but also building those images. They often play a crucial role in their city's hyped rivalry with other major cities around the world, as we witness every four years with cities bidding for the Summer Olympic Games (Short, 2004).

Globalizing cities

Serious concerns have been raised about the almost exclusive focus of global cities research on a handful of large cities, which has resulted in a certain degree of neglect for both smaller cities in developed countries and almost all cities in developing countries (Grant and Short, 2002; Gugler, 2004; Krause and Petro, 2003; Lee and Yeoh, 2004; Öncü and Weyland, 1997; Robinson, 2002; Short, 2004; Simon, 1995). The neglect of so-called non-global cities or ordinary cities is a limit to 'our imaginations about the futures of cities' (Robinson, 2002: 535).

The urge to diversify case study areas has given rise to globalizing processes of other cities, whether they are large or small, whether increasingly integrated into or marginalized from the global economy, whether radically changing with or gradually adapting to, or resisting global trends, and whether benefiting or suffering from the process of globalization. Although it might not mean that all sorts of cities across the globe have finally come to receive their due academic attention, this emerging trend in the global cities literature has encompassed various cities, ranging from Lexington in Kentucky (McCann, 2004), to Moscow (Barter, 2004) and to Accra (Grant, 2002). Many urban journals and conferences have devoted their special issues to an investigation of the distinctive experiences of globalization in globalizing cities (Marcuse and Kempen, 2000; Öncü and Weyland, 1997), forgotten places in the globalization discourse (Lee and Yeoh, 2004), global city-regions (Scott, 2001; Simmonds and Hack, 2000), world cities beyond the West (Gugler, 2004) and cities in the global south and in the mid-range of the global hierarchy (Sassen, 2002).

Extract 7.2: From Short, J.R. (2004) *Global Metropolitan: Globalizing Cities in a Capitalist World*, London: Routledge, pp. 45–6.

[T]o fully understand the connection between globalization and the city, it is important to extend our understanding beyond the narrow focus on the usual suspects of large global cities. While the top level of the global urban hierarchy is an important object of consideration, when it becomes the sole focus of understanding the globalization–city nexus then understanding is skewed and partial. Theories of globalization that only build upon the experiences of a few global cities have a precariously narrow grip on the full range of the urban experience, while the arid search for world cityness dooms a large number of cities to marginality or even exclusion from research on globalization and the city.

Shifting attention away from identifying a narrow range of world cities to a more inclusive concern with globalizing cities extends the range of theorizing on how globalization takes place. Globalizing city is a shorthand term for the idea that many, if not all, cities act as transmission points for globalization and are the focal point for a whole nexus of globalization/localization relationships. We selected seven cities ranging in population size from just over 100,000 to almost seven and half million, and ranging in 'world-cityness' measures from relatively high to not even registering. We purposely selected cities that were not on the usual list of world cities and below the top echelon of the global urban hierarchy. Our case studies could have been different but the general point remains that even small, so called 'non-world' cities can be examined for evidence of globalization. The case studies were brief. Each city could have been the focus of the entire book. However, they were indicative of the rich possibilities of using the globalizing city themes and the selected topics. Globalization is a phenomenon that is occurring around the world in a range of cities. By focusing on the idea and practice of globalizing cities, our understanding of both globalization and the city can only be enriched and deepened. The theme of globalizing cities lays the basis for a sounder theoretical understanding of the impact of globalization on different cities in the world and for a more profound explanation of the connection between urbanization and globalization.

The broadening of study areas beyond the global cities category has prompted a meaningful shift in research focus from categorization to processes in the global cities literature. While existing studies have focused more on the represented aspects of global cities, including large-scale redevelopment, multinational corporations, business services, political projects with global vision and international events, many now turn their attention to delicate changes on streets and neighbourhoods, such as immigrant communities (Allen and Wilcken, 2001; Hamilton and Chinchilla, 2001), architectural designs (King, 2004; Krause and Petro, 2003), land values (Nijman, 2002) and everyday practices in inner cities (Durrschmidt, 2000; Eade, 1997). These micro-scale studies capitalize on the existence of the global (e.g. Wal-Mart stores) within the local, and put a great emphasis on capturing the dynamics of globality and locality in detail (Jordan, 2003). It may seem to be reiterating the oft-cited cliché in the globalization literature that individual places rework globalizing effects, instead of being completely consumed by them. However, the fact that urban places undergo changes in different but comparable ways promotes comparative studies as well as case studies of urban changes at various sites and under various circumstances (Grant and Nijman, 2002). Cities, no matter how localized their cultures and practices are, exist in a globalizing world as differentiated units across a continuous, global grid rather than as isolated, exceptional spheres (Hannerz, 1996; Oren, 2003). Their actions on, and reactions to, global forces could certainly be compared in terms of not only individual distinctiveness and locality but also commonality and globality. The interface of globalization and urbanization in globalizing cities could generate a truly large body of comparative studies based on thorough case studies of various urban changes caused by and causing globalization.

Summary

Urbanization has been closely linked to a series of historical processes, including modernization, industrialization, deindustrialization and capitalism that have restructured, and continue to restructure in some cases, various aspects of human society. Globalization has emerged as another seemingly overbearing mate for urbanization in the current world. While the effects of globalization in all forms and at all levels have been an underlying theme in globalization studies, some urban scholars have successfully conceptualized recent economic, cultural, political, spatial changes in large cities as an enabling factor as well as an inevitable outcome of globalization. This success has been repeatedly celebrated in numerous studies of a handful of global cities where so-called movers and shapers of the current world economy are clustered together.

Despite having created arguably unprecedented excitement in urban studies in the past decade, global cities research has been criticized for both its methodological bias towards categorization, hierarchy and economic analysis, and its geographical bias towards exceptionally large metropolises in the developed world. A globalizing cities approach is a collective term for a wide range of ongoing efforts to fill a gap in the scholarship on the interface of globalization and urbanization across the globe. It attempts to grasp varied aspects of urban change in under-researched, non-global cities as well as non-economic aspects of urban change in global cities. Indeed, a growing number of case studies are being conducted in smaller cities and former socialist cities among other places. It should not be too ambitious to expect that some of these case studies will turn into comparative studies of globalizing cities across the globe, since cities are globalizing in their own ways.

References

Abrahamson, M. (2004) *Global Cities*, New York: Oxford University Press.

Alderson, A.S. and Beckfield, J. (2004) 'Power and position in the world city system', *American Journal of Sociology*, 109 (4), 811–851.

Allen, R. and Wilcken, L. (eds) (2001) *Island Sounds in the Global City: Caribbean Popular Music and Identity in New York*, Urbana, IL: University of Illinois Press.

Barter, J.H. (2004) 'Moscow's changing fortunes under three regimes', in J. Gugler (ed.), *World Cities beyond the West: Globalization, Development and Inequality*, Cambridge: Cambridge University Press.

Budd, L. (1999) 'Globalization and the crisis of territorial embeddedness of international financial markets', in Ron Martin (ed.), *Money and the Space Economy*, Chichester: John Wiley and Sons.

Cai, J. and Sit, V.F.S. (2003) 'Measuring world city formation – the case of Shanghai', *The Annals of Regional Science*, 37 (3), 435–446.

Durrschmidt, J. (2000) *Everyday Lives in the Global City: The Delinking of Locale and Milieu*, London: Routledge.

Eade, J. (1997) *Living the Global City: Globalization as Local Process*, London: Routledge.

Florida, R. (2005) *Cities and the Creative Class*, New York: Routledge.

Friedmann, J. (1986) 'The world city hypothesis', *Development and Change*, 17 (3), 69–83.

Friedmann, J. (1995) 'Where we stand: a decade of world city research', in P.L. Knox and P.J. Taylor (eds), *World Cities in a World-System*, Cambridge: Cambridge University Press.

Friedmann, J. and Wolff, G. (1982) 'World city formation: an agenda for research and action', *International Journal of Urban and Regional Research*, 6 (3), 309–344.

Godfrey, B.J. and Zhou, Y. (1999) 'Ranking world cities: multinational corporations and the global urban hierarchy', *Urban Geography*, 20 (3), 268–281.

Grant, R. (2002) 'Foreign companies and glocalizations: evidence from Accra, Ghana', in R. Grant and J.R. Short (eds), *Globalization and the Margins*, New York: Palgrave.

Grant, R. and Nijman, J. (2002) 'Globalization and the corporate geography of cities in the less-developed world', *Annals of the Association of American Geographers*, 92 (2), 320–340.

Grant, R. and Short, J.R. (eds) (2002) *Globalization and the Margins*, New York: Palgrave.

Gugler, J. (ed.) (2004) *World Cities beyond the West: Globalization, Development and Inequality*, Cambridge: Cambridge University Press.

Hall, T. and Hubbard, P. (eds) (1998) *The Entrepreneurial City: Geographies of Politics, Regime and Representation*, Chichester: John Wiley.

Hamilton, N. and Chinchilla, N.S. (2001) *Seeking Community in a Global City: Guatemalans and Salvadorans in Los Angeles*, Philadelphia: Temple University Press.

Hannerz, U. (1996) *Transnational Connections: Culture, People, Places*, London: Routledge.

Hartley, J. (ed.) (2005) *Creative Industries*, Oxford: Blackwell.

Hiller, H.H. (2000) 'Mega-events, urban boosterism and growth strategies: an analysis of the objectives and legitimations of the Cape Town 2004 Olympic Bid', *International Journal of Urban and Regional Research*, 24 (2), 439–458.

Jenks, M. (2003) 'Above and below the line: globalization and urban form in Bangkok', *The Annals of Regional Science*, 37 (3), 547–557.

Jessop, B. (1998) 'The narrative of enterprise and the enterprise of narrative: place marketing and the entrepreneurial city', in T. Hall and P. Hubbard (eds), *The Entrepreneurial City: Geographies of Politics, Regime and Representation*, Chichester: John Wiley.

Jordan, J. (2003) 'Collective memory and locality in global cities', in L. Krause and P. Petro (eds), *Global Cities: Cinema, Architecture, and Urbanism in a Digital Age*, New Brunswick, NJ: Rutgers University Press.

Keil, R. (1998) 'Globalization makes states: perspectives on local governance in the age of the world city', *Review of International Political Economy*, 5 (4), 416–446.

Keil, R. and Ronneberger, K. (2000) 'The globalization of Frankfurt am Main: core, periphery and social conflict', in P. Marcuse and R. van Kempen (eds), *Globalizing Cities: A New Spatial Order?*, Oxford: Blackwell.

Kim, Y.H. (2004) 'Seoul: complementing economic success with Games', in J. Gugler (ed.), *World Cities beyond the West: Globalization, Development and Inequality*, Cambridge: Cambridge University Press.

King, A.D. (ed.) (1996) *Re-presenting the City: Ethnicity, Capital and Culture in the 21st Century Metropolis*, New York: New York University Press.

King, A.D. (2004) *Spaces of Global Cultures: Architecture Urbanism Identity*, London: Routledge.

Knox, P.L. and Taylor, P.J. (eds) (1995) *World Cities in a World System*, Cambridge: Cambridge University Press.

Krause, L. and Petro, P. (eds) (2003) *Global Cities: Cinema, Architecture, and Urbanism in a Digital Age*, New Brunswick, NJ: Rutgers University Press.

Lee, Y.S. and Yeoh, B.S.A. (2004) 'Introduction: globalization and the politics of forgetting', *Urban Studies*, 41 (12), 2295–2301.

Logan, J.R. and Molotch, H.L. (1987) *Urban Fortunes: The Political Economy of Place*, Berkeley: University of California Press.

Machimura, T. (1998) 'Symbolic use of globalization in urban politics in Tokyo', *International Journal of Urban and Regional Research*, 22 (2), 183–194.

Marcuse, P. and van Kempen, R. (eds) (2000) *Globalizing Cities: A New Spatial Order?*, Oxford: Blackwell.

Martin, R. (1999) *Money and the Space Economy*, Chichester: John Wiley.

McCann, E.J. (2004) 'Urban political economy beyond the "global city"', *Urban Studies*, 41 (12), 2315–2333.

Nijman, J. (2002) 'The effects of economic globalization: land use and land values in Mumbai, India', in R. Grant and J.R. Short (eds), *Globalization and the Margins*, New York: Palgrave.

Öncü, A. and Weyland, P. (eds) (1997) *Space, Culture and Power: New Identities in Globalizing Cities*, London: Zed Books.

Oren, T.G. (2003) 'Gobbled up and gone – cultural preservation and the global city market place', in L. Krause and P. Petro (eds), *Global Cities: Cinema, Architecture, and Urbanism in a Digital Age*, New Brunswick, NJ: Rutgers University Press.

Paul, D.E. (2004) 'World cities as hegemonic projects: the politics of global imagineering in Montreal', *Political Geography*, 23 (5), 571–596.

Robinson, J. (2002) 'Global and world cities: a view from off the map', *International Journal of Urban and Regional Research*, 26 (3), 531–554.

Rutheiser, C. (1996) *Imagineering Atlanta: The Politics of Place in the City of Dreams*, London: Verso.

Sassen, S. (1991) *The Global City: New York, London, Tokyo*, Princeton, NJ: Princeton University Press.

Sassen, S. (1994) *Cities in a World Economy*, Thousand Oaks, CA: Pine Forge Press.

Sassen, S. (ed.) (2002) *Global Networks, Linked Cities*, New York: Routledge.

Schuster, J.M. (2000) 'Ephemera, temporary urbanism, and imaging', in L.J. Vale and S.B. Warner Jr, (eds), *Imaging the City: Continuing Struggles and New Directions*, New Brunswick, NJ: Centre for Urban Policy Research.

Scott, A.J. (ed.) (2001) *Global City-Regions: Trends, Theory, Policy*, Oxford: Oxford University Press.

Shachar, A. (1994) 'Randstad Holland: a "world city"?', *Urban Studies*, 31 (3), 381–400.

Short, J.R. (2004) *Global Metropolitan: Globalizing Cities in a Capitalist World*, London: Routledge.

Short, J.R. and Kim, Y.H. (1999) *Globalization and the City*, New York: Longman.

Short, J.R., Kim, Y., Kuus, M. and Wells, H. (1996) 'The dirty little secret of world cities research: data problems in comparative analysis', *International Journal of Urban and Regional Research*, 20 (4), 697–717.

Simmonds, R. and Hack, G. (eds) (2000) *Global City Regions: Their Emerging Forms*, London: E. and F.N. Spon.

Simon, D. (1995) 'The world city hypothesis: reflections from the periphery', in P.L. Knox and P.J. Taylor (eds), *World Cities in a World-System*, Cambridge: Cambridge University Press.

Smith, D. and Timberlake, M. (1995) 'Cities in global matrices: toward mapping the world-system's city system', in P.L. Knox and P.J. Taylor (eds), *World Cities in a World-System*, Cambridge: Cambridge University Press

Smith, D. and Timberlake, M. (2002) 'Hierarchies of dominance among world cities: a network approach', in S. Sassen (ed.), *Global Networks, Linked Cities*, New York: Routledge.

Taylor, P.J. (2004) *World City Network: A Global Urban Analysis*, London: Routledge.

Taylor, P.J., Walker, D.R.F. and Beaverstock, J.V. (2002) 'Firms and their global service networks', in S. Sassen (ed.), *Global Networks, Linked Cities*, New York: Routledge.

Thrift, N. (1994) 'On the social and cultural determinants of international financial centres: the case of the City of London', in S. Corbridge, R. Martin and N. Thrift (eds), *Money, Power and Space*, Oxford: Blackwell.

Todd, G. (1995) '"Going global" in the semi-periphery: world cities as political projects: the case of Toronto', in P.L. Knox and P.J. Taylor (eds), *World Cities in a World-System*, Cambridge: Cambridge University Press.

Vertovec, S. and Cohen, R. (eds) (2002) *Conceiving Cosmopolitanism: Theory, Context, and Practice*, Oxford: Oxford University Press.

Wilson, H. (1996) 'What is an Olympic city? Visions of Sydney 2000', *Media, Culture and Society*, 18, 603–618.

Wu, W. and Yusuf, S. (2004) 'Shanghai: remaking China's future global city', in J. Gugler (ed.), *World Cities beyond the West: Globalization, Development and Inequality*, Cambridge: Cambridge University Press.

Yeoh, B.S.A. (1999) 'Global/globalizing cities', *Progress in Human Geography*, 23 (24), 607–616.

Yeoh, B.S.A. (2004) 'Cosmopolitanism and its exclusion in Singapore', *Urban Studies*, 41 (12), 2431–2445.

Yeoh, B.S.A. and Chang, T.C. (2001) 'Globalising Singapore: debating transnational flows in the city', *Urban Studies*, 38 (7), 1025–1044.

8   INNOVATION AND CREATIVITY

Andy C. Pratt

This chapter

O Argues that the city has always acted as a site of economic and cultural creativity

O Considers why particular 'cultural industries' are now conventionally described as the driver of post-industrial city economies

O Provides a critical overview of different theories of urban creativity, at the same time noting a tendency to conflate different processes of innovation and creativity

---

Introduction

Innovation and creativity matter, whether in terms of economic opportunity, social problem-solving, the codification of old ideas or simply the generation of new ways of understanding. From an historical point of view, more innovation and more creativity has emerged from cities than from rural areas. Moreover, huge cities appear to be more favoured with innovation and creativity than smaller ones. This suggests cities may have a significant quality that generates innovation and creativity. If more innovation takes place, there is a greater chance that a proportion of it will be translated into novel products and economic growth. The explanation of how, why and where creativity and innovation occurs is complex and contentious, not least because these terms are slippery. For many, the most obvious concern is economic growth, but innovation and creativity may also make cities more 'liveable', either as more interesting and stimulating environments, or as better governed and organized places. Finally, if we want more of these qualities of innovation and creativity, can they be encouraged, or undermined, through policy-making?

The relationship between innovation, creativity, and the city is at once both a simple and a complex one. It is simple in that it trades on a common assumption that cities are the hub of enterprise and culture. It is a complex issue, as with many others, because there is a considerable degree of confusion or fuzziness about what creativity and innovation are. Moreover, picking up on the common-sense notion already mentioned, we need to question whether cities are necessary or sufficient for the promotion, and exercise, of innovation and creativity. The answer to this point is surely that innovation and creativity can be found in rural and peripheral areas as well. What then is the role of cities? We might qualify this question by suggesting that cities simply have more of 'it' and thus set in train a process of cumulative causation: more begets more. In this case, the 'it' may be, among other things, a bigger market or more co-creators that may constitute a critical mass. We may further want to question the precise mechanics of causality: is it a simple diffusion process or one that is structured? Which direction is causality and in what way do, say, innovation and cities relate? It is clear that the innocent assumption of the relationship between cities, innovation and creativity is rather difficult to unravel.

This chapter explores this question by first clarifying what might be understood by innovation and creativity. The former term does have a social science literature related to it, and has long been debated primarily because of its perceived economic importance. However, the term creativity has been less commonly deployed. It is tempting to elide creativity and innovation; however, I want to keep them apart in order to examine the assumptions and tensions within the terms. By examining this tension we will arrive at a more satisfactory understanding of the relationship of both with urbanization.

A Romantic interlude

Romanticism may not seem, at first sight, to have much to do with our topic. However, I want to argue that notions of creativity and innovation are shaped by Romantic ideas. Underpinning much of the debate and potential policy is a question: is creativity a personal trait or a collective outcome? Those who hold the view that the world is made up of random and isolated individuals favour the former and those who argue instead for the structured social nature of action favour the latter. Of course, as always, people have sought to have both sides of the argument, seeing a recursive relationship between individuals and social structures. I want to depart from the usual line of discussion on this topic so as to contextualize these logical points and to positioning them in an intellectual history: a veritable genealogy of creativity and innovation (Osborne, 2003).

A robust notion of creativity can certainly be found in 'Romantic thought' – an artistic and cultural movement prevalent in Europe in the late eighteenth and early nineteenth centuries. The precise interpretation was varied, but in Britain it was positioned as a reaction to rationalism and neo-classical values. The core ideas concern the importance of the individual's subjective experience, which, it was argued, offered unique insight into 'truth' and 'beauty'. Romanticism challenged the dualism of imagination and judgement by proposing that imagination is self-validiating (Welleck, 1963). The conscious rationale of the Romantic artist was to break with accepted forms to gain the freedom of personal expression. It is from this body of thought that we get the notion of 'art for art's sake' (Easton, 1964). Within such a notion, which has come to constitute our lay understanding of the subject, we get the idea of the lone genius who (necessarily) exists on the margins of society, and the mirth that meets any suggestion that art, creativity or innovation can be planned or guided. Nonetheless, practical attempts to harness creativity can be found, for example, in advertising agencies, where the objective has been to acknowledge the otherness of artists/creatives and to incorporate them into a commercial process. Historically, this has been achieved through management of the division by physically separating workplaces, and having different dress, and time-keeping, requirements (Warlaumont, 2001). Society has also managed, and reproduced, this dualism by seeing artists as 'separate' and 'apart', offering financial support that seeks not to taint the art. Art, the output of creatives, is venerated as a thing in and of itself that improves society (by giving us access to 'higher things'). It will be clear that such Romanticism cannot be brushed aside: it constitutes a structure of thought for much Western literature (disclosing an ethnocentrism as well). It is these concepts and values that underpin a notion of individual creativity and innovation.

The Romantic notion also applies to science. Again, the figure of the sole pioneer battling against social norms and structures is presented, as is the 'eureka moment' of discovery, which is, of course, only the beginning of a process. This is brilliantly explored in the work of Latour (1988). We can note in the same way that artists have been constituted by a distinct division of labour, and spatiality, and so have scientists (Latour and Woolgar, 1986; Massey et al., 1992). The classic formulation is the laboratory (or, for artists, the studio) as the site of 'genius'. Whereas creatives are stereotyped for their chaotic genius, scientists are celebrated for their order and logic (as 'men in white coats').

Both cultural and scientific myths are reinforced by social institutions. The two are not equal, but different. In contrast to creativity, innovation is commonly discussed in an instrumental economic framework; the equivalent of 'art' is 'blue skies research'. However, in Western societies the economic value and political legitimacy given to the latter outweighs the former, although both are intrinsically 'useless' until application. In conclusion, the question of whether creativity and innovation are individual or collective enterprises is not simply a choice between alternative logics, but is one imbricated in society. It is important to bear this in mind when discussing these concepts.

Creativity

The term 'creative', as an adjective applied to processes, is relatively new. It has certainly received two quite distinct boosts in the last decade. The first of these occurred as the UK turned its focus on policy-making in relation to the cultural industries (Pratt, 2005). For a variety of reasons it used the term 'creative industries'; the publication of data on employment and output surprised many by the scale of its general contribution to the economy. Searching round for an explanation, many sought to conflate the creative economy with the information/knowledge economy (Garnham, 2005). The latter notion draws upon the work of Bell (1973), in particular his work on 'post-industrial society', which others have sought to restyle as 'the knowledge economy'. Details aside, this has positioned the creative economy as the cutting-edge of the knowledge economy: in other words, as the 'new thing'. The notion travelled around the world quickly as policy-makers sought some of the 'magic dust' of creativity. Bell argued the developed nations were increasingly becoming dominated by people involved in the manipulation of ideas rather than things. Moreover, he argued that scientists, and what others have called the symbolic analysts (Reich, 2000), would add value to products; in fact, they would be the key element in future production. Following this line of argument, 'creativity' is the source of competitive advantage in the post-industrial economy. So, it is understandable that policy-makers should seize upon the 'creative industries' both as a label and as a panacea. Of course, we can find creative activities outside the 'creative economy' (in the car industry, in administration, etc.). This clearly makes the concept unwieldy. At a government level this has led to the exploration of the ways in which creativity can be promoted via the education system (NACCCE, 1999).

A second and significant rise in the popularity of creativity can be noted in the use of creativity as spectacle and entertainment; put simply, as a means of attracting visitors and customers to a place. The notion of creativity as a selling point draws straight from business studies, and the concept of the entertainment (or experience) economy (Hannigan, 1998; Pine and Gilmore, 1999). Retailers and city managers have caught on to the fact that a good experience helps to open people's wallets. In part, this is a response to the fact that shops and malls are increasingly similar. The differentiator proposed here is 'the experience'. Such a notion is very attractive as it arguably requires no latent resources and, with investment in the right labour force and setting, it could succeed anywhere. An early version of this spectacle is embedded in the idea of urban tourism, where the unique assets of a city – its heritage – are the attraction, and hotel bed-nights and consumption are the benefit. In recent years, attempts have been made to attract the 'cultural tourist' to cities with the hope that they will be well off and well behaved, in contrast to the archetypical 'sun and sand' tourism (Pratt, 2000a).

Both cultural tourism and the entertainment economy have been used to justify investment in urban cultural infrastructure. This has extended to the creation of new architectural icons to attract visitors and investors, the classic example being Frank Gehry's Guggenheim museum in Bilbao, Spain (Figure 8.1). The more controversial such schemes are the more they attract publicity. Cities have long competed against one another for foreign direct investment and the nature of environment, or the cultural attractiveness, are heavily implicated in such promotion (Harvey, 1989). Not surprisingly, a number of indices have been developed that rank cities on liveability or even creativity (see Extract 8.1). The latest and most explicit linking of these aims can be found in the work of Florida (2002). Florida's point is not that creativity, and the presence of creative workers, makes cities successful *per se*, but rather that such a

Figure 8.1   Guggenheim museum, Bilbao

setting attracts creative people who like to be there. These workers themselves then become a magnet for high-technology, high growth firms seeking to employ them. The central point about all of these insights is that they are about consuming culture, and not about its production. As such they are not about innovation (product or process), nor are they about cultural production (or creativity).

Extract 8.1: From Mercer Human Resource Consulting

Zurich ranks as the world's top city for personal safety and security, according to a quality of living survey by Mercer Human Resource Consulting (see below). Scores for personal safety and security are based on relationships with other countries, internal stability, and crime, including terrorism. Law enforcement, censorship, and limitations on personal freedom are also taken into account. ... Cities are ranked against New York as the base city, which has a rating of 100. The analysis is part of a worldwide quality of living survey to help governments and major companies to place employees on international assignments. The index is used by companies to judge whether an expatriate is entitled to a 'hardship' allowance.

Mercer Human Resource Consulting Quality of Living: 2007 World Ranking
1. Zurich (Switzerland) 108.1
2. Geneva (Switzerland) 108.0
3. Vancouver (Canada) 107.7
4. Vienna (Austria) 107.7
5. Auckland (New Zealand) 107.3

The top five cities in Asia (world rankings in parentheses):
1. Auckland (tied for 5th)
2. Sydney (tied for 9th)
3. Wellington (12th)
4. Melbourne (17th)
5. Perth (21st)

Osaka (tied for 42nd) was the lowest ranking Asian city in the top 50.

The top five cities in Europe (world rankings in parentheses):
1. Zurich (1st)
2. Geneva (2nd)
3. Vienna (tied 3th)
4. Dusseldorf (tied for 5th)
5. Frankfort (7th)

The lowest ranking European city in the top 50 was Milan (tied for 49th).

The top five cities in the Americas (world rankings in parentheses):
1. Vancouver (3rd)
2. Toronto (15th)
3. Ottawa (tied for 18th)
4. Montreal (22nd)
5. Calgary (24th)

The lowest ranking Americas city in the top 50 was Seattle (tied for 49th).

Source: Mercer Human Resource Consulting (2007)
http://www.mercer.com/referencecontent.jhtml?idContent=1128060#top50all.
This survey should not be seen as 'scientific'; other competitor surveys are carried out, for example the Economist Intelligence Unit's Liveability Ranking, http://www.citymayors.com/environment/eiu_bestcities.html

A trio of less publicized, but nonetheless important, perspectives on creativity and the city have been discussed and deployed as the basis for policy-making. The first of these concerns creativity being used in a socially instrumental manner in cities. The argument here is that the pursuit of creative activities can be distracting and engaging, as well as a means of building understanding and mutual respect. There are many examples of socially innovative projects that use creativity to reinforce social cohesion (Bianchini and Santacatterina, 1997). Those evaluations that have taken place point to significant success (in terms of social cohesion) (DCMS, 1999). Another related use of creativity has been in social problem-solving. The work of Charles Landry (2000) is a testament to the possibility of innovative and socially embedded problem-solving through the use of local social and cultural resources to achieve social cohesion, and sometimes artful outcomes as well.

Thus far we have the notion of creativity as a magic bullet that leads to competitiveness, creativity as a 'honey pot' to boost consumption and attract investment, and creativity as a new cultural resource for problem-solving. A final application draws upon a different conception of creativity, one that concerns the creative industries themselves – cultural production. The work of Becker (1984) and Peterson (1976) challenges the individualist reading of creativity as well as the dominant reading of consumption and culture, and offers an alternative in the identification of an institutional framework that stresses the interconnections and feedback between processes of production, referred to elsewhere as the production system or chain (Pratt, 1997, 2000a, 2000b). Robinson articulates this process well (see Extract 8.2).

Extract 8.2: From Robinson, K. (2001) *Out of Our Minds: Learning to be Creative*, London: Capstone, p. 12.

We all have creative abilities and we all have them differently. Creativity is not a single aspect of intelligence that only emerges in particular activities, in arts for example. It is a systemic function of intelligence that can emerge wherever our intelligence is engaged. Creativity is a dynamic process that draws on many different areas of a person's experience and intelligence. We need to look at what it is in companies and organisations that blocks individual creativity. But this is only half the job. Creativity and innovation must be harnessed and not just released. Creativity is not purely an individual performance. It arises out of our interactions with ideas and achievements of other people. It is a cultural process. Creativity prospers best under particular conditions, especially where there is a flow of ideas between people who have different sorts of expertise. It requires an atmosphere where risk-taking and experimentation are encouraged rather than stifled. Just as individual creativity draws from many different skills and expertise across organisations. Creativity flourishes when there is a systemic strategy to promote it. The cultural environment should be modelled on the dynamics of intelligence. Many organisations stifle creativity in the structures they inhabit and the ethos they promote. If ideas are discouraged or ignored, the creative impulse does one of two things. It deserts or subverts the organisation. Creativity can work for you or against you.

Innovation

The study of innovation is a much more familiar couplet with cities and it is a staple of urban and regional economic analyses, although much of the work is surprisingly aspatial in its expression. The individual–social dualism is also found in the work on innovation. It is considered in the context of diffusion versus more structured processes of transmission of ideas. While first discussed in the economic realm, recent studies have stressed the social dynamics too. An important argument here is that in some cases – in the creative industries in particular – the social dimension is the key to explanation.

The early study of the process of innovation was characterized by a crude linear flow chart that began with an innovation, passed through patenting and ended up with a product in the market-place. From such a perspective one is drawn to 'blockages' to the flow, and to measures such as 'patents' as a surrogate of innovation. Research written from an institutional perspective opened up the importance of the organizational setting of innovation and suggested that it may not be the number of innovations but the means of translating them into products and sales that might be an issue. There is a huge literature on this (see Simmie, 1997, 2003, 2004). Perhaps most significant in this body of work has been that which derives from Lundvall (1992) on National Systems of Innovation (NSI). The point here is that the institutional context of innovation can enable or constrain new ideas turning into successful products in the market-place. A more recursive notion still is provided within the context of the study of the social shaping of technology (MacKenzie and Wajcman, 1999). While the NSI material is focused on the nation-state, it has applications at the urban level as demonstrated by Amin and Thrift's (1994) notion of institutional thickness.

This concept of a dense, variable and interlocking social and economic network has also been used by writers such as Grabher (1993) to explore spaces of innovation; here it is referred to as embeddedness. The notion of embeddedness draws on a legacy of work that can be traced back to Polanyi, who stressed the material interactions, and tacit nature, of economic life. It is not surprising that this approach holds attractions for urban researchers. Grabher (2004) argues that these network interdependencies are exaggerated with groupings of project-based enterprises, that is firms with a limited life, or firms that deal with limited life projects. Classic examples are the film and television industries and advertising.

Analyses of the social dimensions of economic life have pointed to a range of silences in traditional economic accounts of innovation and cities. To understand why, it is important to appreciate that a key strand in this work is that of formal neo-classical accounts of agglomeration. From this perspective, agglomeration happens in part as a consequence of the monopoly advantages afforded by space (many want to be in the same position at once). Moreover, it is argued that close proximity produces externalities or 'spill-over effects'. The problem for neo-classical analyses is that these phenomena are really 'residuals' that are not formally explained by the model. Formally, technology and innovation are also exogenous factors. This is a common trait of formal economic reasoning that draws the limits to explanation so narrowly and proposes unrealistic simplifying assumptions, such that the practical value of explanations are minimal, despite their algebraic sophistication. Arguably, in many cultural and creative industries the 'residual' is of greater explanatory power than the 'core'.

Early classical economic accounts such as those of Marshall (1920) referred to this externality as 'secrets of business in the air', though others have pointed to 'trust' (Gambetta, 1988). This notion has been formalized in the concept of the innovative milieu (Camagni, 1991; Moulaert and Sekia, 2003). Formal economic analyses have sought to account for this within a framework of a minimization of transactions costs (TCA) that occurs under proximity – contacts can be agreed through trust rather than lawyers (Williamson, 1987). Cost savings ensue. Thus innovation is externalized and available to those who are proximate. Such a framework has been deployed by Scott (2000) in his work on the 'image producing industries'.

There is a large debate about the origins and meaning of 'new industrial districts', and yet another body of work on the social and economic transformation of economies that they are indicative of (Amin, 1989, 1994). The key point made in particular by those influenced by Flexible Specialization (FS) accounts of economic development is that the interaction of producers of part-finished goods not only provides supply and demand (Piore and Sabel, 1984; Pratt, 1991), but also allows ample opportunity to experiment with production and switch suppliers to produce novel/innovative items. Storper (1997), building upon both FS, TCA and the embeddedness literature, points to what he terms 'untraded' dependencies. A more recent debate has sought to specify such dependencies more clearly via the notion of 'buzz' (Bathelt et al., 2004; Storper and Venables, 2004).

Yet another large body of work on innovation draws upon the work of Schumpeter. The notion of 'creative destruction' (nothing to do with creativity as discussed above) summed up how firm formation is not a linear or continuous process but one that has distinct upsurges and ruptures. The bunching of new innovations in an economic downturn generates an expanding market, and likewise how many firms go bust as markets overcrowd with 'old' innovations and margins are reduced. Schumpeter saw necessary value in this destruction of obsolete productive capacity as it cleared the way for the new. This cyclical notion of economic processes has been linked to innovation in two ways. First, the idea of business cycles: closer examination of actual production processes has cast doubt on the utility of such an idea to the extent that any product goes from immaturity to maturity without mutating. Second, others have focused on 50-year (Kondratieff) business cycles founded upon transformative technologies: coal, steam, steel, electricity, the internal combustion engine, and semiconductors (Marshall, 1987). It is argued that new technologies (and innovations) create an upsurge of growth. Hall (1985) has argued that different waves produce different regional fortunes. Elaborating, Hall (1998) uses a similar debate to account for the rise and fall of creative cities. Neo-Schumpetarian writers (Dosi, 1983; Freeman, 1986) have sought to pull back from the technological determinism of long waves and have offered institutional arguments for 'lock in' to particular technologies and processes to explain a similar process.

In summary, innovation is linked to various forms of organization. In all processes it involves a division of labour between thinking and doing. The precise nature of this relationship is the nub of the question. This question must be answered in relation to specific industries and forms of production. Arguably, more formal and codified processes can be carried out at a distance or in separate units. However, this does not account for all processes. Critically, many genuinely novel innovations arise not from bureaucratically controlled firms but a looser association. This can be 'internalized' or 'externalized'. In a sense, the ideal type is the mixed-use urban core that facilitates the social 'buzz' and informal interactions. So, cities are innovation hubs not because they have a particular technology or social group, but rather because they offer opportunities for experimentation and the interweaving of production and use. An extreme version of this can be found in the work of 'cool hunters', social anthropologists employed by large corporations both to seek out new trends and to see how existing trends and usage are mutating so that these can be fed back into the next product (Quart, 2003).

We can conclude that there is a strong cultural reading of innovation and creativity that embodies individuality and individualism. Analytically, this is matched by the assumptions of neo-classical economic analysis. Not surprisingly, innovation and agglomeration are some of the most poorly understood areas of neo-classical economics. Approaches that are critical of this norm take a more balanced perspective on both the social and the economic, as well as on the individual and the collective, and have been more successful in offering robust explanations of innovation and creativity. This research tends towards a greater attention to the interrelationships of process and the spatial–temporal embeddedness of action. We should be wary of committing the error of assuming that all processes will be either spatially embedded or disembedded. The latter notion led to a well-known error called the 'death of distance' (Cairncross, 1998), where it was assumed that with the development of electronic communications people would no longer need to be co-present. In many areas of economic activity, just the opposite has happened (Pratt, 2000b). Moreover, we should be sensitive to the fact that different industries have various processes that will be more or less sensitive to these processes. Arguably, the creative industries are more influenced by the need to be socially embedded within a particular milieu than many other activities. One of the insights from work on the cultural industries is that they have an urban focus. Indeed, recent work has pointed to the fact that the creative industries are the third major sector in the London economy (GLA, 2002).

The city

The city is the obvious place to engage with the situatedness of innovation and creativity. It stands to reason that if innovation and creativity are favoured in dense social and economic networks, that cities must be important. But, not all cities, or every part of a city, will benefit or support these activities. As theorists of the global cities have pointed out, the modern corporation usually locates its headquarters and research and development in major cities (Sassen, 2001). On the one hand, activities that can be codified are less sensitive to co-location with other parts of the production chain, or with co-producers. On the other hand, those processes that are more flexible and fluid, and require tacit knowledge exchange, and require diverse and overlapping networks of expertise, are more likely to be locationally limited. Moreover, if they need to be close to final consumption, as 'chart industries' such as music, fashion clothing, film and television, then cities will be a necessity. An interesting example is the relocation of major design studios into urban cultural hubs: Ford and Volkswagen are two recent examples in London's Soho. Whereas the production of designed vehicles can be done remotely as information is sufficiently codifiable, the subtleties of design require a recursive social and physical proximity to other creators, as well as decision-makers.

Research does point to the need for industries relying on a diverse, but free-lance, labour market to be located in a large city where there is a large labour pool; likewise employees need to be in a large city to attain more or less continuous employment. However, it is the need for various forms of social networking that favours places where there are a number of 'neutral' meeting places, such as bars, cafés and restaurants (the situated 'buzz'), or the latest award ceremony. In many cultural industries, information leading to the next job and new ideas circulate through word of mouth and by 'hanging out'. Proximity to other producers and workers, as well as proximity to consumers, is an essential part of keeping 'in the loop'. This is an exceptional group of industries in this respect, but for these the quality of the scanning and processing of knowledge at precisely the right time is the factor that delivers potential success or failure for their products.

The city will inevitably be attractive to those engaged in consumption: either due to proximity or as a consolidated big market. However, this is to assume that mass production takes place in the city. As I have shown elsewhere, production in the cultural sector is moving out of cities (Pratt, 1997), in much the same way that de-industrialization generated out-migration of manufacturing. However, what this leaves in the city is a growing concentration of innovative and productive activities that are heavily co-dependent. This co-dependence may be skill, knowledge, product or market-based. For example, industries that are project-based tend to draw upon a freelance labour market of common occupations.

Consumption does still matter. While we have seen the hollowing out of mass consumption from cities to shopping 'centres' on the peripheries of cities, that retail which remains in the centre tends to be elite and high fashion. The city centre has become an entertainment space – increasingly dominated by eating and drinking. But, above all, it is a performance space for purchasers (either in the street, or in the bars and clubs).

Writers commenting on the birth of the modernist city also discuss the anomie produced by the metropolis by counterpoising it to the community and order of rural areas (Wirth, 1938). We might reinterpret this as a reason why cities are so popular with creators and innovators as cities may free people from some of the limits of social norms (Bradbury, 1991). Here we have the space of production, of new ideas and the possibility of challenging others' views. Of course, this possibility is much enhanced by the wider mixing of backgrounds and experiences that international migration flows afford (Saxenian, 2002). This mix of ideas and investment that constitute migration is a tremendous benefit. Cities, especially trading cities, generate a vital resource for the innovator or creator, a stock of investment and a relatively rich market-place that can act as the financier and first consumer of new products and ideas. Finally, money and power, producers and consumers, operate within a set of legal and regulatory conditions that are subject to various governance structures that are usually located in major cities (see Chapter 9). Innovators need finance, but they also need lawyers and law-makers to protect their inventions, and states to invest in or promote them.

Urban policies

Given the importance of these innovative and creative activities in cities, what has been the policy response? First, as we noted above, there is a dominant Romantic notion suggesting that policy and guidance is not only inappropriate but also inimical to creativity. However, the social scientific understanding of these processes challenges such a notion. It points instead to the value of managing the setting or context, primarily the brokering of relationships. Have policy-makers applied such knowledge? Generally not. There are four dominant strands of policy-making. First, those that focus on cultural consumption have as their objective the generation of tourism and consumption. A recent twist on this is to attract to such consumption sites 'creative workers' who will be the future labour force for innovative industries (Florida, 2002). The policy is establishing 'cultural quarters' or heritage centres (Mommaas, 2004). Second, are those that have as their objective attracting foreign direct investment, to mark out a city from its competitors. Policies generate such resources from scratch, such as large infrastructure projects (new galleries, bridges, buildings, etc.). Third, are those that seek to create idealized physical production spaces, which match creative or innovative environments. Examples are science parks and 'creative hubs'. Finally, a developing area of policy seeks to focus on the strategic assessment of creativity and innovation (Jeffcutt and Pratt, 2002; Pratt, 2005). Examples are policies that seek to promote design, or film and TV production not through simple locational subsidy but through the building of network resources. The hope is that such socio-economic embeddedness will create a degree of inertia, or future proofing, for what are in other senses mobile activities.

Summary

The relationship between cities, innovation and creativity rests on a set of specificities about the production process, the exchange, and the role of tacit and codifiable knowledges. The physical locations and infrastructure are important as places to facilitate interactions. We have noted that some industries, and parts of producers and retailers, have moved out of urban cores, although others – cultural producers – have moved in. Correspondingly, creativity and innovation in the form of the cultural industries have become relatively more important in cities. Finally, not all cities share this re-valuing; the organization of cultural production is structured in a variety of ways. National markets remain important and thus capital cities attract much of the higher value creative activities and secondary cities within the national system tend to act as 'feeder' cities. Likewise, the cultural industries are international, and every nation does not have 'national champions' in all cultural industries. Most, like film, are the preserve of a few select cities in the world (Scott, 2004).

While creativity and innovation are to be found in all areas of social, economic and political life, some of the most intense interactions are found in the relatively new and fast-growing cultural industries. The cultural and economic impact of these activities makes them significant. With the dispersal of manufacturing and much retailing from cities, those activities that require tacit, face-to-face, diverse and uncodified interactions have come to dominate the urban core. The creative industries are one of those industries. They are different from most other industries not because they are 'symbolic', but because they are 'chart industries'. By this I mean that they are very sensitive to what is fashionable at any one time, that the shift of fashion is swift, and that it is a market where the winner takes all. To be a winner one has to release a lot of ideas and on average have hits; success comes from having more hits than average.

This chapter has sought to examine the relationships between the processes underlying creativity, innovation and cities. Cities are neither necessary nor sufficient for innovation and creativity to flourish. Moreover, innovation and creativity are not simple 'magic bullets' that can be added to the mix of cities to deliver competitive advantage. Innovation and creativity are processes; they are ways of doing that are always present. However, for some activities innovation and creativity are the 'core business'; for others they are less so. Innovation and creativity are most intense where there is a flow, and proximity, of challenging ideas and practices. Under such conditions ideas or practices can 'arc' from one area to another in a productive fashion. However, this usually requires close interaction of cognate activities. Critically, the means of transfer is through embodied practice: people moving, talking and doing, learning and misunderstanding. These environments are more often than not found in cities. We should not see an artificial separation between production and consumption. As in the practice of the cool hunters, there is a constant two-way flow. Cities, especially the more avant-garde, fashionable ones, constitute an important space for such learning to take place.

References

Amin, A. (1989) 'Flexible specialization and small firms in Italy: myths and realties', *Antipode* 21, 13–34.

Amin, A. (ed.) (1994) *Post-Fordism: a reader*, Oxford: Blackwell.

Amin, A. and Thrift, N. (1994) *Globalization, institutions, and regional development in Europe*, Oxford: Oxford University Press.

Bathelt, H., Malmberg, A., Maskell, P. (2004) 'Clusters and knowledge: local buzz, global pipelines and the process of knowledge creation', *Progress in Human Geography* 28, 31–56.

Becker, H.S. (1984) *Art worlds*, Berkeley: University of California Press.

Bell, D. (1973) *The coming of post-industrial society*, New York: Basic Books.

Bianchini, F. and Santacatterina, L.G. (1997) *Culture and neighbourhoods*, Strasbourg: Council of Europe Publishing.

Bradbury, M. (1991) *Modernism: a guide to European literature 1890–1930*, Harmondsworth: Penguin.

Cairncross, F. (1998) *The death of distance: how the communications revolution will change our lives*, Boston: Harvard Business School Press.

Camagni, R. (ed.) (1991) *Innovation networks: spatial perspectives*, London: Belhaven Press.

DCMS (1999) *A report for Policy Action Team 10: Arts and Sport National Strategy for Neighbourhood Renewal*, London: Department of Culture, Media and Sport/Social Exclusion Unit.

Dosi, G. (1983) 'Technological paradigms and technological trajectories', in C. Freeman (ed.), *Long waves in the world economy*, London: Butterworths.

Easton, M. (1964) *Artists and writers in Paris: the Bohemian idea 1803–1867*, London: Edward Arnold.

Florida, R.L. (2002) *The rise of the creative class: and how it's transforming work, leisure, community and everyday life*, New York: Basic Books.

Freeman, C. (1986) 'The role of technical change in national economic development', in A. Amin and J. Goddard (eds), *Technological change, industrial restructuring and regional development*, London: Allen and Unwin.

Gambetta, D. (1988) *Trust: making and breaking cooperative relations*, Oxford: Blackwell.

Garnham, N. (2005) 'From cultural to creative industries: an analysis of the implications of the "creative industries" approach to arts and media policy making in the UK', *International Journal of Cultural Policy* 11, 15–30.

GLA (2002) *Creativity: London's core business*, London: Greater London Authority.

Grabher, G. (1993) *The embedded firm: on the socioeconomics of industrial networks*, London: Routledge.

Grabher, G. (2004) 'Learning in projects, remembering in networks? Communality, sociality, and connectivity in project ecologies', *European Urban and Regional Studies* 11, 103–123.

Hall, P. (1985) 'The geography of the 5th Kondratieff', in P. Hall and A. Markusen (eds), *Silicon landscapes*, London: Allen and Unwin.

Hall, P. (1998) *Cities in civilization: culture, innovation and the urban order*, London: Weidenfeld and Nicolson.

Hannigan, J. (1998) *Fantasy city: pleasure and profit in the postmodern metropolis*, London: Routledge.

Harvey, D. (1989) *The urban experience*, Oxford: Blackwell.

Jeffcutt, P. and Pratt, A.C. (2002) 'Managing creativity in the cultural industries', *Creativity and Innovation Management* 11, 225–233.

Landry, C. (2000) *The creative city: a toolkit for urban innovators*, London: Comedia.

Latour, B. (1988) *The Pasteurization of France*, Cambridge, MA: Harvard University Press.

Latour, B. and Woolgar, S. (1986) *Laboratory life: the construction of scientific facts*, Princeton, NJ: Princeton University Press.

Lundvall, B.-Å. (1992) *National systems of innovation: toward a theory of innovation and interactive learning*, New York: Pinter Publishers.

MacKenzie, D.A. and Wajcman, J. (1999) *The social shaping of technology*, Buckingham: Open University Press.

Marshall, A. (1920) *Principles of economics: an introductory volume*, London: Macmillan.

Marshall, M. (1987) *Long waves of regional development*, London: Macmillan.

Massey, D., Quintas, P.D.W. and Wield, D. (1992) *High-tech fantasies: science parks in society, science and space*, London: Routledge.

Mommaas, H. (2004) 'Cultural clusters and the post-industrial city: towards the remapping of urban cultural policy', *Urban Studies* 41, 507–532.

Moulaert, F. and Sekia, F. (2003) 'Territorial innovation models: a critical survey', *Regional Studies* 37, 289–302.

NACCCE (1999) *All our futures: creativity, culture and education, report of the National Advisory Committee on Creative and Cultural Education.* London: Department for Education and Employment.

Osborne, T. (2003) 'Against "creativity": a philistine rant', *Economy and Society* 32, 507–525.

Peterson, R.A. (ed.) (1976) *The production of culture*, Beverley Hills, CA: Sage.

Pine, J.P. and Gilmore, J.H. (1999) *The experience economy: work is theatre and every business a stage*, Boston: Harvard Business School.

Piore, M.J. and Sabel, C.F. (1984) *The second industrial divide: possibilities for prosperity*, New York: Basic Books.

Pratt, A.C. (1991) 'Industrial districts and the flexible local economy', *Planning Practice and Research* 6, 4–8.

Pratt, A.C. (1997) 'The cultural industries production system: a case study of employment change in Britain, 1984–91', *Environment and Planning A* 29, 1953–1974.

Pratt, A.C. (2000a) 'Cultural tourism as an urban cultural industry: a critical appraisal', in Interarts (ed.), *Cultural tourism*, Barcelona: Interarts Turisme de Catalunya, Diputació de Barcelona.

Pratt, A.C. (2000b) 'New media, the new economy and new spaces', *Geoforum* 31, 425–436.

Pratt, A.C. (2005) 'Cultural industries and public policy: an oxymoron?', *International Journal of Cultural Policy* 11, 31–44.

Quart, A. (2003) *Branded: the buying and selling of teenagers*, Cambridge, MA: Perseus Publishing.

Reich, R.B. (2000) *The future of success*, New York: A. Knopf.

Robinson, K. (2001) *Out of our minds: learning to be creative*, London: Capstone.

Sassen, S. (2001) *The global city: New York, London, Tokyo*, Princeton, NJ: Princeton University Press.

Saxenian, A. (2002) 'Brain circulation: how high-skill immigration makes everyone better off', *The Brookings Review* 20, 28–31.

Schumpeter, J.A. (2006) *Business cycles: a theoretical historical and statistical analysis of the capitalist process*, Mansfield, CT: Martino Publications.

Scott, A.J. (2000) *The cultural economy of cities: essays on the geography of image-producing industries,* London: Sage.

Scott, A.J. (2004) 'Hollywood and the world: the geography of motion-picture distribution and marketing', *Review of International Political Economy* 11, 33–61.

Simmie, J. (1997) *Innovation, networks and learning regions?* London: Jessica Kingsley Publishers and the Regional Studies Association.

Simmie, J. (2003) 'Regions, globalisation and the knowledge-based economy', *Urban Studies* 40, 853–854.

Simmie, J. (2004) 'Innovation and clustering in the globalised international economy', *Urban Studies* 41, 1095–1112.

Snow, C.P. (1964) *The two cultures and a second look: an expanded version of 'The two cultures and the scientific revolution',* Cambridge: Cambridge University Press.

Storper, M. (1997) *The regional world: territorial development in a global economy,* New York: Guilford Press.

Storper, M. and Venables, A.J. (2004) 'Buzz: face-to-face contact and the urban economy', *Journal of Economic Geography* 4, 351–370.

Warlaumont, H.G. (2001) *Advertising in the 60s: turncoats, traditionalists, and waste makers in America's turbulent decade,* Westport, CT: Praeger.

Welleck, R. (1963) *Concepts of criticism,* New Haven, CT: Yale University Press.

Williamson, O.E. (1987) *The economic institutions of capitalism: firms, markets, relational contracting,* New York: Free Press.

Wirth, L. (1938) 'Urbanism as a way of life', *American Journal of Sociology* 44, 1–24.

9   STATES AND LAWS

Angus Cameron

This chapter

○ Describes the significance of cities as sites from which different forms of territorial control are exercised

○ Considers the specific role of law, money and taxation in maintaining the power and influence of the city

○ Poses key questions about the current importance of the city as a site of political influence in an increasingly 'footloose' and transnational world

Introduction

Recent literature on cities emphasizes much that is new and exciting about contemporary urban life. Cities are seen as increasingly important sites of an emergent global culture – a culture of global or world cities characterized by new forms of engagement, association, democracy, representation and interaction, all of which are articulated through new and often strange geographies (e.g. Soja, 1996; Amin and Thrift, 2002; Bell and Jayne, 2005). Although there is great variety within this literature, a common theme seems to be that contemporary cities and urban culture more generally are rewriting some of the basic assumptions of social life. The announcement during 2004 that humanity has become, for the first time, a majority urban species suggests that these ideas are both timely and appropriate. While this literature says much about the seeds of the future that can be detected in the contemporary city, it tends to say relatively little about the inheritance from the past. The purpose of this chapter is to highlight just one of these – the enduring and ambiguous relationship between cities and nation-states.

There are two main reasons for focusing on this relationship. The first is simply that it is perhaps the most significant relationship in the long-term development of both cities and states. The ways in which states formed, particularly in Europe, was in part determined by the nature, scale and power of the cities prevailing in particular territories (de Vries, 1976; Tilly, 1990; Reinhard, 1996; Bonney, 1998). There is, for example, a clear correlation between the commercial power of cities such as London and Amsterdam in the later Middle Ages and the subsequent formation of advanced industrial, bureaucratic and imperial states during the eighteenth and nineteenth centuries (Jacobs, 1985; Tilly, 1990). The second reason is that the inheritance of state–city relations is very much still present in the affective capacity of contemporary cities. In practice this means that for all the exciting possibilities that the new urban cultures seem to offer, cities continue to be severely constrained by their long historical subordination to the state (Weber, 1951; Frug, 1999).

The ascendance of state power over the city has been a continuous if not constant process since at least the early nineteenth century. In many ways, the history of the development of advanced capitalist 'national' political-economies – starting in Europe, but spreading worldwide by the late twentieth century – is the history of the assertion of the national state over the earlier disparate social, political and economic geographies of cities (Poovey, 1996; Joyce, 2003). Despite claims to the contrary, this process has not been reversed by the recently ascendant urban order, though the challenges posed to the state by the practices and discourses of globalization have undoubtedly made it more ambiguous. Despite this, the state has largely been written out of recent geographical thinking on cities, producing what can often be a rather one-sided and voluntaristic account of the potential of the city. What I want to argue here is that despite the promises of globalization, urbanization and network spaces, we are still living in an era of what might be called 'state-cities' – cities that remain relatively powerless with respect to the states they inhabit.

This chapter will briefly examine the histories of three overlapping aspects of the relationship between states and cities: taxation, money and law. The history of these three sets of practices and institutions has been an essentially urban one – they are all products of the city and have been in turn instrumental in the development of urban form. All three are also, however, the cornerstones of the theory and practice of national sovereignty, their mutual innovation being integral to the consolidation of the territorial state through a series of shifts in the nature of state/economy relations that took place from around the mid-eighteenth century to the early twentieth. These processes run parallel to and are closely bound up with the development of national cultural identities (Anderson, 1991; Poovey, 1995; Joyce, 2003). Although the idea that states still exercise a meaningful national sovereignty over these matters is increasingly questionable (Beveridge, 1991; Cameron and Palan, 2004; Cameron, 2005), in practice states retain much of the institutional paraphernalia forged during the period of their assertion over and above the city. An awareness of this history, I want to argue here, is essential to understand the limitations as well as the potential of contemporary cities to deliver true social, political, economic and cultural alternatives.

Taxation

In 1835 the British Parliament passed legislation that was to permanently alter the formal fiscal relationship between state and city. The Municipal Corporations Act of that year swept away one of the most significant vestiges of the feudal system by abolishing the bodies that had governed British cities for several hundred years. Created by royal charters starting in the twelfth century, municipal corporations ran cities as corporate enterprises, usually controlled by a small group of powerful individuals, usually leading merchants, who had extensive powers to raise taxes from the city populations, organize militias and police forces, charge tariffs on goods entering the city walls and so on. As Charles Tilly has argued, cities and states, not yet nation-states, but increasingly centralized from the fifteenth century onwards, constituted the sites of very different forms of social power. Cities, Tilly argued, were the natural centres of capital formation in a pre-capitalist world. States, by contrast, which comprised various constellations of monarchy, aristocracy, theocracy and the military, were the sites of coercive power (Tilly, 1990).

The growing significance of mercantile trade and the increasing monetization of inter-city economies meant that city-states developed, often wielding a degree of power and influence comparable to contemporary nation-states (e.g. late-medieval Florence, Venice, Genoa, Amsterdam) (de Vries, 1976; Tilly, 1990). In the context of the more centralized but still very weak patrimonial state of feudal Britain, this dispersed power structure allowed a degree of control to be exercised on behalf of the monarchy without the need for the elaborate and expensive infrastructures now associated with the bureaucratic state. Although taxes were imposed by monarchs and parliaments, their collection was traditionally contracted out to 'tax farmers', who would guarantee a certain revenue to the state in return for a licence to set local, including urban, tax rates (Webber and Wildavsky, 1986). While this system could impose significant burdens during periods of warfare, for the most part the central states of Europe were funded through borrowing from merchant banks rather than through general taxation. When taxes were imposed they were generally on the consumption of commodities (salt, candles, soap, wine, beer, tobacco, etc.), on some aspect of land or housing (window tax, hearth tax, roof tax, etc.), on produce (tithes), or were punitively aimed at particular urban populations (tallages) and therefore well suited to local collection and management (Webber and Wildavsky, 1986). Given that public accounting on a national scale was weak, had a general income tax been proposed it would not have been feasible (Bonney, 1998). This gradually changed throughout Europe – Britain leading the way – as the military and infrastructural demands upon the state grew. As the reading from Pierre Bourdieu (Extract 9.1) suggests, the pressure on states to establish general taxes was in a relation of circular causality with their need to fund the military. The need to create well-equipped standing armies to consolidate and defend both the national territory and growing imperial possessions gave rise to an increasing need to extract a steady supply of money. This, in turn, further stimulated the need to define and control both the territory and population to be taxed (Extract 9.2).

Extract 9.1: From Bourdieu, P. (1998) *Practical Reason*, Cambridge: Polity Press, pp. 42–3 (emphasis in original).

The emerging state must assert its physical force in two different contexts: first externally, in relation to *other actual or potential states* […], in and through war for land […]; and second internally, in relation to rival powers (princes and lords) and to resistance from below (dominated classes). The armed forces progressively differentiate themselves with, on the one hand, military forces destined for interstate competition and, on the other hand, police forces destined for the maintenance of intrastate order.

Concentration of the capital of physical force requires the establishment of an efficient fiscal system, which in turn proceeds in tandem with the unification of economic space (creation of a national market). The levies raised by the dynastic state apply equally to all subjects – and not, as with feudal levies, only to dependants who may in turn tax their own men. Appearing in the last decade of the twelfth century, state tax developed in tandem with the growth of *war expenses*. The imperatives of territorial defence, first invoked instance by instance, slowly become the permanent justification of the 'obligatory' and 'regular' character of the levies perceived 'without limitation of time other than that regularly assigned by the king' and directly or indirectly applicable 'to all social groups'.

Thus was progressively established a specific economic logic, founded on *levies without counterpart* and *redistribution* functioning as the basis for the conversion of economic capital into symbolic capital, concentrated at first in the person of the Prince. The institution of the tax (over and against the resistance of the taxpayers) stands in a relation of *circular causality*, with the development of the armed forces necessary for the expansion and defence of the territory under control, and thus for the levying of tributes and taxes as well as imposing via constraint the payment of that tax. The institution of the tax was the result of a veritable *internal war* waged by the agents of the state against the resistance of the subjects, who discover themselves as such mainly if not exclusively by discovering themselves as taxable, as tax payers […]. It follows that the *question of the legitimacy* of the tax cannot but be raised (Norbert Elias correctly remarks that, at its inception, taxation presents itself as a kind of racket.) It is only progressively that we come to conceive of taxes as a necessary tribute to the needs of a recipient that transcends the king, that is, this 'fictive body' that is the state.

Extract 9.2: From Weber, M. (1958 [1921]) *The City*, Ontario: The Free Press, pp. 186–90.

In England the cities never acquired full tax powers, but for all new taxes the consent of the king was required. ... After the subjection of the cities, the patrimonial bureaucratic state did not reject city economic policy. Quite the contrary. The economic flowering of cities and the conservation of their populations through defence of the sources of their subsistence lay at the very heart of the state's financial interests [...]. However, the city's autonomy in economic regulation was lost. ... Decisive in this was the inability of the city to bring military-political power into the service of its interests in the manner and measure of the patrimonial bureaucratic prince. Only exceptionally were the cities, like the prince, as associations able to take part in the economic opportunities newly opened by patrimonial politics. In the nature of the case that was only possible for individuals, particularly socially privileged persons. Such proto-capitalists included especially many landlords or members of the higher officialdom but in England as well as in France, besides the king himself, relatively few burgher elements participated in the monopolistically privileged domestic and foreign enterprises of patrimonialism. Occasionally ... cities such as Frankfort [*sic.*] participated in a comprehensive manner in risky speculative foreign undertakings. Most cities which did this, however, ran severe risks, for a single failure could destroy them as important political forms.

The economic decline of numerous cities since the sixteenth century ... was due to the fact that the traditional forms of enterprise organized in the city economy no longer represented the activities where the greatest economic gains were to be made. ... Like the revolution at one time worked by feudal military technique so, now, revolutionary changes centred in politically oriented commercial and industrial capital undertakings. Even where these were formally located in the city they were no longer sustained by a city economic policy nor borne by local individual burgher organizations.

The new capitalistic undertakings settled in new locations [...]. The great modern trade and industrial cities of England arose outside the precincts and power spheres of the old privileged corporation. For this reason it frequently displayed archaic elements in its judicial structure such as the retention of the old land courts. The court farm and court leet remained in Liverpool and Manchester until modern times, though the landlord were re-baptized as legal lords.

Although the process of national fiscal consolidation developed haphazardly throughout Europe, indeed it is ongoing in certain parts of Eastern Europe and the former Soviet states, fiscal historians have demonstrated a general drift towards the creation of centrally organized territorial fiscal states roughly from the thirteenth through to the nineteenth century and beyond (Reinhard, 1996; Bonney, 1998). Although the processes in each place differed, by the late nineteenth and early twentieth centuries, most European states had enacted legislation both to create a national fiscal space and, in doing so, to severely restrict the powers of cities to raise taxes outside national frameworks (Frug, 1999).

In fiscal terms, not only did the nation of taxpayers need to be defined in relation to an external world through the consolidation of territory and demography (the nation), the domestic space of the state had to be standardized (Bourdieu, 1998). The introduction of modern accounting standards starting in the fifteenth century (Poovey, 1998), the increasing mathematization of governance through the development of 'political arithmetic' (Frängsmyr et al., 1990) and various treatises on 'national' political economies and taxation (O'Brien, 1999a, 1999b) all presaged moves to strip cities of their independent fiscal powers.

The 1835 Municipal Corporations Act broadly coincided with the introduction of a national income tax in the UK. Although the tax had been brought in before – first in 1799 – this was a reluctant move in response to a particular emergency. Although the 1799 tax became known as 'the tax that beat Napoleon' – since it funded the expansion of the fleet and coastal defences against the threat from Revolutionary France – it was repealed at the first possible opportunity and all records were ordered to be destroyed for fear of creating a precedent (see Figure 9.1). Despite efforts to prevent it, by the 1830s the financial needs of the state made the general introduction of the national income tax inevitable, and alongside it the elimination of competing fiscal jurisdictions (Webber and Wildavsky, 1986; Daunton, 2001). Since then, as recent urban theorists have pointed out, cities have grown and changed both in terms of their internal structures and cultures (economic, political and social) and in terms of their relations with wider national territories. The increasing pace of urbanization, and the increasing pressures on and conflicts over city resources mean that many cities are experiencing 'fiscal crisis' (Joyce and Mullins, 1991; Carmichael and Midwinter, 1999; Frug, 1999; Kincaid, 1999). State control of urban taxation has produced significant problems for some cities where national fiscal norms and allocations do not correspond to the changing needs of growing urban populations (McGee, 1999; Le Galès, 2002). The tensions between fiscal centralism and increasing urbanization are being felt particularly in those parts of the world where cities are growing most rapidly. As McGee (1999) notes with respect to rapidly growing Asian megacities, the issue of *devolution* is of crucial importance to the success of urban governance for it ultimately depends on the willingness of national governments to permit local city governments to assume greater responsibilities for taxation, land management, urban infrastructure and transportation. Within the Asian context the pace of this devolution is very uneven; in comparision, it is non-existent in Western Europe and North America (Frug, 1999). This, ironically, is precisely because of the pressure felt by national governments to impose budgetary restraints. The increasing scale and economic importance of cities relative to the national economy as a whole suggests that, were fiscal control relinquished to urban authorities, the pressure on cities to increase tax rates in the interests of the welfare and competitiveness of their own populations would be immense (Le Galès, 2002).

Money

Although we now routinely associate money with the state – multinationally in the case of the Euro – the innovation of modern forms of money was often carried out within cities. This includes coinage which, although it has existed for many centuries,

Figure 9.1   James Gillray, 1799, *Meeting of the Monied Interest: Constitutional Opposition to the 10 percent: i.e. John Bull's Friends Alarm'd by the New Tex.* London

only became fully controlled and standardized by the major European trading cities from the fifteenth century onwards (Davies, 2002; Helleiner, 2003). Until the nineteenth century money was not standardized or necessarily controlled by the state, but was produced by private banks, merchants, churches, city corporations and others (Helleiner, 2003). That said, the beginnings of standardized currencies began with actions taken by particular cities. During the seventeenth century, for example, the Bank of Amsterdam, founded in 1609, took the unprecedented step of accepting all forms of gold and silver coinage as well as bullion and converted it into a standard coin which it required to be used for all transactions taking place within and through the port. This greatly reduced the administrative cost of money for both the banks and the traders as well as greatly increasing the transparency and security of the coinage (de Vries, 1976: 229). This capacity of cities to manage money and trade in part led to their massive growth during the seventeenth and eighteenth centuries. De Vries (1976) estimates that the populations of the major European trading ports grew by over 250 per cent between 1600 and 1750 as they became the transit points for imperial commodities and for the increasing quantities of gold and silver from the Americas (Davies, 2002). The success of the major cities in standardizing money in this way, however, also brought about their eventual eclipse by national monetary systems. The standardizing logic of the city bank was extended to the creation of national banks towards the end of the seventeenth century, underwritten by guarantees based on government debt and future tax receipts. This in turn spurred the formation of what eventually became state-managed fiduciary money, that is standardized money sanctioned by state authorities and underwritten by state finances, which spread widely from the early nineteenth century to the early twentieth (Helleiner, 2003: 34). At the same time as taxation was being standardized and nationalized, therefore, so the currencies with which it was to be paid were also brought under much tighter state control. Britain created the first national coinage with a value officially fixed to the gold standard in 1816, followed fairly quickly by many others (USA in 1859, Portugal in 1854, Switzerland in 1860, Italy in 1862, France in 1864, Belgium in 1865, Germany in 1873 and Austria-Hungary in 1892) (Helleiner, 2003: 33–4). During the same period paper banknotes, which did not at the time have the same status as metal coins, were also becoming increasingly used and standardized. Helleiner (2003: 35) argues that the most famous example of note standardization was England, where the 1844 Bank Act gradually phased out the notes of various small 'country' banks that had been issuing notes since the mid-eighteenth century. This 'nationalization' of money and banking not only changed the geography of money but changed the nature of money. As Ingham (1999, 2004) has argued, the consolidation and legitimation of paper currency represented a shift of the meaning of money away from being based on precious metals, to being based on the volume of credit circulating in the market. The expansive nature of the debt markets that develop during this period, and which have subsequently grown into the global finance system, were not suited to the relatively restricted circuits of money and goods in cities. Wealth may still be generated in and through urban environments, but the financial system that supports and absorbs that wealth has long since transcended the capacity of individual cities.

Where local communities and third-sector groups within towns and cities have innovated alternative forms of currency – in the forms of Local Exchange Trading Schemes (LETS) – these apply to very restricted circuits of people and place, and in all cases are pegged at a fixed rate to the national currency and have been described as a 'second-class economy' which legitimizes 'the abandonment by mainstream society of the jobless poor, and of the welfare services they depend on' (Bowring, 1998: 107). This can hardly be seen to presage a confident reassertion of urban financial independence.

Despite the relative weakness of such 'alternative' moneys, cities are by no means irrelevant to the (re)production of money. Rather, the geography of urban monetary influence has become highly concentrated in a few global cities such as London, New York and Tokyo (Chapter 6). Those more recent manifestations of money that are not directly regulated by the state – the offshore currency and derivatives markets – occupy a unique legal space which is not contained by any conventional spatial category, urban or national (Palan, 2003; Cameron and Palan, 2004).

Law

Although the various historical processes outlined above vary enormously from city to city and state to state, and while the factors driving the ascendance of the state over the city are extremely complex, one factor underpins them all: law. The location of the law is a matter of considerable controversy. The history of law is one produced by and large in cities, not just because that is where the institutions and personnel were located, but because the need for law itself is partly a product of increasing urbanization. Law develops and evolves as human societies become more and more complex and require ever more elaborate systems of regulation, protection and sanction to maintain justice, equity and the norms of collective social life.

However, law cannot be contained within any single city for all that it might be derived from and articulated with respect to that city's population. Law, like the city itself, leaks out of the physical boundaries and defies containment. This produces the situation wherein law is only ever partially, contingently and temporarily contained by any spatial formation, city, state or any other. The complexities of legal space are illustrated by the eminent jurist William Twining in his account of *Globalisation and Legal Theory*, when he attempts to 'map' the law of Northern Ireland (2000: 138) and concludes that, 'Like it or not, United Kingdom law, the law of the Republic of Ireland, European Union Law, Public International Law, Human Rights Law and developments in the common law world all bear directly on interpreting local legal issues.' Noting that globalization has loosened the association of law, state, and nation, Twining ends his attempt to map law by adopting the metaphor of the 'invisible city' from the Italian novelist Italo Calvino, recognizing simultaneously the localism of legal praxis with the multiplicity and pluralism of law itself. This multiplicity, while less obvious or pronounced during periods of strong national state authority – as asserted by the sorts of laws over taxation and money outlined above – is not new. Indeed, law long predated the nation-state and was in many ways 'globalized' – particularly through its articulation by merchant cities through the *lex mercatoria* (merchant law) – long before the development of any modern sense of territorial jurisdiction (Cutler, 1997; Mertens, 1997).

The problem of the law that we are concerned with here, therefore, is not whether the state retains full legal sovereignty – a concept which jurists and philosophers have argued for many years is extremely problematic (Agamben, 1998; Schmitt, 2003) – but whether its legal precedence with respect to cities remains extant, whatever other jurisdictional scales can be brought to bear. As the excerpt from the US Supreme Court cited by Frug suggests (Extract 9.3), despite the praxis of legal pluralism, the doctrine of state sovereignty strongly prevails, to such a degree that the Federal state reserves the right to override and/or destroy any aspect of a city that it considers a threat. This apparently extreme tension between state and city – in part an artefact of contradictions inherent the formation of the USA as a unified but federal state – has precedents that reach far back into the history of state theory. Hence, Thomas Hobbes' oft-cited warnings over the unchecked power of cities:

> Another infirmity of a Common-wealth, is the immoderate greatnesse of a Town, when it is able to furnish out of its own Circuit, the number, and expence of a great Army: As is also the great number of Corporations; which are as it were many lesser Common-wealths in the bowels of a greater, like worms in the entrails of a natural man. (Hobbes, 1985 [1651]: 374–5)

Extract 9.3: From Frug, G.E. (1999) *City Making: Building Communities without Building Walls,* Princeton, NJ: Princeton University Press, pp. 16–17.

American cities do not have the power to solve their current problems or to control their future development. Cities have only those powers delegated to them by state governments and traditionally these powers have been rigorously limited by judicial interpretation. Even if cities act pursuant to an unquestionable delegation of power from the state, their actions remain subject to state control. Any city decision can be reversed by a contrary decision by the state, a process the legal system calls 'pre-emption.' Moreover, state power is not limited simply to the ability to determine the scope of city decision-making authority or to second-guess the exercise of that authority whenever it seems appropriate to do so. States have absolute power over cities, and the extent of that power has been extravagantly emphasized by the Supreme Court of the US…

In an attempt to limit this subservience to the state, most state constitutions have been amended to grant cities the power to exercise 'home rule.' But cities are free of state control under home rule only on matters purely local in nature. And, nowadays, little if anything is sufficiently local to fall within such a definition of autonomy. As a result, cities are generally treated by American law as 'creatures of the state.'

The Municipal Corporations Act can, in legal terms, be read as a means of curing the nascent state of at least some of the 'worms' in its entrails. The Act, along with numerous other legal, economic, welfare and social reforms around the same period, sought to create what Poovey describes as the 'abstract space' of the nation – disaggregating existing social and spatial practices, primarily those organized at the level of the city, and reconstituting them with respect to a 'conceptual grid that enables every phenomenon to be compared, differentiated, and measured by the same yardstick' (Poovey, 1995: 9). As Patrick Joyce has argued more recently, the Municipal Corporations Act was key to this normalizing process – its decree that justice be strictly separated from municipal government creating a particular type of citizen who was no longer subordinated to the essentially feudal structures of the city.

The partial elimination of older forms of legal relationships in the cities, and the imposition of the emergent norms of nationhood during this period can still be seen in the ambiguities of urban 'citizenship' today. Although analysts can discuss 'effective' citizenship being articulated through the participatory spaces of the city (Holston and Appadurai, 2003), formal legal citizenship still rests with the nation-state, however compromised that concept may be in practice. Despite the pluralism of the law described by Twining above, therefore, and although social life in practice for many urban populations bears little or no relationship to the norms legislated for in the nineteenth-century state, the national legal order maintains a high degree of precedence. Even in the context of the European

Union, which has been marked by an unprecedented internationalization of legal jurisdiction, through the European Courts and human right legislation, we can still see a marked (re)assertion of formal national sovereignty; for example, through Britain's and others' refusal to join the Euro, the Schengen Agreement and other 'superstate' legislation.

Summary

The creation of the social body of the nation from the early nineteenth century to the late twentieth century and beyond has entailed a simultaneous evacuation of difference between places, such as the elimination of the legal, fiscal, monetary and political powers of cities, and imposition of national norms of identity, behaviour, class, gender, etc. All of these were supported and regulated by standardized institutions of taxation, money and law, among others. The process has never been complete, of course. The particularities within and between contemporary cities, for example, are in part a product of the state's inability to fully eliminate what Poovey (1995: 53) evocatively describes as 'old rationalities, like undigestible bits of bone in the craw of modernity'.

By stressing the continuing role of the state in the governance of the city in this chapter, I am not seeking to refute urban geography's current fascination with the city or to deny its importance. I am, however, arguing that awareness of the longer historical relationship between the city and the state has important implications for the contemporary city. If the new spaces of the urban – the networks, the flows, the hyperreal, the mediascape, the glocal – are truly to offer alternatives to the ossified politics of the state, then they will have to overcome the accumulated weight of state power which has dominated the city for nearly two centuries. A politics of the city must also and necessarily be a politics of the state if it is to effect real change for the newly emergent urban culture.

References

Agamben, G. (1998) *Homo Sacer: sovereign power and bare life*, Stanford, CA: Stanford University Press.

Amin, A. and Thrift, N. (2002) *Cities: re-imagining the urban*, Cambridge: Polity Press.

Anderson, B. (1991) *Imagined Communities: reflections on the origin and spread of nationalism*, London: Verso.

Bell, D. and Jayne, M. (eds) (2005) *City of Quarters: urban villages in the contemporary city*, London: Ashgate.

Beveridge, F. (1991) *International Dimensions of Taxation*, Hull: Hull University Law School/Studies in Law.

Bonney, R. (ed.) (1998) *The Rise of the Fiscal State in Europe c.1200–1815*, Oxford: Clarendon Press.

Bourdieu, P. (1998) *Practical Reason*, Cambridge: Polity Press.

Bowring, F. (1998) 'LETS: an eco-socialist alternative?', *New Left Review* 232, 91–111.

Cameron, A. (2005) 'Turning point? The volatile geographies of taxation', *Antipode: Radical Journal of Geography* 38 (2), 236–258.

Cameron, A. and Palan, R. (2004) *The Imagined Economies of Globalization*, London: Sage.

Carmichael, P. and Midwinter, A. (1999) 'Glasgow: anatomy of a fiscal crisis', *Local Government Studies* 25 (1), 84–98.

Cutler, A.C. (1997) 'Artifice, ideology and paradox: the public/private distinction in international law', *Review of International Political Economy* 4 (2), 261–285.

Daunton, M. (2001) *Trusting Leviathan: the politics of taxation in Britain, 1799–1914* Cambridge: Cambridge University Press.

Davies, G. (2002) *A History of Money: from ancient times to the present day*, Cardiff: University of Wales Press.

de Vries, J. (1976) *The Economy of Europe in an Age of Crisis, 1600–1750*, Cambridge: Cambridge University Press.

Frängsmyr, T., Heilbron, J.L. and Rider, R.E. (eds) (1990) *The Quantifying Spirit in the 18th Century*, Berkeley: University of California Press.

Frug, G.E. (1999) *City Making: building communities without building walls*, Princeton, NJ: Princeton University Press.

Helleiner, E. (2003) *The Making of National Money: territorial currencies in historical perspective*, Ithaca, NY, and London: Cornell University Press.

Hobbes, T. (1985 [1651]) *Leviathan*, London: Penguin Classics.

Holston, J. and Appadurai, A. (2003) 'Cities and citizenship', in N. Brenner, B. Jessop, M. Jones and G. Macleod (eds), *State/Space: a reader*, Oxford: Blackwell.

Ingham, G. (1999) 'Capitalism, money and banking: a critique of recent historical sociology', *British Journal of Sociology* 50 (1), 76–96.

Ingham, G. (2004) *The Nature of Money*, Cambridge: Polity Press.

Jacobs, J. (1985) *Cities and the Wealth of Nations: principles of economic life*, Harmondsworth: Viking Penguin.

Joyce, P. (2003) *The Rule of Freedom: Liberalism and the modern city*, London: Verso.

Joyce, P.G. and Mullins, D.R. (1991) 'The changing fiscal structure of the state and local public sector: the impact of tax and expenditure limitations', *Public Administration Review* 51 (3), 240–253.

Kincaid, J. (1999) '*De facto* devolution and urban defunding: the priority of persons over places', *Journal of Urban Affairs* 21(2), 135–167.

Le Galès, P. (2002) *European Cities: social conflicts and governance*, Oxford: Oxford University Press.

McGee, T.G. (1999) 'Urbanization in an era of volatile globalization: policy problematiques for the 21st century', in J. Brotchie, P. Newton, P. Hall and J. Dickey (eds), *East–West Perspectives on 21st Century Urban Development*, Aldershot: Ashgate.

Mertens, H.-J. (1997) '*Lex Mercatoria*: a self-applying system beyond national law?', in G. Teubner (ed.), *Global Law without a State*, Aldershot: Dartmouth.

O'Brien, D.P. (ed.) (1999a) *The History of Taxation, Vol. 1*, London: Pickering and Chatto.

O'Brien, D.P. (ed.) (1999b) *The History of Taxation, Vol. 2*, London: Pickering and Chatto.

Palan, R. (2003) *The Offshore World: sovereign markets, virtual places and nomad millionaires*, Ithaca, NY: Cornell University Press.

Poovey, M. (1995) *Making a social body: British cultural formation 1830–1864*, Chicago: University of Chicago Press.

Reinhard, W. (ed.) (1996) *Power Elites and State Building*, Oxford: European Science Foundation/Oxford University Press.

Schmitt, C. (2003) *The Nomos of the Earth in the International Law of the Jus Publicum Europeaeum* (trs G.L. Ulmen), New York: Telos Press.

Soja, E. (1996) *Thirdspace: journeys to Los Angeles and other real and imagined places*, Oxford: Blackwell.

Tilly, C. (1990) *Coercion and Capital in European States AD 990–1992*, Oxford: Blackwell.

Twining, W. (2000) *Globalisation and Legal Theory*, London: Butterworth.

Webber, C. and Wildavsky, A. (1986) *A History of Taxation and Expenditure on the Western World*, New York: Simon & Schuster.

Weber, M. (1958 [1921]) *The City*, Ontario: The Free Press.

10 PLEASURE AND LEISURE

David Bell, Sarah L. Holloway, Mark Jayne and Gill Valentine

This chapter

○ Argues that the city has always been important as a site of consumption as well as one of production

○ Uses the example of the consumption of food and drink to show that leisure and pleasure have taken different forms over time

○ Suggests that analysis of the changing urban landscapes of eating and drinking reveals important transitions in the way urban space is managed and marketed

---

Introduction

From the rapid growth of the modern city to recent urban restructuring, the production and consumption of spaces associated with pleasure and leisure has been a key feature of urban life (Jayne, 2005). Libraries, museums, sports stadiums, theatres, parks, shopping arcades and department stores are examples of spaces of pleasure and leisure associated with the modern city, while shopping malls, theme parks, cultural quarters, multiplex cinemas, waterfront developments and out-of-town retail parks characterize developments in urban leisure and pleasure over the past twenty-five years.

Urban historians have had a long-standing interest in the provisioning of the city, for example in terms of the retail trade (Fraser, 1981) and of the leisure and pleasure industries (Koshar, 2002). Histories of the commodity culture of the modern city (Richards, 1990) and of 'disreputable' urban pleasures (Huggins and Mangan, 2004) have also been written, drawing attention to the development of both formal and informal sites of leisure and pleasure, and to issues such as regulation and commercialization. Such historical studies have been complemented by analyses of the contemporary leisure industries (Roberts, 2004), informal leisure pursuits (Bishop and Hoggett, 1986), urban retailing and provisioning (Wrigley and Lowe, 1996), and the city's cartography of formal and informal, virtuous and disreputable, 'pleasure zones' (Bell et al., 2001). However, the study of urban pleasure and leisure has been marked by particular inflections and omissions, for example through a focus on the geographies of certain sites – shops, sports stadiums, cinemas – at the expense of others, especially the informal, more mundane pleasures of everyday urban living such as strolling the city streets, for example (though see Edensor, 2000). In particular the focus on sites of so-called 'spectacular' leisure and pleasure, such as the shopping mall, has until recently dwarfed studies of smaller scale, less conspicuous sites and activities. Moreover, pleasure and leisure as urban activities have on the whole been under-researched, reflecting the notion that these are trivial pursuits, far less important to urban form and function (and to urban studies) than productive economic activity – that is, work. Such a bias seems nonsensical now more than ever, given the shift in contemporary cities towards the leisure and experience economies and towards post-industrial consumption and culture (Hall, 1998). Indeed, work for many urbanites today is in the service economies that provide for urban leisure and pleasure, in sectors such as tourism and hospitality, or the cultural industries (Hesmondhalgh, 2002; Hoffman et al., 2003).

Outside urban studies, for example in the expanding fields of leisure and hospitality studies, and in cultural studies, cities have been reconceptualized as key sites for the production and consumption of leisure and pleasure (Lashley and Morrison, 2000; Stevenson, 2003; Roberts, 2004). And leisure and pleasure have become key features in the economic fortunes of cities, as they compete for residents, visitors and capital (Dicks, 2003): the economic value of leisure pursuits such as sport- spectating or attending live music gigs has led cities to reinvent themselves as centres for the production and consumption of leisure and pleasure (see Gibson and Homan, 2004; Gratton et al., 2005).

But what do we mean by urban pleasure and leisure? The problem of delineating and categorizing such diverse pursuits has also hampered the development of a coherent body of theory and evidence. Can formal and informal leisure pursuits, for example, be thought of in the same way, or analyzed using the same tools? Can public and private sector provision be easily bracketed together, or conversely split apart? And what about public space and private space as sites for different forms of pleasure and leisure? Given the enormity and diversity of the topic, we have to narrow down our focus, to sacrifice breadth for depth. So this chapter explores what we consider to be two of the most important, high-profile and (sometimes) conjoined pleasurable and 'leisureable' urban pursuits, eating and drinking. Our focus, moreover, is on commercial spaces of food and drink, largely setting aside both the more private pleasures of cooking, eating and drinking in the home, and the reconstituting of food and drink production as leisure activities (see Bell and Valentine, 1997).

The practices and spaces associated with the production and consumption food and drink have always been significant features of urban form and everyday life. In this chapter, we chart the relationship between the development of cities and political, economic, social, cultural and spatial practices and processes thatsurround commercial food and drink, with a focus on eating out and on alcohol drinking and drunkenness. We use the examples of foodscapes and drinkscapes to show the ways in which geographies of pleasure and leisure can be read as symbolizing broader transformations in the economies of cities and urban living.

Eating and drinking and the growth of the modern city

The organization and formalization of people's work and leisure time is a key feature of both the birth of modern industrial capitalism and the development of modern urban life. The regimentation of time and work along the strict lines of Fordism and Taylorism went hand in hand with the emergence of distinct work-free leisure time and also with the promotion of socially-responsible leisure activities, often referred to as 'rational recreation' (Thompson, 1967). Moreover, attempts at the rational social control of the shopfloor were extended not only to workers' leisure time but also to the physical redevelopment of the city in order to better serve industrial capitalism (Malcolmson, 1973).

One of the key features of this process was a bourgeois attempt to both corral working-class people into particular residential and industrial areas, and also to banish working-class consumption cultures from central urban spaces and places. This was underpinned by rationalized urban master planning, greater state involvement in unruly areas, censorship, licensing laws, planning guidelines and formal policing (Harring, 1983; Cohen, 1997). This led to the criminalization of street pastimes and sports, such as cock-fighting and bear-baiting (Pearson, 1993), and to regulation targeted at working-class drinking cultures, which were seen as a source of social vice, moral decline and as inhibiting industrial productivity (Cunningham, 1980). The reordering of the city extended to practices of food production, distribution and consumption, too, with restriction placed on street trading, regulation of the urban production of food, and the growth of new commercial spaces for food consumption (Burnett, 1999). This reordering was underwritten by the same moral logic of 'civilizing' that marked the regulation of drinking; Valentine (1998) describes how regulation of on-street food production and changing social conventions around public eating worked together to 'civilize' street life in nineteenth-century Britain (see Edensor, 1998, who describes street life beyond Western cities in very different terms).

However, excessive drinking was especially central to bourgeois concern for the health and productivity of the lower classes (see Extract 10.1). One of the most

Figure 10.1   Hogarth's Gin Lane (left) and Beer Street (right)

famous public debates in nineteenth-century Britain regarding alcohol, for example, concerned the relative merits of drinking beer or gin. This debate was underpinned by spatial metaphors of Gin Lane (a place of immoral behaviour, violence, slum life, disorder and potential revolution), and Beer Street (a place of joviality, business success, progress, modern civilized streets, houses and shops) (O'Mally and Valverde, 2004, and see Figure 10.1). Gin was cheap and plentiful and hence was the favoured drink of the working classes. At the time beer was considered to be a healthier alternative, and gin became central to middle-class anxieties over working-class drunkenness, immorality and mob behaviour. Gin thus became a necessary focus of legislation and taxation that ultimately led to reduced usage. Nevertheless, despite drinking establishments being subject to increasing regulation, excessive drinking in the city continued. However, not all observers of the modern period saw the consumption of alcohol in purely negative terms. For example, Mass Observation research (collected from 1937 to 1948 in the UK) was undertaken due to a dissatisfaction with the way in which drinking was reported by official statistics, and also a concern that official measures to curb drunkenness failed to grasp the experience and context of drinking (see Kneale, 2002). In seeking an alternative meaning of drink, *Mass Observations:*

the Pub and the People (1943) provided a method of scrutinizing drinkers not previously attempted, and sought to establish the ways in which drinking and drunkenness were social practices. For example, observers noted that alcohol was not the sole reason for public drunkenness, and that pub sociality encouraged a relaxing of self-control even for those who were not drinking (Kneale, 2002). Mass Observation also looked at spatial factors relating to drinking. For example, pubs in towns and cities in the UK were shown to be concentrated in the old central core and along major roads, largely as a consequence of licensing restrictions related to the newer suburbs. It was also shown that rooms within pubs were strictly gendered; the vault and taproom were masculine spaces, while the lounge or parlour were dominated by couples or mixed groups.

Extract 10.1: From Schivelbusch, W. (1993) *Tastes of Paradise: A Social History of Spices, Stimulants and Intoxicants,* New York: Vintage Books, pp. 147–148.

In the 1840s the young Friedrich Engels reported from industrial areas of Britain:

> It is not surprising that the workers should drink heavily. Sheriff Alison asserts that 30,000 workers are drunk in Glasgow every Saturday night. And this is certainly no underestimate. … It is particularly on Saturday evenings that intoxication can be seen in all its bestiality, for it is then that the workers have just received their wages and go out for enjoyment at rather earlier hours than on other days of the week. On Saturday evenings the whole working class streams from the slums into the main streets of the town. On such an evening in Manchester I have seldom gone home without seeing many drunkards staggering in the road or lying helpless in the gutter. On Sunday the same sort of thing happens again, but with less noisy disturbances. … It is easy to see the consequences of widespread drunkenness – the deterioration of personal circumstances, the catastrophic decline in health and morals, the breaking up of homes.

If one compares this depiction of proletarian drinking in the nineteenth century with similar complaints from the sixteenth century, little enough seems to have changed. … Are we to conclude from this that in these three centuries nothing had changed in the character, quality, quantity, or social import of drinking and drunkenness? Did people in the age of the Industrial Revolution still drink and get drunk in the same way, with the same motives, the same consequences, and the same drinks as people in the sixteenth century?

On the contrary, the success of the new beverages coffee, tea, and chocolate prove that in the interim a quite considerable shift in drinking mores had occurred. … [T]hese hot beverages

(Continued)

deprived alcohol of the status it had once had as the universal drink. Yet the sobriety they established was limited to specific sectors of the population, primarily the middle class. From the seventeenth century on, the bourgeoisie found unrestrained drinking increasingly offensive. Alcohol was not banned, of course, but it was domesticated. The middle-class citizen drank moderately, and he drank in a private circle (at home, in his club, or out amid a table of 'regulars'). ...

Things were quite different, however, for the lower classes. They had never had a share in the coffee culture of the seventeenth and eighteenth centuries. They remain bound to medieval custom in their drinking habits. Alcohol had an incomparably larger place in the lives of the proletariat than it did amongst the bourgeoisie. For the former, drink and drunkenness carried no social stigma; on the contrary, they were almost a symbol of class identity. ... Alongside the motive of drinking to symbolize social fellowship, there is another motive at least as important – escapism. Workers do not drink out of sheer exuberance; they drink to cast off the misery of their lives for a few hours.

Mass Observation also described the symbolic importance of drinking. Observers noted that drinking was a way of creating and transforming social relationships between people, bound up with trust, reciprocity and a relaxing of formal social relations. For example, the working-class drinking practice of 'rounds' and 'treating' one another reproduced social ties and obligations, and hence drinking was seen to have played an important role in making connections between people. Drink was equated with social worth, trust, reciprocity and fraternity. Drunkenness was bound up with group intoxication, and was described by Mass Observation as a social phenomenon of self-liberation from the weekly work routine and time-discipline, which allowed people to break down social inhibitions. It was also argued that sobriety and drunkenness were controlled through social hierarchies and particularly through the guidance of older drinkers. So, even though drunkenness created particular bonds of sociality between drinkers and a temporary suspension of particular social divisions, there were still social norms to which to adhere. Despite the growing popularity of non-alcoholic drinks such as tea, coffee, and milk, and the proliferation of 'drinkable' water supplies for the first time, in working-class areas where excessive drinking was engrained in everyday life and forms of sociability, curbing excessive drinking proved to be a great challenge to middle-class reformists (see Burnett, 1999).

One of the key features of urban change at this time was the commercialization and formalization of spaces to eat and drink. For example, Schivelbusch (1993) charts the emergence of pubs, bars, cafés and restaurants as key and distinct urban landmarks. By the beginning of the nineteenth century, food could be bought at 'eating houses' or 'cookshops', and by those staying overnight in hotels, earlier know as inns. Alcohol was most popularly drunk in pubs and bars, earlier known as ale houses or taverns. However, Schivelbusch shows that before the early nineteenth century all these sites were barely distinguishable from private households that had surpluses of rooms, food and drink, and that made these available to strangers for a price. There was thus a gradual metamorphosis of production and consumption practices that coincided with the industrial revolution and the growth of modern cities. For example, in terms of consumption, changes in the interior of the public house or tavern reflected the requirements of commercial food and drinking cultures and in particular, a physical transformation took place to create a 'hub' in taverns – the counter.

Schivelbusch shows that food and drink were once served from the kitchens of houses opened up to the public. The kitchen was an all-purpose room centred on an open hearth where customers would eat and drink at a price. In the largest inns, and for upper-class travellers only, there may have been a separate dining room. However, Schivelbusch suggests that around 1800, restaurants began to develop as separate from inn-keepers' private rooms, becoming properly commercialized spaces where the clientele were served. A counter now marked the boundary between production and consumption areas. Moreover, the counter-turned-bar took on further significance as the 'typical' way of having a drink: due to the physical proximity to the tavern-keeper and to the drink itself, customers could stand at the bar and engage with the tavern-keeper. The bar took on particular significance in big-city drinking houses in the USA and the UK. In the UK, for example, the so-called gin palaces were a genuine product of the industrial revolution, where industrialized production techniques allowed high-alcohol content distilled spirits to be served to customers, shortening the inebriation process as well as the time drinkers stayed at the bar (see McAndew and Edgerton, 1969).

However, the importance of the 'counter' to eating and drinking cultures is variable across the world. For example, while the US bar developed along similar lines to the UK's pubs, in Europe the bar never acquired the same significance. For example, Oldenberg (1989) shows that in France drinks were most often poured at the bar/counter but served at tables. Similarly, drinking cultures in German taverns were focused around service at large tables, with only favoured regulars occupying the privileged location, standing at the bar.

In a similar vein, the proliferation of public eating habits was also place-specific. For example, Mennell (1985) argues that affluent aristocrats and those made rich through the industrial revolution were able to entertain more at home, so fewer restaurants opened in the UK than in mainland Europe and North America. Mennell also suggests that appreciation of more elaborate cuisine further down the social scale explains the larger number of restaurants in France. The preparation of fine food required expensive ingredients and a large kitchen staff, and the production of emerging French 'cuisine' was difficult for the less wealthy to achieve at their home. Mennell argues, moreover, that in post-revolutionary France, all social classes had come to aspire to a cuisine more elaborate than they could prepare at home, and eating out began to take on the cultural connotations it carries today.

Warde and Martens (2000) write that the very idea of paying to dine out for pleasure was only gradually accepted in Britain, starting in the late eighteenth century. However, restaurants as we understand them today were largely a creation of the late nineteenth century, and specialized places to eat and drink, with a choice of food and drink rather than a set menu, were essentially a commercial innovation of the twentieth century. Importantly, however, the emergence of pubs, bars, restaurants and cafés as key urban landmarks contributed, along with the proliferation of shopping arcades, department stores, boulevards and plazas, to the development of agglomerated entertainment and retail quarters, and ensured that the city centre was a hub of mass consumption in the modern city. This concentration proved problematic to the ongoing bourgeois project to impose order and control on urban life (Marcuse, 1964; Weber, 1967; Sennett, 1977; Frisby, 2001). For example, Monkkonen (1981) shows how attempts to 'civilize' the early-modern American city were undermined by the availability of public transport. Public transport allowed easy and relatively cheap access to the city centre. Organized in order to take industrial workers via central connecting stations to industrial areas and office and service sector workers directly to the central business districts, as well as middle-class shoppers to central entertainment and retail quarters, public transport ensured that people of all social groups could travel to centrally agglomerated drinking establishments.

Monkkonen (1981: 550) also suggests that the city of the late nineteenth and early twentieth centuries was characterized by 'relatively intense use of public space by people of all classes' and that bourgeois control, particularly at night time, was never fully established in city centres. Thus, while depictions of the rationalization of industrialized modern urban life – the development of city centres, spatial segregation and a 'civilizing' process, relating specifically to urban drinking and eating in public space – provide a valuable vision of urban change, there is undoubtedly a tendency to oversimplify this process and its outcome. Moreover, the social relations and cultural practices and processes that surround urban eating and drinking have often been ignored, despite the class-based readings of differential consumption patterns outlined above. Moreover, it is important to remember that industrialized capital accumulation, bourgeois political control, the suppression of unruly working-class leisure activities, and the redevelopment of the city were differentially constructed and experienced in different places at different times (Miles and Paddison, 1998).

Eating and drinking in the contemporary city

Over the past thirty years, many cities throughout the world have re-invented themselves as sites of consumption. However, some of the failings that surrounded the conceptualization of the production and consumption of food and drink in the modern city have not been fully redressed. For example, changes in urban life have been explained in terms of political and socio-cultural change associated with

Fordism, Post-Fordism and neo-Fordism (Lash and Urry, 1987; Amin, 1994; Kumar, 1995) and changes in the local state, such as the rise of the 'entrepreneurial city' (Hall and Hubbard, 1998). Hand in hand with these changes has been the decline of heavy industry and manufacturing and a move towards a more service-based, cultural and symbolic economy. Cities have become sites of consumption rather than production (Jayne, 2005). A key part of this transformation has been the growth of the leisure, night-time and experience economies (Lash and Urry, 1994). Nevertheless, the tensions between generalized visions of urban change and more nuanced understandings of local specificities or social and cultural differences have been hard to reconcile.

Over the past three decades increased competition has arisen around the efforts of cities to create new images in order to attract speculators, businesses and consumers. Buildings and venues such as theatres, art galleries, convention and exhibition centres, as well a supporting cast of café bars, restaurants, fashion boutiques, delicatessens and other cultural facilities, have been shown to help generate the buzz of creativity, innovation and entrepreneurialism that is seen as crucial to contributing to the competitiveness of cities (Florida, 2002). Key features of the contemporary city thus include market segmentation, gentrification and branding, as well as increased globalization and corporatization. Bars, cafés and restaurants are now seen as playing an increasingly significant role in gentrification and urban regeneration (Latham, 2003). However, much work on gentrification fails to account fully for local specificity, and in particular depictions of food and drink spaces have failed to fully reflect complexity and diversity.

Despite this contention, there are nonetheless useful writings that have helped to illuminate the relationship between pleasure and leisure through analysis of foodscapes and drinkscapes in our cities. For example, in *Landscapes of Power*, Zukin (1991: 206) draws a parallel between the transformations to New York's cityspace brought about by processes of gentrification and the rise of *nouvelle cuisine*: 'gourmet food – specifically, the kind of reflexive consumption beyond the level of need that used to be called gastronomy – suggests an organization of consumption structurally similar to the deep palate of gentrification'. For Zukin, this is exemplified in how both urban gentrification and *nouvelle cuisine* appropriate and subvert 'segmented vernacular traditions', including building styles or cooking styles, leading to the serial reproduction of a narrow range of key elements and reflecting new regimes of the production and consumption of cultural value. Both gentrification and *nouvelle cuisine* ambivalently combine tradition with innovation, or authenticity with novelty. This ambivalence is symbolized for Zukin in the chasing out of 'other' occupants of space or providers of food, such as homeless people or 'downtown cafeterias'.

Zukin (1991: 201–203) also emphasizes the role of the critical infrastructure: 'men and women who produce and consume, and also evaluate, new market-based cultural products', and who have instigated 'not just a shift in taste, but in the way taste is produced'. This critical infrastructure, made up of people sometimes referred to as 'cultural intermediaries' (see Featherstone, 1991), actively work, through both their jobs and through their leisure activities, to set the boundaries of legitimate taste, and to embody and perform taste through their crafting of a lifestyle. The commercial hospitality sector is a vital space in which taste is produced and consumed, through food and drink, music and décor, ambience and service style. Restaurants and bars are therefore very important in producing and continually reproducing the 'feel' and the 'buzz' of a particular neighbourhood, and in keeping it 'hip' – thereby fuelling the area's ongoing gentrification (see Extract 10.2).

Extract 10.2: From Bell, D. (2002) 'Fragments for a new urban culinary geography', *Journal for the Study of Food and Society*, 6 (1), 10–21, pp. 14, 17–18 (original emphasis).

The contemporary city is a space of consumption and a site of spectacle. It is also a site of contestation, a site of refusal. Played out on the streets and plazas, the political dramas of everyday life are materialized in the practices of city living. The sociological transformations of the late modern age – all that postmodernization, deteritorialization, globalization – get worked through at the level of the everyday, the commonplace, the banal. As a *concentration* of these processes, the buying, cooking and eating of food gives us a way into thinking through the city as a node in the disjunctive flows of contemporary culture. … Nothing sums up the postmodern metropolis better than the frantic commingling of cultures and cuisines – making fusion food a culinary cipher for *multiculti cosmo-metro* life. Instead of preserving distinct ethnic cultures, they are here mixed, or rather *allowed to collide*: not blended in some melting pot out of which comes an indistinct mélange, but cultures rubbing up against each other, jostling, making new and surprising juxtapositions. Syncretic combinations emerge that hybridize and creolize diverse ingredients, playfully pick-and-mixing (James 1997). Here is Elizabeth Miles on one star of fusion food: 'Wolfgang Puck creates cuisine that both expresses his own identity(ies), and mirrors what he perceives as the identity(ies) of his customers. … [T]hese identities reflect the multi-ethnic, multi-cultural, multi-gendered, nomadic paradigms of postmodernism' (Miles 1993: 193) – the dishes articulate 'points of identification' (195), complexly and playfully combining ingredients and cooking practices: 'No "real" nor "intended" meaning emerges from this accretion of foods, vocabularies and techniques, but rather a pastiche of possible readings' (196).

Of course, it isn't very far from fusion to confusion, and reactions to fusion food can play up the boundary-blurring dangers of over-fusioning. This is often disguised as a clash of taste, but can be read as distaste about the clash of cultures, too. … Miles includes another useful dimension in her reading of Puck: that his dishes also speak of the relationship between the city and nature: 'These recipes presume a great, diverse natural bounty magically melting from specific farmland and ocean locations into the decentred city. This is the postmodern landscape, where nature meets city in a seamless continuum of goods and commodities' (Miles 1993: 199). Abundance and availability are here factored in, domesticating nature by cooking and eating it. The city is the stage for this process: the place where nature turns to nurture.

Eating out has become a central part of the experience economy of cities. It is not just the food we are consuming, but eating out can be considered as entertainment itself, and importantly visiting restaurants is bound up with knowing about and tasting the world through experiencing different cuisines, ambiances and themed environments. For example, Finkelstein (1999) renames eating out as 'foodatainment', emphasizing that it is about so much more than just eating. Foodatainment is regularly conscripted into the place promotion techniques central to urban regeneration, with parts of the city sold on the basis of their food offer, especially, perhaps, in the case of ethnic foods, as in Chinatowns (Bell, 2002). The form of foodatainment emphasized by Finkelstein – high-style restaurant dining, like Wolfgang Puck's playful fusion food – is accompanied in the city today by other forms of food-related entertainments, from the pleasures of wandering around a sumptuous food hall or deli, visually consuming the produce on display, to the equally pleasurable but more everyday experiences of coffee shops, take-aways and local bars.

And, of course, the experience economy of cities or districts also has parallels in what we might call 'drinkatainment' – the production of themed bars and pubs, ranging from the staged authenticity of Irish theme pubs to soviet-styled vodka bars. Both foodatainment and drinkatainment have become cornerstones of the urban regeneration script, which increasingly emphasizes the value of the night-time economy to cities seeking to improve their fortunes. Chatterton and Hollands' *Urban Nightscapes: Youth Cultures, Pleasure Spaces and Corporate Power* (2003) and Hobbs et al.'s *Bouncers: Violence and Governance in the Night-time Economy* (2003) explore in detail many, but not all, debates and theoretical concerns regarding the night-time economy and urban drinking cultures. These writers ask important questions, such as who and what is involved in producing nightlife spaces (designing, planning, building, marketing, selling); who and what is involved in regulating them (legislation, surveillance, gatekeeping, policing); and who consumes what in these spaces.

Both accounts highlight an increasing corporatization of urban consumption and culture, enduring social and spatial inequalities, and the changing role of the state as key features shaping cities at night in the UK. Chatterton and Hollands, for example, show that increasingly internationalized production and distribution systems have led to the proliferation of large-scale drinking establishments (Extract 10.3). Moreover, while city-centre regeneration has created opportunities for diverse groups to visibly participate in this new urban economy, they argue that these groups often become commercially incorporated into the mainstream. Moreover, they write that historic, residual and alternative cultures are often excluded in cities where the zoning or quartering of lifestyles and identities into particular spaces makes the city more socially and spatially polarized (Bell and Jayne, 2004).

Extract 10.3: From Chatterton, P. and Hollands, R. (2003) *Urban Nightscapes: Youth Cultures, Pleasure Spaces and Corporate Power*, London: Routledge, pp. 235–7.

Downtown nightlife, then, is being sanitized and cleansed of 'undesirable' elements through continued gentrification of housing and leisure markets and the growth of central urban professional service class. Smith's (1996) analysis of the revanchist city – a vengeful programme aimed at displacing certain types of activity and people, especially non-consumers, the homeless, the urban poor, punks and skaters – has much relevance here. Non-consumers in the corporate-dominated city are cast as deviants – if you're not buying, why are you here? (Atkinson, 2001). While the night does retain some fluidity, there are clear trends towards demarcation, sanitization and privatization of nightlife spaces, not to mention marginalization of alternatives. ...

In terms of urban nightlife, we are clearly only at the tip of the gentrifying iceberg. Brain (2000), echoing Bauman (1998), highlights that while those in stable employment (the puritans by day and hedonists by night) are seduced by the delights of pleasurable consumption, there are many who are excluded from such fun. The exclusive nature of recent developments in the night-time economy is simply out-pricing many social groups and reinforcing perceptions that it is not a place for them (Chatterton and Hollands, 2001). There are groups of young people living in outer estates or ghettos who have never felt enfranchised by the bright lights of the urban core, for reasons of price, access, racism, safety or style. What is clear is that current waves of restructuring is likely to further disenfranchise such groups, as well as some current traditional users with only minimum resources.

Moreover, younger city-centre nightlife consumers are exposed to little choice and have few opportunities outside a narrowly defined role as a 'consumer'. As a result, rather than finding a wealth of opportunities for alternative or independent cultural styles, most young people simply seek escapism in the less risky world of corporate-packaged nightlife on the mainstream. This perhaps says more about the lack of actual choice downtown, and the difficulties involved in locating and travelling to alternative and fringe nightlife spaces. ... In the context of nightlife, then, consumption options continue to be curtailed. A further aspect of this non-local branded space is a clear functional separation between spheres of consumption and production. ... Larger non-local venues do little to promote or connect with existing cultural practices, and in general they are directed from remote head offices and have scant interest or knowledge of local musical tastes, styles and habits. Corporate entities are given reign to produce, market and sell nightlife and alcohol, and accrue vast profits. Yet they are largely left free to regulate themselves, taking little responsibility for nightlife problems such as violence, noise and social segregation.

Despite contemporary corporate success in dominating urban life, conflicts around the role which drinking plays in urban life are as high-profile today as they were in the nineteenth century. Currently, for example, a 'moral panic' has grown up in the UK around city-centre drinking cultures (Lister et al., 2000). In the midst of urban regeneration initiatives that have developed night-time economies based around new consumption landscapes characterized by hybrid corporate café–bar–club venues, concerns about alcohol-fuelled disorder, drunken brawling, public sex acts, and litter from take-away food wrapping strewn across streets have emerged (Jayne et al., 2006). Such disorder, of course, would appear at face value to be at odds with urban renaissance, as well as being reminiscent of bourgeois fears over drinking in the modern city.

The UK Government's White Paper, *Time for Reform* (Department of Culture, Media and Sport, 2001), is an attempt to give the local state new powers to tackle alcohol-related problems by taking a more active role in shaping landscapes of drinking. At the heart of this political response has been a growing concern over drink-fuelled violence and vandalism among young adults – representations of youth that produce a whole raft of restrictive regulations, ranging from CCTV surveillance (Toon, 2000) to curfews (Collins and Kearns, 2001) and attempts to curb underage drinking. Policy debates regarding drinking are also currently taking place in Spain, the USA and Australia (Chatterton and Hollands, 2003). In the UK, alongside new legislation to crack down on violence and crime relating to unruly nights out, there are contradictory calls and new legislation aimed at speeding up economic development and the further deregulation of the night-time economy. The economic value to cities of drinkatainment and its role in regeneration is yet to be fully squared with public order concerns over the effects of over-indulgence: the pleasures of 'European-style' pavement cafés are at odds with tightened regulation of on-street drinking.

Beyond the West, other writers have highlighted the different roles spaces of food and drink can have in transforming urban cultures. For example, Edensor (2006) describes the development of the waterfront area in the city of Port Louis in Mauritius, noting how the 'Keg and Marlin' pub has provided a new place for local people to mingle, overcoming strict racial and religious segregation. The pub opens up the possibility of communal participation and is 'a convivial venue for alcohol-fuelled loud talk and expressive behaviour, modes of conduct that transgress ordinary constraints of family and community oriented social practice' (Edensor, 2006: 17).

Similarly, Thomas (2002) shows how economic and social transformations in public spaces in Hanoi, Vietnam, have paved the way for a dramatic change in the way in which streets, pavements and markets are experienced and imagined by the populace. She argues that an emerging street culture has further destabilized state control of public space. The consumption of alcohol, young people racing motorbikes, religious meetings, football crowds, and an outpouring of grief during the funeral possession of national pop stars are just some of the new forms of behaviour in public space that are causing political concern. Thomas argues that this new use of public space for everyday activities has been a catalyst for crowd formation, and utilizes the work of Habermas (1974), who argued that the growth of urban public culture in eighteenth-century Europe, centred on eating, drinking, leisure and meeting places, fuelled the development of the public sphere, to suggest that gathering together to exchange information and ideas has allowed a 'public sphere' to develop in much the same way.

This line of argument is followed in more detail by Latham's (2003) ethnographic study of a neighbourhood in the Western Bays area of Auckland, New Zealand, where he detects and tracks a new form of public culture, based around cafés, bars and restaurants. Crucially, he notes that 'what is happening there is about more than an aesthetics of consumption': the cafés and bars have 'acted as a key conduit for a new style of inhabiting the city' (Latham, 2003: 1706–1710). 'Consumption has quite literally helped to build a new world' (Latham, 2003: 1713), he writes, highlighting the importance of studying how people make use of bars and cafés in their everyday lives, but also the importance of looking outwards, from the micro-practices, to witness their broader impacts.

Latham argues for a more contextualized understanding of the role these sites play in new patterns of urban living. He notes the key role of entrepreneurs in developing these spaces, showing that many of these key players saw what they were doing as 'a kind of socio-cultural project' (Latham, 2003: 1717); they were invested in producing new ways of living and not just new markets for their food and drink. The entrepreneurs creating hospitality spaces in this neighbourhood in Auckland have played a crucial role in defining the 'feel' of the neighbourhood, and in consciously shaping their bars and cafés to promote particular kinds of conviviality. Importantly, Latham notes that the two streets he examined are not purified spaces of gentrification in which older, conflicting uses of space were chased or crowded out. In place of the chasing out of pre-existing vernacular traditions described in downtown New York by Zukin (1991), therefore, a 'convivial ecology' was seen by Latham to spill out into the streets, generating new solidarities and new collectivities, and a greater sense of belonging (see also Laurier and Philo, 2004).

Summary

The studies outlined in this chapter, and others like them, point to the importance of understanding urban eating and drinking spaces as more than sites of economic exchange or social problems; what's at stake is a collective, creative endeavour to produce and reproduce social and cultural practices and concrete enactments of a new way of living in cities (see Bell and Binnie, 2005). New geographies of hospitality, yet to be fully mapped, are emerging as cities transform and are transformed by changing consumption cultures.

Merrifield (2000) argues that urban space is kept alive by conflict. Indeed, as Laurier and Philo (2004: 4) suggest, the 'rumbustiousness of the crowd, as [well as] the self-interest of the individual are transformed into a new kind of sociality' – in places such as pubs, bars, cafés and restaurants. Latham and McCormack (2004: 716–717) have drawn attention to the many ways that 'alcohol plays a part in the eventful materiality of the urban', and they too track a tension between the promotion of cities as places of pleasure and the regulation of those pleasures or, as they put it, a clash between 'different modes of governmentality and economy'. And so it is with countless other pleasures of the city, formal or informal, planned or spontaneous, individual or collective. But we should be wary of romanticizing some pleasures and demonizing others, or seeing transgression and normalization too straightforwardly or self-evidently.

Miles' (2000) argument concerning young people's occupation of public space is useful here. Miles argues that young people's seemingly transgressive practices of hanging out at the mall, or on street corners, or skateboarding in public spaces, have little in fact to do with transgression and conflict; rather they represent an attempt by young people to carve out a place for themselves in the city, to assert their identities as active participants in consumer society. Eating and drinking should perhaps be considered in terms of the connectivities and belongings generated in public space – even in so-called commercial hospitality spaces – and as being grounded in the pleasures, enjoyment and 'riskiness' of heterogeneous groups of people mixing in city spaces (see Jayne et al., 2006). Going out for dinner, having a few drinks – these are simple pleasures, but their simplicity should in no way eclipse their importance. Mapping the geographies of these simple pleasures in the city is a similarly important task.

References

Amin, A. (1994) *Post-Fordism: a reader*, Oxford: Blackwell.

Atkinson, R. (2001) 'Domestication by cappuccino or a revenge on urban space? Control and empowerment in the management of public spaces', *Urban Studies*, 40(9), 1829–1843.

Bauman, Z. (1999) *Consumerism, Work and the New Poor*, Buckinghamshire: Open University Press.

Bell, D. (2002) 'Fragments for a new urban culinary geography', *Journal for the Study of Food and Society*, 6 (1), 10–21.

Bell, D. and Jayne, M. (eds) (2004) *City of Quarters: urban villages in the contemporary city*, Aldershot: Ashgate.

Bell, D. and Binnie, J. (2005) 'What's eating Manchester? Gastro-culture and urban regeneration', *Architectural Design*, 75(3), 78–85.

Bell, D. and Valentine, G. (1997) *Consuming Geographies: you are where you eat*, London: Routledge.

Bell, D., Binnie, J., Holliday, R., Longhurst, R. and Peace, R. (2001) *Pleasure Zones: bodies, cities, spaces*, New York: Syracuse University Press.

Bishop, P. and Hoggett, P. (1986) *Organizing around Enthusiasms: mutual aid in leisure*, London: Comedia.

Brain, K.J. (2000) *Youth, Alcohol and the Emergence of the Post-modern Alcohol Disorder*, Occasional Paper 1, London: Institute of Alcohol Studies.

Burnett, J. (1999) *Liquid Pleasures: a social history of drinks in modern Britain*, London: Routledge.

Chatterton, P. and Hollands, R. (2001) *Changing Our Toon: nightlife and urban change in Newcastle*, Newcastle: University of Newcastle.

Chatterton, P. and Hollands, R. (2003) *Urban Nightscapes: youth cultures, pleasure spaces and corporate power*, London: Routledge.

Cohen, P. (1997) *Rethinking the Youth Question*, London: Macmillan.

Collins, D.C. and Kearns, R.A. (2001) 'Under curfew and under siege? Legal geographies of young people', *Geoforum* 32 (3), 389–404.

Cunningham, H. (1980) *Leisure in the Industrial Revolution, 1780–1880*, New York: St Martin's Press.

Department of Culture, Media and Sport (2001) *Time for Reform: proposals for the modernisation of our licensing laws*, London: HMSO.

Dicks, B. (2003) *Culture on Display: the production of contemporary visitability*, Maidenhead: Open University Press.

Edensor, T. (1998) 'The culture of the Indian street', in N. Fyfe (ed.), *Images of the Street: planning, identity and control in public space*, London: Routledge.

Edensor, T. (2000) 'Moving through the city', in D. Bell and A. Haddour (eds), *City Visions*, Harlow: Pearson.

Edensor, T. (2006) 'Caudan: domesticating the global waterfront', in D. Bell and M. Jayne (eds), *Small Cities: urban experience beyond the metropolis*, London: Routledge.

Featherstone, M. (1991) *Consumer Culture and Postmodernism*, London: Sage.

Finklestein, J. (1999) 'Dining out: the hyperreality of appetite', in R. Scapp and B. Seitz (eds), *Eating Culture*, Albany, NY: SUNY Press.

Florida, R. (2002) *The Rise of the Creative Class*, New York: Basic Books.

Fraser, W.H. (1981) *The Coming of the Mass Market, 1850–1914*, Basingstoke: Macmillan.

Frisby, D. (2001) *Cityscapes of Modernity*, Cambridge: Polity Press.

Gibson, C. and Homan, S. (2004) 'Urban redevelopment, live music and public space: cultural performance and the re-making of Marrickville', *International Journal of Cultural Policy* 10 (1), 67–84.

Gould, K. (2005) 'Eating out', *The Guardian Weekend*, 2 July, p. 95.

Gratton, C., Shibli, S. and Coleman, R. (2005) 'Sport and economic regeneration in cities', *Urban Studies* 42 (5/6), 985–999.

Habermas, J. (1974) *The Structural Transformation of the Public Sphere: inquiry into a category of bourgeois society*, Cambridge: Polity Press.

Hall, T. (1998) *Urban Geography* (2nd edition), London: Routledge.

Hall, T. and Hubbard, P. (eds) (1998) *The Entrepreneurial City: geographies of politics, regimes and representation*, London: John Wiley.

Harring, S. (1983) *Policing a Class Society: the experience of American cities, 1865–1915*, New Brunswick, NJ: Rutgers University Press.

Hesmondhalgh, D. (2002) *The Cultural Industries*, London: Sage.

Hobbs, D., Hadfield, P., Lister, S. and Winslow, S. (2003) *Bouncers: violence and governance in the night-time economy*, Oxford: Oxford University Press.

Hoffman, L., Fainstein, S. and Judd, D. (eds) (2003) *Cities and Visitors: regulating people, markets, and city space*, Oxford: Blackwell.

Huggins, M. and Mangan, J. (eds) (2004) *Disreputable Pleasures: less virtuous Victorians*, London: Frank Cass.

James, A. (1996) 'Cooking the books? Global or local identities in contemporary British food cultures', in D. Howes (ed.), *Cross-cultural Consumption: global markets, local realities*, London: Sage.

Jayne, M. (2005) *Cities and Consumption*, London: Routledge.

Jayne, M., Holloway, S.L. and Valentine, G. (2007) 'Drunk and disorderly: alcohol, urban life and public space', *Progress in Human Geography* 30 (4), 451–468.

Kneale, J. (2002) 'The place of drink: temperance and the public, 1956–1914', *Social and Cultural Geography* 2 (1), 43–59.

Koshar, R. (ed.) (2002) *Histories of Leisure,* Oxford: Berg.

Kumar, K. (1995) *From Post-industrial to Post-modern Society*, Oxford: Blackwell.

Lash, S. and Urry, J. (1987) *The End of Organised Capitalism*, Cambridge: Polity Press.

Lash, S. and Urry, J. (1994) *Economies of Signs and Space*, London: Sage.

Lashley, C. and Morrison, A. (eds) (2000) *In Search of Hospitality*, Oxford: Butterworth Heinemann.

Latham, A. (2003) 'Urbanity, lifestyle and making sense of the new urban cultural economy: notes from Auckland, New Zealand', *Urban Studies* 40 (9), 1699–1724.

Latham, A. and McCormack, D.P. (2004) 'Moving cities: rethinking the materialities of urban geography', *Progress in Human Geography* 28 (6), 701–724.

Laurier, E. and Philo, C. (2004) 'Cafés and crowds', published by the Department of Geography and Geomatics, University of Glasgow at: http://www.geog.gla.ac.uk/olpapers/elaurier004.pdf.

Lister, S., Hobbs, D., Hall, S. and Winslow, S. (2000) 'Violence in the night-time economy – bouncers: the reporting, recording and prosecution of assaults', *Policing and Society* 10 (4), 283–402.

Malcolmson, R. (1973) *Popular Recreations in English Society, 1700–1850*, Cambridge: Cambridge University Press.

McAndrew, C. and Edgerton, R.B. (1969) *Drunken Comportment: a social explanation*, London: Nelson.

Marcuse, H. (1964) *One Dimensional Man*, London: Abacus.

Mass Observation (1943) *The Pub and the People*, London: The Cresset Library.

Mennell, S. (1985) *All Manners of Food: eating and taste in England and France from the middle ages to the present*, Oxford: Blackwell.

Merrifield, A. (2000) 'The dialectics of dystopia: disorder and zero tolerance in the city', *International Journal of Urban and Regional Research*, 24 (2), 473–489.

Miles, E. (1993) 'Adventures in the postmodern kitchen: the cuisine of Wolfgang Puck', *Journal of Popular Culture*, 27 (2), 191–203.

Miles, S. (2000) *Youth Lifestyles in a Changing World*, Buckingham: Open University Press.

Miles, S. and Paddison, C. (1998) 'Urban consumption: an historical note', *Urban Studies*, 35 (5–6), 815–832.

Monkkonen, E.H. (1981) 'A disorderly people? Urban order in the nineteenth and twentieth centuries', *The Journal of American History*, 68 (3), 539–559.

Oldenburg, R. (1989) *The Great Good Places: cafés, coffee shops, bookstores, bars, hair salons, and other hangouts at the heart of community*, New York: Marlowe and Company.

O'Mally, P. and Valverde, M. (2004) 'Pleasure, freedom and drugs: the use of "pleasure" in liberal governance of drugs and alcohol consumption', *Sociology* 38 (1), 25–42.

Pearson, G. (1993) *Hooligan: a history of respectable fears*, Basingstoke: Macmillan.

Richards, T. (1990) *The Commodity Culture of Victorian England: advertising and spectacle, 1851–1914*, London: Verso.

Roberts, K. (2004) *The Leisure Industries*, Basingstoke: Palgrave.

Schivelbusch, W. (1993) *Tastes of Paradise: a social history of spices, stimulants and intoxicants*, New York: Vintage.

Sennett, R. (1977) *The Fall of Public Man*, Cambridge: Cambridge University Press.

Smith, N. (1996) *The New Urban Frontier: gentrification and the revanchist city*, London: Routledge.

Stevenson, D. (2003) *Cities and Urban Cultures*, Maidenhead: Open University Press.

Thomas, M. (2002) 'Out of control: emergent cultural landscapes and political change in urban Vietnam', *Urban Studies*, 39 (9), 1611–1624.

Thompson, E.P. (1967) 'Time, work discipline and industrial capitalism', *Past and Present*, 38 (1), 56–97.

Toon, I. (2000) 'Finding a place in the street: CCTV surveillance and young people's use of urban public space', in D. Bell and A. Haddour (eds), *City Visions*, Harlow: Pearson.

Valentine, G. (1998) 'Food and the production of the civilised street', in N. Fyfe (ed.), *Images of the Street: planning, identity and control of public space*, London: Routledge.

Warde, A. and Martens, L. (2000) *Eating Out: social differentiation, consumption and pleasure*, Cambridge: Cambridge University Press.

Weber, M. (1967) *The Protestant Ethic and the Spirit of Capitalism*, London: Routledge.

Wrigley, N. and Lowe, M. (eds) (1996) *Retailing, Consumption and Capital: towards the new retail geography*, Harlow: Longman.

Zukin, S. (1991) *Landscapes of Power: from Detroit to Disney World*, Berkeley: University of California Press.

SECTION THREE
COMMUNITIES AND CONTESTATION

From the earliest times, cities have been noted for their capacity to encapsulate social diversity and difference. In contrast to the rural, where villages often remain predicated on the myth – if not always the reality – of kinship and stability, the city has always been known as a melting-pot of peoples from different backgrounds. This social mixity has ambivalent outcomes. One consequence is the splintering of cities along the fault-lines of race, sexuality, class, political affiliation or religion. In many contexts, these divides have taken concrete form (as in the 'peace lines' that long marked the sectarian divides in Belfast); in others they exist only in the imaginations of citizens (but are none the less significant for all that). A consequence is that some urban dwellers find their occupation and use of space routinely challenged by those social groups who depict them as troublesome Others (and hence as 'out of place'). In the urban West, for example, white, middle-class, heterosexual values reign supreme, and those who challenge hegemonic values may be pushed to the metaphorical (and sometimes literal) margins of cities.

The city may therefore be described as a highly contested social arena, riven by tension and fear. The counterpoint to this argument, however, is that the city is a space where exposure to difference breeds a cosmopolitan and tolerant outlook, and where people from different backgrounds get along. Nowhere is this more evident than in the city's public spaces – its streets, squares, parks and open spaces. Theoretically open to all, public spaces provide opportunities for different social groups to pursue a rich variety of activities. Public spaces are thus arenas where difference is on display. Gay, lesbian and bisexual groups, for example, have announced their presence in city space through a variety of parades, love-ins and carnivals, while religious festivals (Hanukah, Christmas, Diwali, Eide, etc.) are often celebrated on the city streets in non-exclusive ways. The city unavoidably exposes us to different values, memories and ideologies, and part of the urban experience is to accept and celebrate this diversity. One outcome is the emergence of *hybrid* cultures and practices. For instance, cities constantly generate new styles of cooking, music, fashion and literature that combine elements from traditionally separate cultures. This *fusion* of cultures is thus crucial in maintaining cities' roles as a centre of innovation and creativity (see also Chapter 8).

Yet while some cities appear crucibles of creativity and cosmopolitanism, not all of those who live in cities respect difference to the same degree – and some differences are clearly more tolerated than others. Racism, homophobia, sexism and snobbery unfortunately still thrive in many cities. Some commentators go so far as to suggest that urban dwellers are actually become less rather than more tolerant over time, with post-millennial tensions about crime and terrorism fuelling a diverse range of exclusions. Islamaphobia, for instance, has been evident in attacks on Muslim communities and opposition to mosque building in UK cities, while asylum seekers are also often regarded as unwelcome visitors to cities, stereotyped as malingerers by an aggressive media. Street homeless populations, prostitutes, buskers, beggars and those simply 'hanging out' on the street have also become figures of suspicion, subjected to increasingly punitive forms of policing and surveillance. In the words of US urbanist Mike Davis, Burgess's urban ecology of concentric zones of settlement has been replaced by a pervasive 'ecology of fear'.

Far from buying into a relentlessly negative prognosis of the current social state of cities, our contributors in this section aim to offer a more balanced appraisal. In Chapter 11, for example, Marie Price considers how both forced and voluntary migration is focused around a set of leading world cities, producing vibrant ethnoburbs and diasporic spaces where immigration is most pronounced. Such spaces become the launching pad for new cultural and affiliational identities, and are often foci of tourism and investment. Likewise, such city spaces become woven into a complex skein of familial, social and economic networks that have global reach. As such, senses of belonging are created in and through cities, with specific cities prospering through this incorporation into a global space of flows (as Price shows with reference to Birmingham's ethnic business networks). Yet Price also reveals the precarious circumstances facing many migrants, with insecurity and prejudice often forcing immigrants to take marginal or piecemeal work. As she relates, the gendering of home and work life means this tendency is particularly apparent among female migrants. Consequently, many immigrant women have a somewhat tenuous hold on the city, and lack the citizenship rights others may take for granted.

David Wilson picks up on issues of immigration in a somewhat different manner in Chapter 12, exploring the way that successive generations of non-white migrants, in the USA and UK especially, have become increasingly segregated within urban social space. While first-generation migrants often cluster for mutual support and the maintenance of cultural values, Wilson alludes to the forms of racism that serve to displace non-white populations from spaces of privilege and power. The notion of the 'ghetto' is itself loaded with cultural assumptions and stereotypes of criminality and despair, but in this chapter Wilson uses the term advisedly to scrutinize the processes conspiring to corral non-white populations into some of the least prosperous census tracts. His analysis draws on a rich tradition of critical urban research to expose the role of urban space in reproducing notions of racial difference and, given that he spends little time describing how some non-white migrant groups have been assimilated into Western cities successfully, his review reminds us that there is much that still needs to be said and done about the pernicious racialization of the city.

Issues of race have also been explored in the literatures on urban public space, where the need for immigrant and newly-settled groups to become 'streetwise' because of discrimination has often been noted. Yet it is not only those marked as racially Other who have to be resourceful in their use of public space. Failure to adapt the social mores and codes of civility which adhere to the street can mark out individuals as discrepant, prompting the disapproval of dominant social groups who may summon the forces of law and order. Yet learning how we should act and appear on the streets is a complex matter. Joyce Davidson and Mick Smith's chapter accordingly considers how notions of comportment, culture and civility have changed over time as the distinction of culture/nature has become more deeply embedded in urban life. As they show, there is a mesh of written and unwritten rules about what behaviours are acceptable in public and private space. Non-conforming groups have often been described as belonging to the 'dangerous classes', with their failure to conform to notions of civility often a justification for their displacement. Gentrification – the widely noted phenomenon where working-class populations are displaced by upper-class incomers – is thus often predicated on notions that the inner city is uncivilized and needs to be 'reclaimed' by decent citizens. As Davidson and Smith show, questions about manners and comportment are consequently tied into all number of urban policy debates.

Sarah Holloway's chapter on living spaces picks up on some of these themes about social conformity, not least when she considers the way housing style and decoration communicates our identity and class. There is clearly a wide literature in urban studies about housing, but Holloway argues that much of the literature ignores social issues of occupation, and instead treats housing as an abstract commodity. As she demonstrates, the home is a space of work, rest and play, and may be a space in which dominant orders are resisted and recast. For instance, collaboration and cooperation begins within the home, with economic, social and political networks typically taking shape in domestic space before spreading across different neighbourhoods, cities or even nations (and here one can think about the diverse range of e-businesses, support groups and societies which are run out of homes). By the same token, however, homes may be spaces of oppression, and Holloway's chapter also explores how the home is bound into power relations that serve to oppress (for example) women, people of colour or sexual minorities. For some, the home may feel more like a prison than a haven.

Exploring the social relationships played out in the home, Sarah Holloway's chapter reminds us that the social geography of the city is not just about the life on its streets. Yet for some urban dwellers, the sad truth is that they know of no other life. Gerald Daly's chapter accordingly explores the multifarious urban geographies of homelessness. Long associated with the city, homelessness takes a number of different forms, the most discussed of which is the visible concentration of homeless peoples on the street, rough-sleeping. For the homed population, the visible presence of homelessness on the streets tends to provoke a variety of emotions. Notable here is the disgust that is provoked when street dwellers disturb the established boundaries between public and private spaces by, for example, washing, urinating, eating or sleeping in public spaces. The general sympathy expressed for those who have no home thus often dissipates when homelessness is experienced at first hand, triggering a *turf politics* in which the homeless are usually the losers. Documenting the marginalization of the homeless in urban social space, Daly thus outlines the contribution that urban studies has made to understanding the processes that lead to and exacerbate homelessness. Yet even if homeless populations have just a precarious hold on urban space, Daly also shows how they use the city as a productive space, a space of encounter and a space of sociality. In the final analysis, he shows that the city often provides routes out of homelessness, as well as ways in. It is this ambivalence that is at the heart of the urban condition, and which remains central to urban social studies of all kinds.

11 MIGRATION AND SETTLEMENT

Marie Price

This chapter

O Outlines the importance of cities as foci for immigration and emigration, with a particular focus on the 'gateway' cities through which a high proportion of the world's migrants and refugees pass

O Details some of the ways in which migration transforms the economic, social and political life of cities

O Notes the high proportion of migrants working in city economics in the least skilled, lowly paid and precarious jobs, and highlights an important gendering of urban migration

---

Introduction

The impact of migrant flows on cities is a long-standing concern in urban research. Much of this work has focused on the flow of internal migrants to cities throughout the world; a process that has been underway for decades, with over half the world's people now living in urban areas (PRB, 2005). This chapter focuses on the impact of immigrants, that is to say the foreign-born, on cities. One of the most frequently noted aspects of globalization is the rise in the number and diversity of foreign labourers working and living in those world cities which lie at the heart of the networked global economy (such as London and New York) (Sassen, 1994, 2002a; Samers, 2002). Collectively, millions of immigrants are redefining global networks by creating ethnic enclaves in distant countries, returning billions of dollars in remittances to countries of origin, and creating new levels of cultural diversity. Yet this is not just the case in a few world cities, but in scores of new immigrant destinations. In the process, urban spaces are being created which are celebrated as cosmopolitan centres of diversity and pluralism and condemned as localities of displacement with heightened polarization along racial and class lines.

Immigration is a lens through which we can view the reconfiguration of urban social space as millions of economic immigrants settle in select cities across the globe. By focusing on immigrant labour and settlement in the city it is possible to see how global processes become localized, from highly visible Chinese commercial districts, to the enclaves of residential guest workers, or the barely visible lives of foreign nannies in scattered suburban settings (Cox, 2006). In contrast to internal migrant flows, the growth in the number and percentage of foreign-born in cities raises important questions about citizenship, identity, race, and the integration or exclusion of distinct groups of people. This issue is of increasing importance to policy-makers at the urban and national scales, especially in the cities of Western Europe, Japan, North America, and the Persian Gulf, where foreign labourers are increasingly relied upon due to unmet demands for both highly skilled and low skilled labourers, as well as the overall ageing of the population.

This chapter develops several themes with regards to the socio-spatial issues surrounding immigrants in cities. First, the range of immigrant destinations will be explored, from the archetypal gateways of New York City and London to the less recognized destinations of Dubai and Johannesburg. Second, the legal status in which immigrants enter the city will be addressed, noting the significant effects that temporary worker programmes have had on the increase of foreign labour. Third, the feminization of migration, especially as it relates to the care industry, has resulted in circuits of women from developing countries flowing into major cities to care for children, and to provide domestic and healthcare services. With regards to the settlement of immigrants in cities, there are diverse trends from highly centralized enclaves in both city centres and suburbs to increasingly dispersed patterns of settlement across metropolitan areas. This chapter will conclude with a discussion of how immigrants claim space, both literally and symbolically, through the formation of commercial spaces to the maintenance of transnational social networks.

Traditional gateways and emerging destinations

The term 'gateway' is typically used in North American and Australian contexts to designate a major metropolitan area where large numbers of immigrants have settled (Ley and Murphy, 2001; Clark and Blue, 2004; Gozdziak and Martin, 2005). Cities such as New York, Toronto, Los Angeles, Vancouver, Miami, Sydney and San Francisco are the chosen destinations for many of the world's immigrants (Foner, 2000; Benton-Short et al., 2005). These are cities with a long history of immigrant settlement, but even these cities are witnessing greater flows of people from more diverse countries coming with a higher level of skills. For example, in metropolitan Sydney (an area of some four million people) one-in-three residents are foreign-born, and the top ten 'sending countries' include traditional and newer countries such as the UK, China, New Zealand, Vietnam, Lebanon, Italy, Philippines, India, Greece and South Korea (Burnley, 2001; Voigt-Grad, 2004).

Sociologist Audrey Singer's (2004) typology of immigrant gateways for US cities over the past century underscores the temporal aspects of a particular city's appeal to immigrants. Her six-fold gateway typology includes Former (Buffalo or Cleveland), Continuous (New York City or San Francisco), Post-Second World War (Los Angeles or Miami), Emerging (Washington, DC or Atlanta), Re-Emerging (Denver, or Portland, Oregon) and Pre-Emerging (Salt Lake City) gateway cities. Analyzing over a century of census data, Singer demonstrates that very few US cities continually attract immigrants – even in the context of a nation which has received more official immigrants than any other country. The economic context that made Cleveland or St Louis attractive cities for immigrants in the 1920s did not hold for the 1980s and 1990s. Moreover, cities that attracted very few immigrants in the early twentieth century, such as Washington or Atlanta, are now major destinations. Typically, cities that have experienced sudden spikes in immigrant populations experience the most friction as local governments, churches, schools and healthcare providers respond to the particular needs of a relatively new and diverse population that requires services. Yet for Canadian, US and Australian cities, the idea of an urban immigrant gateway that receives diverse groups of people is an integral part of interpreting the social and economic geography of cities (Extract 11.1).

Extract 11.1: From Smith, M.P. (2001) *Transnational Urbanism, Locating Globalization*, Oxford: Blackwell, pp. 76–8.

The most obvious dimensions of the story of Mexican migration to Los Angeles are its timing and vast scale. More than two thirds of all Latinos living in Los Angeles are of Mexican origin. Many of these are longtime Angelenos who are Mexican-American US citizens. They are the offspring of earlier generations of Mexican migrants, whose migration preceded the epoch of global economic restructuring and was driven more by the ebbs and flows of US agricultural policy and its permissions and constraints than by global developments. ...

A more recent dimension of Mexican migration to Los Angeles is rooted in the operation of another dimension of US state policy, the Immigration Reform and Control Act (IRCA) of 1986. The amnesty provisions of IRCA allowed undocumented Mexican workers who had been living in California for four years preceding the amnesty to obtain green cards (legal permanent residence). Although not envisaged in the IRCA amnesty, many of the hundreds of thousands of male workers who had been living in transnational households then chose to bring their wives and children to live with them in California, thus dramatically altering the gender composition of more recent Mexican migration. ...

(Continued)

The law had several unintended consequences. Low wages in the agricultural sector drove many male farm workers who had brought additional household members from Mexico to California to move to cities like Los Angeles in search of higher-paying urban service and manufacturing jobs, in the state's then still robust metropolitan economies. ... Increasing numbers of the children of Mexican migrants entered the public schools, setting the stage for one of the most punitive, exclusionary, and unconstitutional provisions of Proposition 187.

As can be seen from these developments, the IRCA-driven expansion of recent Mexican migration is an interesting phenomenon both within California and transnationally. It has changed the demographics and expanded the transnational social networks operating in Los Angeles as well as many other cities and towns in California. Yet it too may be viewed as a historical extension of long-term bi-national relations of production, labor supply, household survival, and social network formation between California and Mexico, rather than as either a singular manifestation of economic globalization or a proof of Los Angeles's emergence as a 'global city.'...

The most recent period of Mexican transmigration to Los Angeles (late 1980s–1990s) was a politically produced outcome of historically specific policy initiatives. The newest wave of transnationalization of Los Angeles was driven by declining living standards and growing income polarization within Mexico, produced by a combination of the austerity policies of international banks, the Mexican debt crisis, and the conscious decision of Mexico's ruling political elites to meet this crisis by export-oriented neoliberal economic policies. ...

These developments, exacerbated by NAFTA [North American Free Trade Agreement], have made it difficult for many households in Mexico but the elite level to survive on the income they could generate in Mexico's borders. This, in turn, led to the formation of ever more transnational households capable of tapping into the income-producing possibilities of cities like Los Angeles.

Many European cities, such as London, Paris and Amsterdam, also have complex immigration histories that have accelerated in the post-colonial period (Sassen, 1999; Hagendoorn et al., 2003; Foner, 2005). Yet in the European context, the idea of gateway has been less enthusiastically embraced, in preference for a vision of foreign workers as labourers rather than permanent settlers and citizens. As Europe was rebuilt after the Second World War, programmes were created that encouraged guest workers from (for example) North Africa and Turkey. These labour migrations inevitably led to the formation of new immigrant communities in cities throughout Europe. Since the 1990s, many European cities have experienced a steady rise in their foreign-born populations, both from flows within an enlarged European Union and arrivals from throughout the world. By 2000 several European cities, such as Amsterdam, Geneva, Brussels, Frankfurt, London, Munich and Zurich, had populations of at least 20 per cent or more foreign-born residents. Other major cities, such as Cologne, Paris, Stockholm, Birmingham, Dusseldorf, Vienna,

Cities with over 250,000 and 1,000,000 Foreign-born Residents

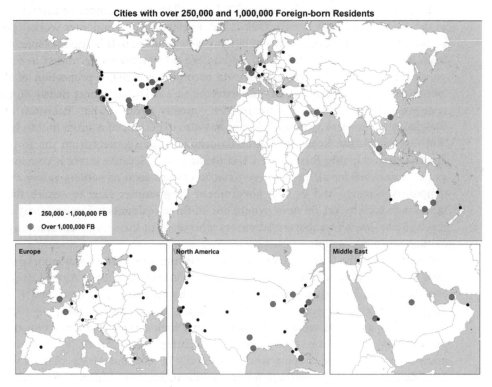

Figure 11.1 Global urban immigrant destinations, based on data from the Globalization, Urbanization and Migration website

Hamburg, Bonn, Berlin, Rotterdam, Dublin, Copenhagen, Lyon, Athens and Marseille, experienced steady increases in immigrants so the foreign-born accounted for 10 to 20 per cent of their populations. Some European cities, such as Lisbon, are both sending native-born people to work in other EU countries while receiving foreign labour from Africa, South America and the former Soviet Union (see www.gstudynet.com/gum for data on the foreign-born by city).

There is tremendous variation in immigration policies in Europe, and laws are constantly being changed to address new and larger flows of immigrants. In some countries, such as the UK and France, immigrants often become full citizens. Yet in other countries, such as Germany, the path to citizenship is less assured. These newer immigrant gateways should perhaps be viewed more as 'turnstiles' than as settings for permanent settlement, where foreign-labourers work for varying periods of time before returning home or moving on to another city in the transnational network (Castles and Davidson, 2000).

Generally speaking, contemporary urban immigration flows are largely directed to North American, European and Australian cities. Yet as Figure 11.1 demonstrates, there are also immigrant destinations in the Middle East, notably in Dubai, Riyadh, Mecca, Jiddah and Tel Aviv. The large number of immigrants in the Arab cities of the Persian Gulf is due to established temporary worker programmes encouraging thousands of labourers emigrating to this region, especially from North Africa and South Asia. Seldom are these workers permitted to settle permanently in these cities, but they do account for a major proportion of the labour force and the population. The extreme case is Dubai, where nearly 80 per cent of the population is foreign-born, mostly from India, Pakistan and Bangladesh. The Israeli city of Tel Aviv has also experienced a surge in immigration, especially due to the arrival of thousands of newcomers from the former Soviet Union (notably Russian and Ukrainian Jews). Because Israel is conceived as the homeland for all Jews, these newcomers are seen as settlers rather than temporary workers and explicit government programmes exist to absorb them into Israeli society. Yet Tel Aviv (where one-in-three residents is foreign-born) also receives non-Jewish temporary labourers who work in the service economy, agriculture and construction.

There are relatively few urban immigrant destinations in the developing world, save from former immigrant gateways in South America (Buenos Aires, Caracas, São Paulo) and emerging gateways such as Johannesburg, Singapore and Hong Kong. Due to the dire economic problems facing South America since the 1980s, these former gateways are now the sites from which former immigrants (or their descendents) leave to carve out new lives in Europe, the USA or Japan. But for many regions in the developing world, especially in Asia and Africa, there is very little data gathered on the foreign-born at the urban scale. In fact, for sending countries such as the Philippines, national censuses are more interested in questions about emigration (people leaving) than immigration.

The data problems for assessing the foreign-born in African cities are also significant. The South African case is noteworthy because excellent data do exist. With the end of Apartheid in 1994, Sub-Saharan Africans (both skilled professionals and labourers) began to migrate to South Africa, which boasts the continent's strongest economy. From having very few foreign-born, metropolitan Johannesburg is now home to at least 400,000 immigrants. Many from the neighbouring countries of Mozambique, Zimbabwe and Zambia have settled among black South Africans in the former townships. Immigrant flows are also changing the demographic and ethnic composition of Cape Town. The intensity and size of these new flows has spurred an international effort to document migration flows throughout southern Africa (see the Southern African Migration Project, www.queensu.ca/samp).

Singapore is one of the most prosperous cities in the developing world. This city-state was a former British colony, and as such it has a complex history of colonial and labour migration. Yet it was the city's economic growth in the 1980s that encouraged a steady rise in foreign labourers. Estimates in 2005 revealed that 18 per cent of the city-state's population was non-native, yet foreign labour accounted for up to 30 per cent of Singapore's labour force. The majority of this foreign labour comes from neighbouring Malaysia, followed by ethnic Chinese from China, Hong Kong and Taiwan and then much smaller numbers of immigrants from South Asia and Indonesia. Increasingly, the Singaporean government relies upon foreign temporary workers to provide needed labour, but at the same time access to citizenship is strictly limited and undocumented workers and over-stayers are deported (Yeoh, 2006).

There are notable absences in Figure 11.1 that indicate the avoidance of the some of the world's largest cities, especially in South and East Asia. The world's two largest countries, India and China, have relatively few foreign-born labourers in their cities, nor do they apparently need them, given their abundant labour supplies. Even in India's celebrated Silicon Valley – Bangalore – the number of foreign-born workers is relatively low. The robust economies of Tokyo and Seoul have not relied upon large numbers of foreign workers (although even in these relatively homogeneous cities the numbers of foreign-born workers is on the rise). In the Tokyo case, the 'return' migration of ethnic Japanese to the mother coun-try (especially from Brazil and Peru) has resulted in a significant increase in foreign-born residents. So, too, have 'trainee' programmes that let in Chinese workers for set periods of time.

The place of settlers, temporary workers and the undocumented

There is an inherent tension between national immigration policies, and the localities – typically cities – receiving large numbers of immigrants. What may be a workable policy at the national scale, say allotting temporary worker visas for low-skilled labourers, can become a local crisis when such workers need health-care, housing and education for their children. In a global age when capital and information freely crosses borders, the movement of people is still highly regu-lated by the state and state policies can shift dramatically in response to the needs of the labour market and sudden shifts in national security concerns.

City governments and their residents do respond to the various opportunities and issues resulting from new and diverse immigrant flows. Typically, urban plan-ners and various non-governmental associations respond to needs in housing, transportation and education. For example, the city of Vienna created a public housing complex in the 1990s called 'Global Yard' which mixes native-born Viennese with foreign-born residents at a fifty–fifty ratio to encourage greater social integration (Figure 11.2). Resettlement centres are created and maintained by governmental and non-governmental actors to deal with refugees and economic migrants. Cities can also make public space available for festivals and other events that display the cultural diversity of residents. In some cases, urban governments can be proactive in formalizing the political status of its non-citizen residents, as in the case of Berliners in the 1990s with regards to the citizenship status of its immigrant-derived ethnic population (Ireland, 2004). Or city officials can provide locations where immigrant day labourers can register and wait for

Figure 11.2   The 'Global Yard' – a public housing project in Vienna, Austria, that is home to equal numbers of foreign-born and native-born residents

employers. Cities are often sites of protest against ethnic immigrant populations (and their descendents), as witnessed in the anti-Arab immigrant protests by white youths on a Sydney beach in December 2005. As cities are the places where foreign-born and native-born are most likely to come into contact, they become vital settings for both tolerance and intolerance of ethnic and cultural difference.

Immigrants can hence be roughly divided into three main categories: settlers (with legal permanent residence or citizenship), temporary workers (legally allowed to work and reside for a set period of time), and undocumented labourers. Major immigrant destinations typically contain all three groups, although in varying proportions. The look and feel of an immigrant gateway is shaped by the major immigrant source countries and the legal status of the newcomers.

Metropolitan New York has foreign-born residents from all over the world with no one country dominating the flow (Foner, 2000). Recognized as one of the world's traditional gateways for immigrant settlers, New York also has many temporary workers (notably in the high-tech fields) as well as undocumented migrants who easily blend into the hyper-diverse city with its well developed, albeit segmented, labour markets that rely on foreign labour. Over time, however, many immigrants do naturalize and thus are able to vote. Consequently, New York City politicians have always been very conscious of the diverse immigrant communities, actively courting their support and attending immigrant-organized events (Guarnizo et al., 1999).

The ability of temporary workers to assert themselves in the life of a city is less assured. Cities in the Middle East, such as Dubai and Riyadh, insist that their foreign-labour is admitted on a temporary basis and it is often segregated from the native population in discrete residential zones. Seldom do these immigrants receive full citizenship, nor the rights and resources available to the native-born population. Yet even in these cases, where such a substantial demand for foreign-labour exists, undocumented populations exist and have an impact on the city. In the more liberal Persian Gulf city of Dubai, one hears a myriad of languages on the streets as well as diversity in dress and appearance due to its enormous (and largely male) foreign labour force, especially from South Asia (Castles and Miller, 2003).

A basic role of the state is to regulate who can legally enter its borders; thus one of the most debated political issues concerns the appropriate role of the state with regards to the rising number of undocumented migrants. There are two basic approaches that state governments can take, both of which directly impact cities. One strategy is to regularlize (i.e. make legal through amnesty programmes) the status of illegal immigrants. The second is to detain and repatriate those immigrants who are deemed illegal. Both of these policies are costly, although regularization of migrants transfers more costs to local governments whereas repatriation does not. In the USA, the debate regarding the undocumented is especially strident as it is estimated that the USA has the largest undocumented population in the world – 10–14 million immigrants (mostly from Mexico and Central America). While only 3–4 per cent of the total population, undocumented migrants are a disproportionate part of the labour market in major metropolitan areas of the country, and especially in the sunbelt cities of Los Angeles, Phoenix, Houston, San Diego and Dallas-Fort Worth. In the spring of 2006 there were over a dozen marches in cities across the country protesting a punitive federal law that would make undocumented immigrants felons, thus forcing massive deportations. The largest march was in Los Angeles, where over 500,000 protestors (mostly immigrants from Mexico and Central America) peacefully rallied for recognition of their work and a path to citizenship. But undocumented migrants can also be a concern for developing counties, such as Costa Rica, with its large numbers of undocumented labourers from Nicaragua entering the country (and its principal city of San Jose) each year to work in industry, agriculture and services jobs.

The feminization of urban migration

Global migration trends based on United Nations data show that the proportion of female immigrants is growing, accounting for nearly half of all immigrants. In the developed countries in Europe and North America, women immigrants actually outnumber male ones. Even in regions such as the Middle East, where male immigrants have traditionally dominated the flow, the ratio of male to female immigrants is less skewed than in past decades. Some female migration is due to family reunification policies or dependent spouses accompanying workers. Yet greater attention is being paid to the number of female immigrants filling employment needs throughout the world.

The growing movement of women across international borders, and especially into cities, is associated with the globalization of the care industry – employment in child- and elder-care, housekeeping, nursing and hospital support. In cities throughout the world it is increasingly common to see immigrant women caring for children and the elderly, cleaning homes and tending to the sick. Ironically, the gains that many women have made in the formal workforce have led to the commodification and denigration of traditional 'home-making' tasks. Much of this reproductive labour has become 'low-skilled' and 'low-status', and jobs are more often than not filled by immigrant women (Extract 11.2).

Extract 11.2: From Sassen, S. (2002) 'Global cities and survival circuits', in B. Ehrenreich and A.R. Hochschild (eds), *Global woman: Nannies, Maids and Sex Workers in the New Economy*, New York: Metropolitan Books, pp. 257–60.

Globalization has greatly increased the demand in global cities for low-wage workers to fill jobs that offer few advancement possibilities. ... The fact that these workers tend to be women and immigrants also lends cultural legitimacy to their non-empowerment. In global cities, then, a majority of today's resident workers are women, and many of these are women of color, both native and immigrant. ...

A substantial number of studies now show that regular wage work and improved access to other public realms has an impact on gender relations in the lives of immigrant women. Women gain greater personal autonomy and independence, while men lose ground. More control over budgeting and other domestic decisions devolves to women, and they have greater leverage in requesting help from men in domestic chores. Access to public services and other public resources also allows women to incorporate themselves into mainstream society; in fact, women often mediate this process for their households. Some women likely benefit more than others from these circumstances, and with more research we could establish the impacts of class, education and income. But even aside from relative empowerment in the household, paid work holds out another significant possibility for women: their greater participation in the public sphere and their emergence as public actors.

Immigrant women tend to be active in two arenas: institutions for public and private assistance, and the immigrant ethnic community. The more women are involved with the migration process, the more likely it is that migrants will settle in their new residence and participate in their communities. And when immigrant women assume active public and social roles, they further reinforce their status in the household and the settlement process. Positioned differently from men in relation to the economy and state, women tend to be more involved in community building and community activism. ...

And so two distinct dynamics converge in the lives of immigrant women in global cities. On the one hand, these women make up an invisible and disempowered class of workers in the service of the global economy's strategic sectors. Their invisibility keeps immigrant women from emerging as the strong proletariat that followed earlier forms of economic organization, when workers' positions in leading sectors had the effect of empowering them. On the other hand, the access to wages and salaries, however low; the growing feminization of the job supply; and the growing feminization of business opportunities thanks to informalization, all alter the gender hierarchies in which these women find themselves.

The book *Global Woman* (Ehrenreich and Hochschild, 2002), illustrates the international circuit of care-related migration as well as the trafficking of women in the global sex industry. Mexican and Central American women fill much of the demand in the USA and Canada. Women from India, Sri Lanka, Malaysia, the Philippines and Indonesia provide much of domestic labour in the Middle East. In Western Europe, domestic immigrant labour comes from the Philippines, Sub-Saharan Africa and Eastern Europe. Lastly, in the wealthy Asian cities of Singapore, Taipei, Hong Kong and Kuala Lumpur, immigrants from the Indonesia and the Philippines are especially common. The spectacle of thousands of Filipina domestic women gathering in Hong Kong parks to socialize on their one day off a week is a regular sight. In terms of the international sex trade, Thai women are often recipients of 'entertainment' visas for employment in the wealthy East Asian cities of Tokyo and Seoul. There is much written about the forced trafficking of women for sex work, which does occur but probably not at the rates suggested by some human rights reports. What is clear is that both immigrant and native-born women are involved in sex work, but we still know very little about whether this population migrated voluntarily or were coerced.

The role of women in the global care industry has many conflicting consequences (Huang et al., 2006). In the Philippines, female overseas contract workers are celebrated as national heroines through their vital economic contributions to a remittance-based economy. Yet, they are often sharply criticized for social ills resulting from their inability to care for their own children while they toil for months at a time in Singapore, Riyadh or Rome as temporary workers. In the USA, where many care providers enter illegally from Mexico and Central America, undocumented women often do not see their children for many years until they are able to regularlize their own status or find a way to have their children join them (either legally or illegally). Interestingly, research of the long-term impact of female immigration have observed that as women become an important part of the immigrant flow, the immigrant community itself is more likely to gain permanency. This is often at odds with the desires of the host country officials.

Due to the nature of domestic work in people's homes, there are many opportunities for abuse and few ways to regularize or unionize this sector of the economy (Cox, 2006). Examples of women living in slave-like conditions in urban and suburban households are commonly reported in newspapers and pursued by various government and non-governmental agencies. The Philippines government and NGOs, for example, have enacted laws and advocated for the better protection of the rights of overseas workers, especially women contracted to do domestic work. Yet a consequence of Filipino workers asserting their rights has led recruiters to turn to other Southeast Asian countries (notably Indonesia or Vietnam) where female labour is considered more 'docile' and less expensive. Given the economic needs of many women in the developing world and the labour demands of urban residents in both developed and developing countries, it is likely that immigrant women will continue fill the gaps in urban care-related and commercial sex sectors for decades to come.

Patterns of urban immigrant settlement

The dominant narrative explaining the settlement and assimilation of immigrants in cities was first developed by American sociologists in the early 1900s and has often been applied to urban immigrants in traditional settler societies in North America and Australia. The Chicago School model assumes that immigrants with limited means initially settle in poorer quality inner-city neighbourhoods, most often amidst other co-ethnics. These neighbourhoods are often viewed as problematic settings where social services are limited and ethnic populations isolated. Yet over time, with greater social and economic attainment, immigrants will leave the urban core for the suburbs, indicating their integration into the host society (Waters and Jimenez, 2005).

This spatial assimilation model maintains the socioeconomic status of immigrants and is reflected in their residential distribution. Under the assumptions of the model, *space* is the critical variable, with the locational concentration of immigrants a surrogate measure of social and economic isolation and the dispersion of immigrants among the native-born as an indication of social integration into the dominant society. Yet the model definitely falls short when explaining the immigrant settlement patterns of visible ethnic minorities (especially Hispanics, Asians and African-ancestry immigrants in the US context) who for various reasons may maintain ethnic enclaves in urban and suburban settings rather than disperse themselves. Geographers Richard Wright and Mark Ellis (2000) also critique the model as applied to the USA for assuming spatial dispersion implies social integration, arguing that many dispersed suburban immigrants live in relative social and linguistic isolation. Even with its faults, the spatial assimilation model is the baseline for questions of immigrant integration into urban space. Yet as evidence from new immigrant destinations accumulates, the inadequacies of the model are significant.

While cities such as New York still have a large proportion of its immigrants in the urban core, the post-industrial nature of many other cities means more employment opportunities may be found in the suburbs and in edge cities. Consequently, many immigrants by-pass the city centre entirely, settling directly in the suburbs. In a study of recent immigrants to Metropolitan Washington DC, the researchers found that newcomers were highly dispersed and preferred the suburbs to the city centre, with only 12 per cent of new arrivals intending to reside in the District of Columbia (the inner city) and nearly 60 per cent residing in two large and extremely affluent suburban counties (Fairfax County, Virginia and Montgomery County, Maryland) (Price et al., 2005). Research on Sydney's immigrant settlement patterns shows that 11 per cent of immigrants settled in the inner city, whereas the rest of the foreign-born were found in older and newer suburbs (Ley and Murphy, 2001).

Different patterns emerge in cities where a large percentage of residents live in public housing. In French cities, and especially in Paris, foreign-born and non-white ethnic groups (especially from North Africa) tend to be concentrated in housing projects on the suburban fringe. The inner city tends to be a much more expensive, and more ethnically French, area of settlement. Immigrants tend to be highly concentrated in suburban housing projects in which they are isolated from many of the educational and employment opportunities found in the city. This frustration turned into protest in the autumn of 2005 when ethnic and immigrant North Africans rioted in the streets, burning cars and protesting against their sense of disenfranchisement from the dominant white French society.

The Portuguese capital of Lisbon offers another intriguing example of immigrant settlement patterns. This historic capital has long been a site of emigration, rather than a destination for immigrants. With Portugal's entrance into the European Union in 1986 as well as ongoing economic problems in the former Portuguese colonies, the number of immigrants began to soar in the 1990s. New immigrants from the formers colonies of Cape Verde, Brazil, Mozambique and Angola are adding to the ethnic and racial diversity of Lisbon. In addition, large numbers of Ukrainians are also settling in Lisbon. In the Lisbon case, more immigrants are settling in the suburbs than the city centre, and levels of residential segregation of immigrants and ethnic minorities are lower than in other European and North American cities. Still, due to the housing shortage, many African immigrants are living in shanties (self-built housing) that lack plumbing and electricity. This familiar pattern in the developing world is not common in Europe; the Portuguese government launched a special housing programme in 1993 to try to eradicate the shanties in greater Lisbon. As a result of these efforts, roughly half of the shacks were eliminated (Fonseca et al., 2002).

A core issue surrounding urban immigrant settlement is the issue of spatial segregation, be it voluntary or forced. The assumption is that segregation is generally an indicator of immigrant marginalization and discrimination, although ethnographic work on suburban enclaves paints a more complex picture in which self-segregation is often conducive to building solidarity and an ethnic economy. Geographer Ceri Peach's exploration of 'good' and 'bad' segregation contrasts the 'ghetto' model with the 'ethnic village' model of immigrant settlement in North American and European cities (Peach, 1996). While the former is seen as discriminatory and socially destructive, the latter is seen as voluntaristic and protective. Thus by mapping immigrant settlement patterns we can see areas of concentration but the processes by which these concentrations formed (be it structural discrimination, immigrant agency or a combination of these and other factors, such as access to housing or transportation) can only be determined through studying social practices and attitudes on the ground.

Take, for example, the phenomenon of Chinatowns, which are found in many cities throughout the world and are representative of the diverse flows of Chinese immigrants beginning in the nineteenth century (Anderson, 1987; Craddock, 1999). Some Chinatowns began as efforts from host societies to segregate the Chinese population and force them to live in particular areas of the city. This was certainly the case for the origin of many early Chinatowns in North America. Yet, Chinese ethnic and commercial centres sometimes owe their origin to the social and economic advantages of forming clustered settlements. Thus there are Chinatowns found in many Southeast Asian cities that do not owe their origin to segregationist policies. The advantages of Chinese ethnic clustering are still evident today in the phenomenon of the *ethnoburb*.

Geographer Wei Li (1998) defines *ethnoburbs* as suburban ethnic clusters of residential and business districts in large metropolitan areas. Ethnoburbs are typically multi-ethnic communities in which one ethnic community has a significant influence but may not be the majority. They function in many ways like an inner-city enclave but they also tend to have better schools, cheaper housing, lower crime and greater anonymity associated with the suburbs. In the case of the San Gabriel Valley in Los Angeles County, ethnic Chinese immigrants (from China, Taiwan, Hong Kong and Southeast Asia) rely on the ethnoburb to make a living and conduct business using various ethnic social networks. It is also a setting for shopping, eating Chinese foods, speaking in one's native language and acquiring news from home. While Li's ethnoburb was developed in the U.S. context, similar formations are noted in Canada and Australia. The ethnoburb, like its inner-city immigrant enclave, is also an important locality for examining the role of immigrant social networks in remaking the social spaces of cities.

Urban immigrant networks

The distribution of immigrants throughout cities in the world is not random but typically results from a mix of structural forces, historical geopolitical linkages between places (colonies with former colonizers) and a complex web of immigrant social networks. Through contacts with family, friends, community members or recruiters, immigrants usually arrive in destinations where they know other co-ethnics. The phenomenon of chain migration – the staged movement of people from one location to another – is well documented for both internal and international migration from rural areas to cities and from cities to cities. Such networks have potential to both facilitate and impede integration into the host society. Similarly, while immigrant social networks are often celebrated as important buffers from the difficulties of settling in new societies, these same networks can also exploit fellow co-ethnics into usurious employment and housing arrangements.

The ethnic economy often relies upon trust (a trust sometimes enforced because of the lack of formal financial services available to newly-arrived populations). Many agreements are carried out on a handshake and the knowledge of social and economic isolation if an agreement is not honoured. Immigrant groups form associations that support both the ethnic community and their economic interests in a host city. Thus, in a gateway city there might be an association of Indian computer programmers or Korean dry-cleaners. Equally important are hometown associations created by immigrants in destination cities to support communities back home by collecting funds to build schools, sports facilities, churches, clinics or provide electricity. Mexican immigrant hometown associations in the USA play a vital role in the flow of capital, technology and information returning to sending communities in Mexico (Orozco and Lapointe, 2004).

In multi-ethnic cities with significant immigrant populations, the role of the ethnic economy is often noted. This economy serves both newcomers as well as ethnic groups established for several generations. In São Paulo, Brazil, ethnic Japanese entrepreneurs have played an important role in that city's development since they began to arrive in large numbers in the early twentieth century. Initially brought over to work in the surrounding coffee plantations, Japanese immigrants acquired their own lands, moved to the city, and became an important force in commerce and the agro-industrial economy of Brazil. This same ethnic community was mobilized in the early 1990s. As the Brazilian economy continued to struggle, many Japanese Brazilians took advantage of changes in Japanese immigration law that allowed people of Japanese ancestry to legally enter Japan as permanent residents. Tens of thousands left Brazil, but many have also returned finding that they were more Brazilian (and less Japanese) than they realized (Roth, 2002).

Since the 1990s, many social scientists began to write about the phenomenon of transnational migrants (Basch et al., 1994). Transnationalism results in a multi-directional construction of social networks or fields between host and home communities that cross international boundaries. In the process, meaningful linkages between distant places are formed that force consideration of how human mobility influences livelihood systems, place-making and identity. Transnationalism, over long periods of time, creates a sense of 'in-betweeness' where an immigrant may not fully belong to either his or her country of origin or host country. Put differently, how long can a Bangladeshi immigrant live in Birmingham and still be considered Bangladeshi? Likewise, when will that same immigrant be seen as fully English? (Extract 11.3).

Extract 11.3: McEwan, C., Pollard, J. and Henry, N. (2005). 'The global in the city economy: Multicultural economic development in Birmingham', *International Journal of Urban and Regional Research*, 29 (4), pp. 917–18, 928–9.

Birmingham has received little attention as a global city, yet its place in the global economy is evolving in new ways that are often related to its multiculturalism, post-colonialism and the transnationalism of many of its residents. The 2001 census revealed that 26.4 per cent of the city's population categorized themselves as other than 'White', with 19.5 per cent of the population identifying as Asian/Asian British … and above average proportions of White Irish, Mixed White and Black Caribbean, Indian, Pakistani, Bangladeshi and Black Caribbean people. The argument here is that diversity is recognized as a strength and, indeed, as a route to economic development.

Birmingham's economic position is constructed through, and interwoven with, numerous minority ethnic economic networks, some more visible than others. Birmingham's overseas Chinese population, for example, is prominent in property development within the city. The areas of Sparkbrook and Sparkhill include Pakistani banks and the city will shortly become home to the first branch of the new Islamic Bank of Britain plc to be located outside of London. The formation of Britain's first Irish Business Association in the city is another example of how some of Birmingham's economic networks are identified primarily by their ethnicity. … The Irish business community in Birmingham is constructing an ethnic identity based on 'minority white' experiences. It is explicit about Irishness and economic advantage, and includes any business of Irish origin or with trading or cultural links with Ireland.

The production of particular commodities has also become associated with (minority) ethnicity within the city, for example, ethnic foods and particular musical styles (Bhangra). Birmingham's multicultural and transnational economic activities extend across a range of economic operations and products. The Sparkhill area is the centre of the South Asian jewellery quarter, the retailing of clothing, sarees and other textiles. Perhaps less well known is the Greek-Cypriot fish-frying network, which, as early as 1989, constituted 25 per cent of Birmingham's 300 fish and chip shops. Today the city's halal butchers number more than 50; Birmingham's National Halal Centre is growing through demand from non-Muslims and is exporting goods such as halal baby food throughout Europe. ...

Although data is scarce, some figures suggest that up to 33 per cent of Birmingham's business activity occurs within minority ethnic owned enterprises. The possibilities for continued development of this business activity are myriad.

Birmingham is a distinctive global city, but this distinctiveness is not based on its recent and very substantial development of prestige urban regeneration projects or even its marketing slogan as 'The Meeting Place of Europe'. Rather, its global distinctiveness is, in part, based on its residents' diasporic and transnational (economic) roots, which extend across Europe into Asia and the Pacific Rim.

There is still much discussion about how long a community of immigrants can maintain a transnational status. Much depends on the continuing flow of new immigrants as well as return flows to the countries of origin. With regards to cities in particular, the concept of *transnational urbanism* was coined by Michael Peter Smith (2001) as a metaphor to capture the importance of cities in the maintenance of immigrant-derived transnational networks. Smith's work emphasizes that translocal connections are forged and maintained through two approaches: (1) transnational actors are materially connected to socioeconomic opportunities, political structures, or cultural practices *found in cities* at some point in their transnational communication circuit or (2) they maintain transnational connection by using advanced means of communication and travel that have been historically connected with the culture of cities. The value of articulating transnational urbanism is that it carves out an agency-oriented space for multiple actors, in this case immigrants, to use both cultural and economic capital to assert their place in the global economy.

Transnational urbanism is a direct result of the widespread growth of immigrant populations in cities, and as such it has made cities the centre-stage for transnational players, be they multinational corporations or immigrant hometown associations. How cities accommodate this diversity without being deeply fractured by it is a difficult balancing act. It is important to see integration as not a unidirectional phenomenon of immigrants adapting to the host society. Inevitably, with substantial immigrant numbers, the host society itself must adapt and

Figure 11.3 A welcome sign in 11 languages at a Montgomery County library in Maryland, USA

change, as symbolized by a welcome sign at a public library in suburban Washington, DC (Figure 11.3). It is the two-directional aspect of immigrant integration that is necessary but also threatening to native-born populations in major immigrant destinations.

Summary

In cities around the world, but especially in Western Europe, Australia, the Persian Gulf and North America, immigrants play a fundamental role in the labour force and the social life of the cities. True, the vast majority of people on this planet live in their county of birth, but more and more, a constellation of cities in developed economies depend upon a large immigrant population to function. For North American and Australian cities the numbers are reminiscent of the early twentieth century, although the diversity is far greater. For many European cities, both the scale and diversity of the flows are a relatively new phenomenon. For all the major receiving countries, immigration has become one of the central political issues of our time and is likely to remain so given the demographic and economic realities driving foreign labour flows.

A constant tension in labour-related immigration is the desire for labour but not necessarily a desire for more settlers or new citizens. Many immigrants arrive in cities as part of temporary worker programmes. Yet, as has been shown time and time again, and despite the considerable efforts of state and local authorities, foreign labour is made up of real people with needs and reasons to form attachments to places and communities. Regardless of status or intent, and even where citizenship or permanency is aggressively denied, the tendency is to settle. Thus cities are the crucibles for this experiment in globalization from below. Immigrants contribute to the hyper-diversity and cosmopolitanism that makes transnational urbanism palpable. But the difficulties associated with adjusting to such rapid social change are significant for immigrant and native-born alike.

References

Anderson, K.J. (1987) 'The Idea of Chinatown: The Power of Place and Institutional Practice in the Making of a Racial Category', *Annals of the Association of American Geographers* 77 (4), 580–598.

Basch, L., Glick Schiller, N. and Szanton Blanc, N. (1994) *Nations Unbound: transnational projects, postcolonial predicaments and the deterritorialized nation-State*, New York: Gordon and Breach.

Benton-Short, L.M., Price, M. and Friedman, S. (2005) 'Globalization from Below: The Ranking of Global Immigrant Cities', *International Journal of Urban and Regional Research* 29 (4), 945–959.

Burnley, I.H. (2001) *The Impact of Immigration in Australia: a demographic approach*, Oxford: Oxford University Press.

Castles, S. and Davidson, A. (2000) *Citizenship and Migration: globalization and the politics of belonging,* New York: Routledge.

Castles, S. and Miller, M. (2003) *The Age of Migration,* (3rd edition), New York: Guilford.

Clark, W.A.V. and Blue, S.A. (2004) 'Race, Class and Segregation Patterns in US Immigrant Gateway Cities', *Urban Affairs Review* 39 (6), 667–688.

Cox, R. (2006) *The Servant Problem: domestic employment in a global economy,* London: I.B. Tauris.

Craddock, S. (1999) 'Embodying Place: Pathologizing Chinese and Chinatown in Nineteenth-century San Francisco', *Antipode* 31 (4), 351–371.

Ehrenreich, B. and Hochschild, A.R. (eds) (2002) *Global Woman: nannies, maids and sex workers in the new economy,* New York: Metropolitan Books.

Foner, N. (2000) *From Ellis Island to JFK: New York's two great waves of immigration,* New Haven, CT: Yale University Press.

Foner, N. (2005) *In a New Land: a comparative view of immigration,* New York: New York University Press.

Fonseca, M.L., Malheiros, J., Esteves, E. and Caldeira, M.J. (2002) *Immigrants in Lisbon: routes of integration* (Estudios de Regional e urbano 56), Lisbon: Lisbon University.

Gozdziak, M. and Martin, S.F. (eds) (2005) *Beyond the Gateway: immigrants in a changing America,* Lanham, MD: Lexington Books.

Guarnizo, L.E., Sánchez, A.I. and Roach, E.M. (1999) 'Mistrust, Fragmented Solidarity, and Transnational Migration: Colombians in New York City and Los Angeles', *Ethnic and Racial Studies* 22 (2), 367–396.

Hagendoorn, L., Veenman, W. and Vollebergh, J. (eds) (2003) *Integrating Immigrants in the Netherlands,* Aldershot: Ashgate.

Huang, S., Rahman, N.A. and Yeoh, B.S.A. (2006) *Asian Women as Transnational Domestic Workers,* London: Marshall Cavendish Academic.

Ireland, P.R. (2004) *Becoming Europe: immigration, integration, and the welfare state,* Pittsburgh, PA: University of Pittsburgh Press.

Ley, D. and Murphy, P. (2001) 'Immigration in Gateway Cities: Sydney and Vancouver in comparative perspective', *Progress in Planning* 55 (3), 119–194.

Li, Wei. (1998) 'Anatomy of a New Ethnic Settlement: The Chinese Ethnoburb in Los Angeles', *Urban Studies* 35 (3), 479–501.

McEwan, S., Pollard, J. and Henry, N. (2005) 'The Global in the City Economy: Multicultural Economic Development in Birmingham', *International Journal of Urban and Regional Research* 29 (4), 917–929.

Orozco, M. and Lapointe, M. (2004) 'Mexican Hometown Associations and Development Opportunities', *Journal of International Affairs* 57 (2), 31–52.

Peach, C. (1996) 'Good Segregation, Bad Segregation', *Planning Perspectives* 11, 379–398.

PRB (2005) *World Population Data Sheet,* Washington, DC: Population Reference Bureau.

Price, M., Cheung, S., Friedman, S. and Singer, A. (2005) 'The World Settles In: Washington, DC, as an Immigrant Gateway', *Urban Geography* 26 (1), 61–83.

Roth, J.H. (2002) *Brokered Homeland: Japanese Brazilian migrants in Japan*, Ithaca, NY: Cornell University Press.

Samers, M. (2002) 'Immigration and the Global City Hypothesis: Towards an Alternative Research Agenda', *International Journal of Urban and Regional Research* 26 (2), 389–402.

Sassen, S. (1994) *Cities in a World Economy*, Thousand Oaks, CA: Pine Forge Press.

Sassen, S. (1999) *Guests and Aliens*, New York: The New Press.

Sassen, S. (2002a) *Global Networks, Linked Cities*, New York: Routledge.

Sassen, S. (2002b) 'Global Cities and Survival Circuits', in B. Ehrenreich and A.R. Hochschild (eds), *Global Woman: nannies, maids and sex workers in the new economy*, New York: Metropolitan Books.

Singer, A. (2004) 'The Rise of New Immigrant Gateways', in *The Living Cities Census Series*, February 2004. Washington, DC: The Brookings Institution.

Smith, M.P. (2001) *Transnational Urbanism: locating globalization*. Oxford: Blackwell.

Voigt-Grad, C. (2004) 'Towards a Geography of Transnational Spaces: Indian Transnational Communities in Australia', *Global Networks* 4 (1), 25–49.

Waters, M.C. and Jimenez, T.R. (2005) 'Assessing Immigrant Assimilation: New Empirical and Theoretical Challenges', *Annual Review of Sociology* 31, 105–126.

Wright, R. and Ellis, M. (2000) 'Race, Region and the Territorial Politics of Immigration in the US', *International Journal of Population Geography* 6, 197–211.

Yeoh, B.S.A. (2006) 'Bifurcated Labour: The Unequal Incorporation of Transmigrants in Singapore', *Tijdschrift Voor Economische en Sociale Geografie* 97 (1), 27–36.

12   SEGREGATION AND DIVISION

David Wilson

This chapter

O   Reviews ideas concerning the division of urban space on ethnic and racial lines

O   Compiles evidence from the USA and UK to show that ethnic segregation remains a significant phenomenon – and a major problem

O   Argues that current processes of residential segregation are informed by a racism informed by negative media representations of 'non-white' spaces

---

Introduction

Cities continue to be haunted by the spectre of massive residential segregation and the marginalization of minority ethnic populations. This trend, accelerated since 1980, is exemplified by the spatial concentration of African Americans in US inner cities, migrants from North Africa in the French *banlieue*, Turkish families in specific German housing estates or the clustering of Commonwealth groups in British inner cities. What is clear from such examples is that the spatial form of ghettoization varies markedly between nations. In the UK, for example, segregation occurs mainly at the level of the street or block; in the USA this involves entire neighbourhoods. Irrespective, we can now talk of a new kind of ghetto terrain having formed in both places since 1990, forming what I describe as 'the glocal ghetto'. In these spaces, poverty and hopelessness have deepened, welfare safety nets have disappeared, and social stigma has escalated (see Wacquant, 2002; Ehrenreich, 2003). And state schools, a traditional support for this youth, further deteriorate. Deep cuts in funding and growing rates of impoverished children damage the role of many of these schools as learning centres.

The forces creating this new afflicting ghettoization are complex. First, much from the recent past continues to operate. Business alliances and their web of institutions (realtors and estate agents, banks, developers, city government) still pursue profit and tax revenues that serve to isolate the poor in forbidding terrains through processes of financial and social exclusion. But there is also much that is new. Now, the sense of a powerful economic globalization dramatically sweeps across Urban America and Urban Britain. Business alliances, sternly invoking this reality, galvanize government and populations to transform cities with one goal in mind: to attract resources. While cities scramble to physically and socially upscale, poor neighbourhoods and citizens are left to rot. Tied to this, a new kind of local politics, termed neoliberalism, further pushes individual actions and private markets as determinants of social welfare. Programmes like New Labour's welfare-to-work 'New Deal,' the Crime and Disorder Act, Workfare, No Child Left Behind, and Faith-based Service Provision sear these cities and their poor.

This chapter examines this intensified ghettoization, focusing on the experience of UK and US cities. Reviewing literatures on segregation in the 'global' and 'neoliberal' era, I suggest, we see the rise of a new post-1990 regime of ghettoization that embodies the interplay of evolving economies, political formations, constructions of culture, and production of space. This new regime is shown to be rooted in the ascendant sense of ominous global times and the insistence that cities must enhance their economic competitiveness. At the heart of this, ghettoization is the new normalized reality to stash populations and land-uses deemed detrimental to 'the global competitiveness project'.

A new ghettoization and marginalization

Urban studies has a long-standing interest in segregation and marginalization. The term 'ghetto' is, accordingly, a key notion in the literature, describing the creation of landscapes exclusive to a particular minority social group. But who is being ghettoized today? In the USA, poor African Americans continue to suffer most severely. This trend was initiated in the 1920s with booming industrialization in cities and displacement of labour from southern farms that compelled rural blacks to enter the Fordist industrial economy. Ghettoization intensified after 1945: stepped-up exclusionary zoning, realtor steering, and bank redlining cordoned off this population as cities struggled to redevelop decaying downtowns. While 1920s newly arriving blacks to cities found segregated restaurants, stores, beaches, and schools, post-war migrants found burgeoning and isolated ghettos. By 1950, America's most notorious black ghettos – New York's Bedford-Stuyvesent and Harlem, Cleveland's Hough and Glenville, Philadelphia's Tioga and Allegheny West, and Chicago's South Side Black Belt had taken shape.

In the UK, principal subjects have been Black Africans, Black Caribbeans, Bangladeshis, and Pakistanis. Like the USA, Britain has experienced a sizable (but smaller) minority population migration to its urban cores since the Second World War. Many arrived from former colonies, motivated by perceptions of greater job and educational opportunities in Britain (Connelly, 2004). Newly-settled 'thick ethnic' blocks in neighbourhoods, typically old and dilapidated, became Britain's most pronounced zones of economic struggle. Time-tested mechanisms of exclusion – stigma, residential steering, and exclusionary planning – continued unabated in these cities. These blocks, dotting districts like Tower Hamlets and Hackney in London, Cheetham Hill and Gorton in Manchester, and inner Birmingham, quietly emerged as Loïc Wacquant's (2002) deepened prisons without walls. Minus bars and steel, these spaces now foreclose afflicting activity spaces of residents in an intensifying ghettoization.

A deepened ghettoizing and marginalizing of these low-income neighbourhoods since 1990 is difficult to refute (see Extract 12.1). Two developments indicate this: worsened living conditions in these ghettos and the widespread practice of the media and policy community describing more problematic neighbourhoods and people. These trends may be surprising given the pre-1990 realities. Reagan and Thatcher served up virulent anti-poor programmes and rhetoric in the 1980s that notoriously deepened ghettoization and marginalization. Both spoke of the characteristics of what Thatcher infamously termed 'those inner cities' – 'irrepressible welfare mothers, footloose fathers, welfare chisellers, and rancid ghettos' – while systematically dismantling welfare programmes. But now, I suggest, this rhetoric has shifted to the local level and is still intense and more institutionalized in policy. Unlike the 1980s regime of segregation, this latest one leads with a new cast of characters – local reporters, politicians, planners, and editorialists – who colour and embellish the anti-poor rhetoric with local flavour.

Extract 12.1: From Marcuse, P. (2000) 'Cities in Quarters', in G. Bridge and S. Watson (eds), *A Companion to the City*, Oxford: Blackwell, pp. 270–82, 273–4.

One may speak of separate residential cities. … The gentrified city serves the professionals, managers, technicians, yuppies in their twenties and college professors in their sixties: those who may be doing well themselves, yet work for and are ultimately at the mercy of others. The frustrated pseudo-creativity of their actions leads to a quest for other satisfactions, found in consumption, in specific forms of culture, in 'urbanity' devoid of its original historical content and more related to consumption than to intellectual productivity or political freedom. The residential areas they occupy are chosen for environmental or social amenities, for their quiet or bustle, their history or fashion; gentrified working-class neighbourhoods, older middle-class areas, new developments with modern and well-furnished apartments, all serve their needs. Locations close to work are important, because of long and unpredictable work schedules, the density of contacts, and the availability of services and contacts they permit.

The suburban city of the traditional family, suburban in tone if not in structures or location is sought out by better-paid workers, blue- and white-collar employees, the 'lower middle class', the petit bourgeoisie. It provides stability, security, the comfortable world of consumption. Owner-occupancy of a single family house is preferred (depending on age, gender, household composition), but cooperative or condominium or rental apartments can be adequate, particularly if subsidized and/or well located to transportation. The home as symbol of self, exclusion of those of lower status, physical security against intrusion, political conservatism, comfort and escape from the work-a-day world (thus often substantial spatial separation from work) are characteristic. ... The tenement city must do for lower-paid workers, workers earning the minimum wage or little more, often with irregular employment, few benefits, little job security, no chance of advancement. Their city is much less protective or insular. In earlier days their neighbourhoods were called slums; when their residents were perceived as unruly and undisciplined, they were the victims of slum clearance and 'up-grading' efforts; today they are shown their place by abandonment and/or by displacement, by service cuts, deterioration of public facilities, political neglect. ...

The abandoned city, economic and, in the US, racial, is the place for the very poor, the excluded, the never employed and permanently unemployed, the homeless and the shelter residents. A crumbling infrastructure, deteriorating housing, the domination of outside impersonal forces, direct street-level exploitation, racial and ethnic discrimination and segregation, the stereotyping of women, are everyday reality. The spatial concentration of the poor is reinforced by public policy; public (social, council) housing becomes more and more ghettoized housing of last resort (its better units being privatized as far as possible), drugs and crime are concentrated here, education and public services neglected.

In the USA, worsened living conditions are indicated in a sample of recent changes in six black ghettos (Table 12.1). The numbers are ominous. The intensity of deprivation, measured by ratios of high- and medium-poverty to low-poverty people, significantly deepened for all six neighbourhoods (an average of 71.7 per cent). For example, Cleveland's Fairfax and Hough neighbourhoods had gains in this of 87.2 per cent and 193.9 per cent. In this context, both neighbourhoods (like all six) experienced substantial population loss. Changes in level of squalor, measured by proportion of people living below the poverty line, also increased (an average of 4.8 per cent). Thus, Chicago's Englewood and Woodlawn registered 43.8 per cent and 41.2 per cent of their populations living below poverty, increases of 2.9 and 3.9 per cent from ten years earlier.

Renditions of lived realities by writers in these ghettos further supports the notion of worsened living conditions. In the USA's quintessential urban laboratory, Chicago, Lealan Jones (1997) notes increased poverty and deprivation in

Table 12.1   Changes in black ghetto neighbourhoods in major US cities

	Cleveland		Philadelphia		Chicago	
	Fairfax	Hough	Fairhill	Hartranft	Englewood	Woodlawn
% population change, 1990–2000	–13.1	–19.2	–22.8	–7.0	–16.8	–14.3
% below poverty level, 2000	35.7	41.3	57.1	33.9	43.8	41.2
% below poverty level change, 1990–2000	–0.9	+14.7	+12.6	+1.4	+2.9	+3.9
% of housing units substandard, 2000	21.0	20.9	22.0	21.8	23.7	22.9
% change in ratio of high and medium to low poverty level, 1990–2000	+87.2	+193.9	+52.3	+26.7	+41.9	+28.2

Unit of measurement: census tract.
High poverty measured by people with incomes two times or greater below the poverty level.
Source: US Census Bureau

Southside neighbourhoods: Grand Boulevard, Oakland, and North Lawndale. To Jones, decrepit shops and stores fill out marginal retail strips anchored by fast-food restaurants and liquor stores. Isolation is exacerbated as gentrified zones, just north and west, separate themselves via gating, and obsessive surveillance and monitoring. Similarly, writer and photographer John White (2004) narrates a growing pronounced hopelessness of residents in these areas, frustrated by few decent paying jobs and youth opportunities. To White, residents keenly feel a sense of forgotten and abandoned communities. As neighbourhood populations thin, those left behind suffer.

At the same time, these black ghettos have also been widely re-represented since 1990 in media, policy and planning pronouncements. I elaborate on this in the next section and simply identify the trend here. Data from a sample of five prominent US daily newspapers (Table 12.2), for example, shows that these ghettos have been increasingly cast as pathological consumptive spaces whose populations are lost in a haze of cultural aberrance. This representation differs markedly from the 1980s period, when the Reagan and Thatcher 'national regime of representation' widely cast black ghettos and populations as spiralling downward and productively dysfunctional (Mercer, 1997; Wilson, 2005, 2007). This seemingly subtle difference is significant: a shift from the production to the consumption metaphor suggests that spaces and people are now infected by an uncontrollable 'eating', devoured by pathology (Norton, 1993).

In UK cities, neighbourhoods with clusters of Black Caribbean, Bangladeshi, Pakistani, and Black African populations have often experienced similar declines in quality of life. This is most pronounced in London, which contains more than

Table 12.2   Percentage of stories presenting black ghettos as pathologically consumptive

	1985–91	1992–97	1998–2003
Philadelphia Inquirer	4 (8%)	6 (12%)	9 (18%)
Kansas City Star	2 (4%)	3 (6%)	5 (10%)
Cleveland Plain Dealer	1 (2%)	5 (10%)	6 (12%)
St. Louis Post-Dispatch	1 (2%)	6 (12%)	7 (14%)
Indianapolis Star	3 (6%)	7 (14%)	6 (12%)

Derived partially from Wilson (2007)

Table 12.3   Ghetto districts in the UK

District	City	Ethnic or racial minority (non-white) (%)	Population living in one of the 10% most deprived wards in England (%)
Hackney	London	40.6	100
Tower Hamlets	London	49.0	97
Newham	London	60.6	95
Manchester Borough	Manchester	12.1	79
Knowsley	Knowsley (Merseyside)	4.0	79
Easington	Easington (Durham)	3.8	79
Liverpool	Liverpool	5.2	72

Source: UK Census

40 per cent of these populations nationally (see UK Census, 1991 and 2001). As in Birmingham, Liverpool and Manchester, these London populations disproportionally occupy inner-city deprived neighbourhoods (55 per cent of black children and 73 per cent of Bangladeshi and Pakistani children in London currently live in poverty). In Tower Hamlets, for example, 33.4 per cent of the population is Bangladeshi, and 97 per cent of the area's population lives in one of the 4 per cent most deprived English wards (Table 12.3). In London's Hackney, Black Caribbeans and Africans make up over 20 per cent of the population; 100 per cent of this area's population lives in one of the 4 per cent most deprived wards in England (Table 12.3). In less ethnically diverse Manchester and Liverpool, these small minority populations (less than 3 per cent of city populations) cluster in highly isolated and poor Cheetham Hill and Gorton, and West End communities, respectively.

A closer look at the statistics amplifies the deprivation. In Tower Hamlets and Hackney, the two most deprived areas in the UK in 2001, 35.6 and 28.0 per cent of populations had no qualifications for work (Table 12.4). Only 51 per cent and

Table 12.4   Statistics on select marginalized ghettos in the UK

	No qualification for work (%)	Level 1 qualification (%)	Proportion of working age population employed (%)	Deprivation rank in the UK
Tower Hamlets	35.6	20.1	51	1
Hackney	28.0	14.0	60	2
Newham	28.2	22.3	60	3

Source: UK Census

60 per cent of people of working age in these areas were employed, the lowest proportion of any areas in England. In the Newham, East London ethnic 'ghetto', where poverty has dramatically deepened, the economic inactivity rate for adults was 40 per cent: 28.2 per cent of adults had no qualifications for work, and 22.3 per cent had only level one qualifications (Table 12.4). In all of these terrains, physical landscapes reflect this vividly: dilapidated houses and buildings line the streets and thoroughly dominate the physical fabric (see Atkinson, 2003; Connelly, 2004). The scars of deterioration, like poverty, spread inexorably across these blocks.

Expositions of living conditions in these communities rival the grim narrations of the US ghettos. For example, Thompson et al. (2002) discuss an increased decrepitness in London's Peckham district. They depict a downtrodden landscape riddled with multiplying numbers of angry and discordant youth with few serious job opportunities and little hope. Once solid blocks have unravelled in the face of two forces: increased poverty and misery. Similarly, writer Paul Harrison (1992) chronicles Hackney's physical decline and grinding poverty. In his narrative, crack houses, abandoned buildings, and eroded streets dot forgotten, disinvested blocks. Like a bad dream to Harrison, Hackney's wretchedness is poignant and overpowering. In sum, these economic and physical indicators reveal much: the UK's minority ghettos, like their US brethren, continue to spiral downwards.

The new representation of ghettoization and marginalization

The everyday processes that have historically promoted segregation and ghettoization in US and UK cities continue to operate. Thus, realtors and estate agents in business alliances are understood to 'steer' the racial and ethnic poor to inner-city blocks. In Chicago, Cleveland, and Indianapolis, realtors staunchly protect and buffer downtown gentrified neighbourhoods from 'ethnic and racial invasion'. In Bradford, UK, one site of the infamous 2001 race riots, residents widely decry the creation of isolated spaces established for ethnic minorities through a process of social 'dumping'. At the same time, banks continue to *redline* low-income communities, meaning that they are unwilling to make loans to those who live in high-risk areas. In Milwaukee, Pittsburgh, St Louis and Detroit, inner cities experience paltry flows of credit and leaders struggle with banks to attract more. In London and Newcastle, low-income areas are denied credit in a new high-tech twist: banks use computerized credit-scoring and risk assessment tools that rely on spatially referenced postcode data (Leyshon and Thrift, 1995).

City governments, equally important, continue to structure city morphology through time-tested programmes and policies. At the core of this, segregative zoning ordinances further institutionalize patterns of segregation. Zoning in London, Birmingham, Manchester, Philadelphia and Cleveland meticulously articulates and protects 'healthy' (i.e. investment leveraging) districts. At the same time, established city programmes and policies like historic preservation, tax increment financing, and public–private partnerships channel new investment to targeted residential and commercial spaces. In Chicago, St Louis, Cleveland, Newcastle and London, historic preservation designates select neighbourhoods for tax abatements and enhanced prestige. In Leeds, Manchester, Indianapolis and Baltimore, public–private partnerships focus 'regeneration' efforts on residential areas deemed 'private-sector salvageable' and attractive to relatively affluent households.

But a dizzying arsenal of new programmes and policies has recently unfolded (in the spheres of city growth, labour regulation, economic development). Their origins, in changed circumstances, reflect the recent obsession with new global times and the need for cities to uncompromisingly respond (i.e. 'go global'). City planning and media pronouncements trumpet this rhetoric and the need to physically and socially restructure (i.e. build attractive cores, construct upscale gentrified communities, aestheticize public spaces, discipline the poor for being a burden on the local state). Now, these ghetto spaces are more virulently denigrated and accepted as warehousing zones for 'anti-growth' land-uses and populations. Many of these spaces, of course, have always warehoused the racial and ethnic poor and been seared by negative representations. But today these processes deepen, with the isolation more institutionalized and accepted.

Why the obsession with globalization in these cities? Because there is truth to the general notion of a more hypermobile world; its luminous images are compelling, and business alliances can benefit economically from propagating this reality. In this context, globalization that seamlessly emerged from the 1980s as the new policy buzzword in the USA and UK has been jumped on. The proof of globalization's scripted rise? Mentions of 'globalization' or 'global economy' in *Business Week* escalated from 160 in 1990 to over 290 in 2000 (Wilson, 2007, p. 35). Moreover, 40 US and UK local and national media sources in 1991, 1995 and 2000 had 158, 2,035 and 12,636 references respectively to the notion of globalization. To the venerable *New York Times* (in Wilson, 2007), a new global era was simple fact: it has set in as a 'fluid, infinitely expanding and highly organized system that encompasses the world's entire population, but which lacks any privileged position or places of power'.

The post-1990 rhetoric of an ominous globalization – and the need for cities to respond – has been thick. For example, Chicago planner C. Redd (personal communication, 2004) calls globalization 'the new world that Chicago and other cities in America now have to respect. ... It's the new epoch we find ourselves in, there's no turning back to a past, a different time.' Redd's globalization ensnares Chicago like 'a vice that now squeezes it'. To London Mayor Ken Livingstone (cited in GLA, 2003), 'the globalization of the world economy is increasing rapidly ... the economic success of the whole UK depends on London as a gateway to the international economy. ... This is exactly the danger we face ... there should be no illusions about this: any brake on London's success would have a disproportionally severe impact on the whole national economy.'

In this frame, city restructuring has been widely presented as both a physical and social undertaking. London Mayor Ken Livingstone, in a White Paper (in London Europe, 2003), called globalization a test on 'sustaining London['s] [physical and social] fabric ... as a leading global city'. To Livingstone, 'London's sustained economic growth depends upon' re-doing city forms and 'finding employees with relevant skills and with offsetting the migration from London to the rest of the UK'. Similarly, to Baltimore Mayor Martin O'Malley (2004), the city now must forcefully embed itself in the new global reality: 'we are at another stage in our history, where we are changing from an economy of the past to an economy of the future. ... [W]e need to attract ... information and biotech industries ... [and play to our assets], historic buildings, historic neighbourhoods, a great waterfront ... that's the future.'

In this context, US and UK ghetto spaces have widely become accepted terrains to capture and retain 'city-damaging people'. This acceptance in planning and policy communities has been at the core of this deepened ghettoization. Comments by planners reflect this. To Weir (personal communication, 2003) in Chicago, where the city actively promotes gentrification and 'cultural upgrading' across the city, 'we need to identify and cultivate redevelopment that will elevate Chicago's culture and standing – this is what the public wants and would benefit from ... the South Side [ghetto] is there, it looms and is a problem ... it's unfortunate but it's true.' To St. Louis planner Cole (personal communication, 2004), where public–private partnerships re-make downtown and extend gentrification, 'ghettos and upgrading in St. Louis are a striking juxtaposition ... one's got to be encouraged, the other eliminated. For what purpose? It's clear we [St. Louis] have to be able to compete in the global economy ... like Chicago, Indianapolis, and other Midwest cities.'

Planners in the UK articulate the same sentiments. One central role, now, is to strategically partition cities and regenerate city centres to attract what Zukin (1995) calls 'trophy investments'. In Manchester, where the local state actively restructures the core (Peck and Ward, 2002), Chris Brown (2003, p. 6) of Igloo Regeneration notes: 'And finally, urban life. That's what people live in cities for. It's Love Saves the Day, Café Pop, Earth the Market Restaurant, Affleck's Place, the Farmers Market, Church Street Market, Urbis, the Gallery, Chinatown, the Curry Mile ... all things that make Manchester different, and better ... .' Brown's vision of a re-made Manchester obliterates the plight of ghetto terrain and focuses on sites of consumption (clubs, bars and museums). A ruthless prioritization annihilates the existence of ghetto space. In this context, such spaces can continue to warehouse undesirables as rhetoric celebrates the displacement of 'problems' rather than resolving them (see Atkinson, 2003). Planners now typically dream and plan according to the neoliberal ideal: an opulent and aestheticized city fabric that pushes anything else to the distant margins (see Extract 12.2).

Extract 12.2: From Short, J. and Kim, Y.-H. (1999) *Globalization and the City*, New York: Addison Wesley Longman, p. 97.

Urban representation has two distinct discourses. The first is the positive portrayal of a city; the city is presented in a flattering light to attract investors, promote 'development' and influence local politics. But every bright light casts a shadow. The second discourse involves the identification of the shadow, the dark side that has to be contained, controlled or ignored. This discourse works through silence, as some issue and groups are never mentioned, and through negative imagery, as some groups and issues are presented as dangerous, beyond the confines of civil debate. ...

The official web sites of cities across the world are good examples of the bright-sided portrayals. A large number of cities have constructed their web sites to provide information on investment, tourism, affairs, education and community. The city of Detroit web site, for example, contains, among others, the facts that the city has hosted dramatically increasing inward investment and that it also has diverse cultural resources (http://www.ci.detroit.mi.us). The city's slogans represent an economically promising and culturally rich Detroit: 'Making it better for you,' 'New city for a new century,' 'It's a great time in Detroit' and '1998 best sports city.' The opposite representation of Detroit is founded in numerous images of America's poorest in its inner city ... Neill's (1995a, b) disturbing metaphor, 'lipstick on the Gorilla,' critically insinuates Detroit's failure of image-driven development projects. The promising web sites and the derelict photo images show two different Detroits

---

Not surprisingly, much mainstream thought in US and UK cities now accepts these ghettos as cast-off zones. This isolating has become more normative, something civically pragmatic. For many, of course, these ghettos have been deeply ambivalent entities for a long time, accepted and avoided but believed to be unfortunate and tragic. Now, these benevolent feelings have thinned in a new functional and pragmatic mindset. In this context, harsher anti-poor rhetoric builds on a history of socially constructing these spaces and populations. Unlike the 1980s virulent rhetoric, when these spaces and people were widely scripted as culturally unravelling and increasingly different from the mainstream (see, for example, Sowell, 1985; Williams, 1987), they now have supposedly come completely apart. This fear, loathing, paranoia and bewilderment has been assisted by the media's coverage of the 2001 race riots in Burnley, Oldham and Bradford (UK) and the 1993 Los Angeles riot (USA).

Contemporary representations of racialized space are, however, not the incendiary and shocking depictions of spaces and populations that were evident in the 1980s. Reaganite and Thatcherite oratory offered stunning renditions of a people and place that supposedly had to be told. Post-1990 renditions are, rather, more localized, 'factual' depictions of hopelessly lost and pathological spaces and people. In the 1980s, urban non-whiteness – a space melding with a skin tone – was firmly established as a proxy for dangerousness in common thought (Wacquant, 2002; Wilson, 2005). The vestiges of a disadvantaged people, an imprint from the US civil rights era, were ravaged by this 1980s regime of representation. The post-1990 project builds on this and more profoundly normalizes processes of isolation and segregation. In this context, an offered sense of dangerousness intermixes with an animated sense of irretrievably pathological and unrecoverable people. Moral degeneracy has supposedly become dominant and calcified; non-white ghettos have become profoundly *sub-cultural*.

Yet the specifics of the representing differ in the USA and the UK. In the USA, these spaces and populations are widely portrayed as 'living things' infected by a consumptive degeneracy. This kind of representing – nothing new in itself – has become more profoundly used. It features ghettos plagued by a relentless impulse to devour societal resources as out-of-control 'eaters'. They ravenously consume subsidies, public goodwill and consumer goods in a frenetic devouring. The public now is to see these areas and populations as irresponsibly and unremorsefully engulfing. Comments by Chicago politician N. McGrath (cited in Wilson, 2007) reflect this: [Chicago's State Street Corridor] drains the city's resources and goodwill. ... It ... asks for more and more [resources] as it struggles ... subsidies and expertise are channeled to help the area but seem to have little effect.' To McGrath, this black ghetto bluntly and unapologetically demands and eats in a mindless frenzy. This consumption, a ravaging metaphor, speaks volumes as a communicative device about the character of these spaces and populations.

Populations are especially vilified by this metaphorical presentation. Their neighbourhoods and terrains are served up as dominated by a listless and sordid cast of characters: hustling men, dimly-cognitive youth, roaming kids, discordant teachers, cowering elderly men and women, and turf-obsessed gangs. A similar cast of characters animated 1980s discourses, but they tended merely to be falling rapidly into disrepute. Now, the offering is a total plunge into discordance. Each is either out of control–ravenous or suffering from the dramatic reverberations of an afflicted community. The narrated action is perversely colourful. The broader society's morals and values, once dimly seen, have disappeared in this ascendant cultural wasteland. To imagine this group of people is to fathom the worst of what can happen when people lose sight of a supposed binding societal culture (see Extract 12.3).

Extract 12.3: From Lott, T.L. (1999) *The Invention of Race: Black Culture and the Politics of Representation*, Oxford: Blackwell, pp. 120–1.

The voice of black urban poor people is best represented in black popular culture through rap music. As a dominant influence on black urban youth, rap music articulates the perspective of a black lumpenproletariat. For this reason, class lines have been drawn around it within the black community. This 'underclass' status of rap, however, tends to conceal the fact that it has certain social and political dimensions that suggest that something other than pathology is occurring in black youth culture.

Black urban males have been depicted in mass media as the number-one criminal threat to America. In George Bush's presidential campaign, Willie Horton was used as a cultural signifier to perpetuate the time-honoured myth of the black male rapist who deserved to be lynched. The social and political function of mass media's image of the so-called 'underclass' is to routinely validate this claim, a practice that reached a pinnacle at the time of the Central Park gang rape when the demand for a lynching was published by Donald Trump in a full-page ad in the *New York Times*. Trump's ad is a noticeable manifestation of the mass media's general tendency to employ a Girardian script when dealing with such potent racial fears. News reporting of incidents such as the Central Park Rape and the Charles Stuart Murder Case were cultural spectacles that validated the consciousness responsible for racist lynchings, a consciousness already prepared by history. The mass media's labelling of black men as criminals serves in the consciousness of many whites as a justification of anti-black vigilantism, including the Goetz-styled slayings and the police beating of Rodney King.

The response of the black community to this media has been divided along class lines. The black middle class denounces as a negative image any association of black people with crime by media. In hip-hop culture, however, crime as a metaphor for resistance is quite influential. Unfortunately, the generation of this new meaning frequently has been misunderstood. The point of the rap artist embracing the image of crime is to recode this powerful mainstream representation.

In their reflections on the social and political significance of their music, rap artists have sometimes likened their cultural practice to a kind of alternative media, that is, 'black America's TV station'. What this means is that rap music is a cultural reaction to the hegemony of television's image of black people. The crime metaphor in rap culture serves notice on mass media's ideological victimization of black men. Rap artists have declared war against the dominant ideological apparatus. Their purpose is to invalidate, on a constant basis, the images of black men in mass media with various recoding techniques that convey other meanings to their largely black audience.

Land uses are also widely coded to reflect the ghetto's descent into consumptive degeneracy. Spaces are ravaged in the form of deteriorated parks, dangerous and foreboding streets, boarded-up houses, scorched and vacant lots, and seamy play-grounds. Patterns of land, like transparent mirrors, illuminate a problematic group blinded by cultural dysfunction. A seemingly endless eating of crack and other drugs, civic energies, government subsidies, and blaring rap music takes its toll on this community's physiognomy. A scarred physicality, in communication, follows from devastating actions: families with resources and pride fleeing, investors staying away, and government continuing to ineffectually pump subsidies into these terrains. With this impulsive acting and eating, community physical fabrics further tear.

Alternatively, UK ghetto spaces and their populations are commonly represented somewhat differently: as living beings infected more by a productive rather than a consumptive degeneracy. This kind of representing features people and spaces as underachieving and inept in the sphere of work and civic contribution (see also Helms and Cumbers, 2006). At its core, both remain pathologically idle and shiftless in a debilitating malaise that ensnares them. The theme is resonant: while consumption commonly invokes sense of aggressive and expansive desires, production illuminates a sense of people's core worth: their work habits, motivation to achieve, and work ethics in families. Many depictions, of course, include both kinds of signifiers. But often leading with a notion of productive degeneracy, they configure something crucial to defining individual citizenship and neighbourhood worthiness: a community's sense of family dynamics, social norms and institutional fabric.

This scripting of productive degeneracy now widely permeates common media reportage of these neighbourhoods. Often, narratives 'voice' local politicians and/or offer 'factual' accounts of neighbourhood conditions to do the deed. For example, London's Muslim leader Lord Ahmed of Rotherham (in BBC News, 2004) is quoted to verify the supposed poor work ethic and joblessness that dominates the city's East End. As Ahmed says, parents have lost control of younger generations who largely don't work and are violent. A productivist ethic has broken down, and civic responsibility is shunned. Similarly, Beju (2004) depicts Manchester's Cheetham Hill as unsafe for whites, a ghetto stockpiled with unemployed and aimless men. To Beju, a dominating aimlessness now dictates new social ways and mores. The supposed missing ingredient in these areas – an entrepreneurial and civic spirit – saps the lifeblood of these terrains. Lord Ahmed's and Beju's ghettos, seen through the lens of production, have hit rock bottom.

These renditions of consumptive and productive dysfunctionality ultimately criminalize these ghettos. Spaces are communicated as deeply disordered and malfunctioning. Spaces of dysfunction, reflecting an unravelled social fabric, trap a cast of beings who become its unwitting subjects. These people now know only to mindlessly plunge ahead and ravenously consume or productively malinger. A new normalized ethic – living for the moment – cannot be easily shed. This offering, a potent politics, proposes an inanimate, concrete setting as a breathing, brooding, irresponsible being. A place has evolved as a kind of beast, lurching forward and unaware of its shortcomings and destructive ways.

Workfare and the regulation of the new poor

New programmes equally mark this new regime of ghettoization. This arsenal, initiated at diverse governmental levels, is increasingly implemented in local strategies (Helms and Cumbers, 2006). It emphasizes private markets and individual actions to regulate numerous things: work habits of the poor, uses of land, users of public spaces and services, and spatial patterns of investment. These schemes, supported by the rhetoric of globalization and city need to respond to this, are seen to lash into shape faulty people and land-uses. Yet a conspicuous silence looms in public discourse across these cities: the lucrative repercussions for business interests that arise from this partitioning of urban space. Leading programmes, in the US context, are Workfare, Downtown Revitalization Strategies, Faith-based Initiatives, and No Child Left Behind. Downtown Revitalization is the stepped-up strategy to further re-make city downtowns for discerning commercial and visual consumption. Planning apparatuses from New York to Los Angeles codify this policy and use enabling tools like historic preservation, tax abatements, block grants and tax increment financing. Mayors, City Councilpersons and legislators follow the supposed imperative of new global times and build on efforts to transform downtowns inaugurated during the 1980s brief real-estate boom (after the painful 1980–81 recession). The ideal prize: 'trophy centrepiece' investments which can discursively and materially anchor city transformation (e.g. elite hotels, large conference centres, ritzy malls, glittering theatre and entertainment centres).

The Cleveland case is illustrative. Since 1990, mammoth public resources have helped 'culturally globalize' the city. Its shiny 'subsidy lane' – anchored by the Rock and Roll Hall of Fame, Browns Stadium, the Old City historic area, and swaths of gentrified neighbourhoods – received public expenditures of more than $350 million. Mayors Michael White and Jane Campbell subsidized this by pulling $60 to $70 million from its low-income neighbourhoods (reducing expenditures for housing improvement, job creation and training initiatives, and social service provision) (Hennepin, personal communication, 2004). A mix of fanaticism and glibness has characterized this global drive. The Civic Task Force on International Cleveland (2003), formed by Mayor Campbell in 2001, immediately and rancorously debated the best slogan: Wake Up the Sleeping Giant: International Cleveland; One World Cleveland; or Cleveland: the World's Hometown. Discussion had already dismissed a finalist for best slogan: Where Global Opportunities Don't Knock – They Rock-N-Roll.

But in the shadows of the sloganeering, its two poorest black ghettos, Hough and Collinwood, have become more impoverished (more than 70 percent of households unofficially live below the poverty level). Perhaps most tellingly, cash-starved and struggling services have severely contracted or disappeared. Indicative of this, the Collinwood Community Services Center, a major community resource, now struggles to provide basic necessities to residents: meals, day care, and housing assistance (Naymik, 2003). On top of already severe cutbacks, a crushing debt, $300,000, threatens facility closure. A critical question is the agency's very survivability. 'At this point, I don't know', stated South Collinwood councilperson Roosevelt Coats in an interview. Cash is needed to cover basic operating costs: outstanding utility and food bills, salaries, and maintenance. Desperate measures to stay afloat are now routine. 'I offer creditors fifty bucks, whatever they will take, even a dollar', say Executive Director Wallace Floyd. 'I'm honest and tell them that we just don't have the money.'

Workfare, now policy in America's 40 largest cities, compels welfare recipients (disproportionally black and female) to work. Animated by the presentation of new global times, it pushes welfare recipients into low-wage jobs, 'getting tough' on the unproductive (Peck, 2001). The mandate: no government benefits without work. The dilemma: most workfare participants find themselves in circumstances where financial resources acquired remain pathetically low, work is demeaning and discouraging, and scant opportunities exist to conduct job search or attend school. Its latest provisions, draconian in tenor, tighten eligibility requirements: states no longer have to require workfare participants to receive minimum wages and anyone convicted of a drug offence is forever barred from obtaining welfare.

New York City's use of workfare is illustrative. Over 40,000 people currently work in workfare assignments in the country's largest undertaking. Workers clean toilets, pick up trash, tend to grounds, and chase off homeless people (Piven, 2005). The average pay for workers is $5.80 per hour and $3,600 per year, substantially lower than the $18–22,000 per year for the average entry-level unionized worker. Assignments are increasingly in the rapidly expanding fast-food sector of the city: Burger King, Hardees, McDonald's, Wendy's, Arby's, and Sams. Workers cook, clean floors and bathrooms, do counter work, and scrub machinery. In this context, many workers are compelled to hold down multiple jobs, often working 50 to 60 hours per week to scrape by (Ehrenreich, 2003). Entrapment in this dead-end, low-wage service economy is extremely destructive: it too often saps energy, hope, and the ability to search for better jobs or enroll in education programmes.

Faith-based provision of government resources, another new tool in this regime, rapidly multiplies in America's cities. It is said to unleash 'armies of compassion' to help those with inadequate commitment to waged work and community upgrade. Mayors and planners across Urban America advertise it as shaping something now essential: a more productive and culturally unifying citizenry in global times. These things need to be shored up, it is said, to bolster the city's entrepreneurial ethos in these scary days. Increasingly, faith-based organizations provide resources and services that the federal government once provided: drug and job counselling, family planning, youth mentoring, food distribution, affordable housing provision and homeless shelters. In the process, these organizations (unlike government providers) can constitute their boards on a religious basis, display religious symbols and hire religiously compatible workers.

But much evidence suggests that this poses new problems to the black poor. The dilemma is the use of volunteers, use of workers exempted from state and local regulations, and the tendency of many workers to proselytize religious beliefs (Piven, 2005). Now, social service provision increasingly features religious tenets, reduced regulations and cost-cutting of general operations. Neoliberal values anchor this offering: privatize public resources, reduce wages of workers (or use volunteers) to 'streamline' organizations, and re-make the poor with spiritual and entrepreneurial values. In this context, the poor in New York, Chicago and LA report a litany of problems: the need to masquerade as good Christians to receive food and shelter, client prioritization by religious conviction rather than need, poorly-trained and poor quality counsellors, and religion often pushed as a cure-all panacea. Black ghetto populations, hurting and vulnerable, become test sites for a new kind of cultural engineering that exacerbates affliction.

In the UK, welfare-to-work is a leading new programme that afflicts ghetto spaces. This initiative, like the US workfare, is presented as a progressive measure to liberate the jobless from incapacity and unemployment (see Hyland, 2002). But the reality, again, is different: the poor are typically pushed into low-wage jobs that frequently deepen poverty and marginality. Central to this, a continued national deindustrialization and post-2000 economic malaise have created few moderate- to decent-paying jobs. Many welfare-to-work participants thus become removed from the welfare rolls but stay mired in poverty. Yet enormous political mileage is gained by this welfare roll reduction – public expenditures are reduced, politicians appear tough on the non-working poor – and the programme continues. The result: the programme now spreads to almost unimaginable venues. Most notoriously, welfare-to-work's recent 'factories within fences' scheme now brings workfare to prisons across the UK. Inmates in four prisons now do telesales work for private companies. From jail cells, they market furniture, home security equipment and airline tickets (see Rufford, 1997). Business efficiency, these subcontracting-to-prison companies report, has never been better.

Welfare-to-work has been shown to be especially debilitating in Inner London's ghetto terrain. Poverty among Black Caribbean, Black African, Pakistani and Bangladeshi households in East London deepens (see London Poverty Working Group, 2003). Here men and women trudge off to typically poorly paying, demeaning work (often in the city's burgeoning fast-food economy) with little opportunities for advancement. Perhaps unsurprisingly, this economy now relies on welfare-to-work to staff stores. Burger King, a UK-run franchise, has an elaborate hire and train programme that uses welfare-to-work as its centrepiece. Training and re-training of these workers is done at many store sites depending upon openings and needs. The other fast-food chains (McDonald's, KFC) have forged a similar dependency on this programme. Welfare-to-work now feeds the downtown economic tiger, evolving as a reliable and subsidized source of cheap labour for local business. The programme, in this sense, is seen by growing numbers of planners and politicians to assist local economic development (Hersh and Johs, 2003).

Neoliberalizing UK cities also correspond to US cities in another way: policies increasingly seek to cleanse downtowns as spaces for investment (Atkinson, 2003;

Raco, 2003). Ghetto spaces, in the process, feel the brunt of this. But the UK and US experiences in this regard are not identical. Unlike many US cities, this policy in UK cities often conflicts with still heavy doses of social welfare rhetoric. This is articulated especially by local politicians sensitive to and carrying the legacy of a national social welfare tradition. But these politicians still speak fervently of new, ominous global times that acutely disciplines cities. In this context, aggressive policy across UK cities routinely clears streets of the poor, sex workers, and minority youth that are taken to 'infect' middle-class consumption and habitation zones. This restructuring is fuelled by national legislation like the Crime and Disorder Act (1998) and its Anti-Social Behaviour Orders Initiative that locally institutionalizes this drive (Atkinson, 2003).

The results blanket UK cities from London to Newcastle. In London's gentrifying West End, purifying space features police banishing sex workers under zero tolerance policies. At the same time, a community-targeted rhetoric trumpets a new family-oriented and wholesome terrain (Hubbard, 2004). In Glasgow, the city centre under 'Operation Spotlight' is purified by arresting or removing panhandlers and minority children (the often stated goal is to eliminate 'yob culture'). Bolstering this, police monitor vendors for distribution of 'homeless' or 'minority youth'literature; this material is deemed by the local police as 'land-use trashing' (MacLeod, 2002; Atkinson, 2003). In Middlesbrough, police sweeps across the core area (sanctioned by former policeman turned get-tough mayor Ray 'Robocop' Mallon) interrogate anyone seemingly 'out of place' and detrimental to 'sense of community'. Mallon is uncompromising in his strategy to 'protect purity and order' (Butler and Simon, 2002). The police, according to Hetherington (2003), are roaming harassers who seek to rid the streets of any sign of class 'otherness'.

As these examples illustrate, regeneration agencies and the police now widely act in partnership to develop these purifying strategies. The recent rash of Anti-Social Behaviour Orders in the UK reflects this. A breach of an ASBO (e.g. excessive begging, verbal assault, flyposting, scrawling graffiti) is a criminal offence that is both arrestable and recordable. The police, following the legislation, are the lead agency in investigation and prosecution. Furthermore, the designation of more than 140 dispersal zones identifies areas with excessive 'anti-sociality'. In all cases, regeneration agencies and the police collaborate to plan and implement eradication of 'disorder and social dysfunction'. In many of the spaces, including those in London, Birmingham, Newcastle and Manchester, these spaces have been identified as core city locations having clusters of begging, the homeless and the poor (see Craig and Dillon, 2005).

Summary

Today, in the shadows of sparkling downtown high-rises and gentrified neighbourhoods, impoverished 'coloured' spaces in US and UK cities have worsened. This chapter argues that these spaces are now cast as and widely treated as 'contaminants' to a new civic imperative: globally-compelled' city restructuring. But these terrains in London, Manchester, Cleveland, Detroit, St. Louis and Pittsburgh have for decades warehoused the racial poor and been assigned negative representations. Yet this ghettoization and marginalization is more virulent today under hardened neoliberalism and its expansive city restructuring. Now, a deepened poverty and deprivation marks these areas, and their residents are widely and punitively presented as being profoundly and destructively consumptive or productive.

The role of global rhetoric has been crucial in this process. It has been a kind of trigger in local settings to mobilize key ghetto-afflicting forces ('global' redevelopment, destruction of the welfare state, attack from a new low-income punishing government rhetoric). This rhetoric, anchoring a new regime of ghetto creation, has been a perceptual apparatus that has penetrated the common consciousness and planning practice. It has served up a digestible reality that, following Robin Wagner-Pacifici (1994), guides the construction of programmes and policies by making certain actions thinkable and rational and others not. As this apparatus has resisted and beaten back competing visions of city, it has ensnared many non-white populations within a web of signifiers and meanings with profound repercussions. In this setting, I term these changed spaces 'global ghettos'. Spurred by this global rhetoric, these spaces have devolved into one-dimensional instruments for naked isolation and confinement. In the name of civic survival (i.e. cities becoming more taut entrepreneurial places), ghetto spaces have become more calcified zones of entrapment. The public, even as they ambiguously accept integration in abstract principle, now sanctions a social and symbolic denigration of these non-white communities codified in planning and policy initiatives. A normative category, the unworthy ghetto space, is defined and infused with meanings in relation to global discourses of business and competition. An important part of the new global reality, I thus suggest, is the hype of globalization, whose deft deployment todaycontinues to ravage ethnic ghetto spaces and their people.

References

Atkinson, R. (2003) 'Domestication by Cappuccino or a Revenge on Urban Space: control and empowerment in the management of public spaces', *Urban Studies*, 40 (7), 1829–1843.

BBC News (2004) 'Met Targets South Asian Criminals', website news.bbc.co.uk/1/hi/England/London/3807349.stm

Beju (2004) 'About', essay on life in London and Manchester, website topshelf.8m.net/about_me.html

Brown, C. (2003) 'Daring to be Different: What Do We Want from Urban Living?', *Manchester Forum*, Winter, 6–7.

Butler, P. (2002) 'Victory for Ray "Robocop" Mallon Adds to Labour's Mayoral Misfortunes', *Guardian Unlimited*, Friday, 3 May, p. 2.

Civic Task Force on International Cleveland (2003) 'Globalizing Cleveland: The Way to Go', unpublished document, available from City of Cleveland.

Cole, B. (2004) Discussion with Planner, City of St. Louis, 4 November.

Connelly, C. (2004) *Fair Housing in the US and the UK*. Paper presented at the Environment and Housing Conference, 3 July, Cambridge, UK.

Craig, I. and Dillon, M. (2005) 'ASBOs Do Work, Says Research', *Manchester Evening News*, 19 February.

Ehrenreich, B. (2003) *Nickled and Dimed*. New York: Owl.

GLA (2003) 'Investing in London: The Case for the Capital', website www.london. gov.uk.mayor/case_for_london2001/mayor-intro/sp.

Harrison, P. (1992) *Inside the Inner City*, London: Pelican.

Helms, G. and Cumbers, A. (2006) 'Regulating the New Urban Poor: Local Labour Market Control in an Old Industrial City', *Space and Polity*, 10 (1), 67–86.

Hennepin, B. (2004) Discussion with Planner, City of Cleveland, 4 March.

Hersh, J. and Johs, D. (2003) *Globalization, New Technologies, Inequalities and the Commodifiction of Life and Well-Being*. London: Pluto.

Hetherington, T. (2003) 'Robocop Mallon Gets on the Case as Mayor', *Guardian Unlimited*, website http://society.guardian.co.uk/mayorquestion/story/O,81,50, 957833.00.html.

Hubbard, P. (2004) 'Cleansing the Metropolis: sex work and the politics of Zero Tolerance', *Urban Studies* 41, 1687–1702.

Hyland, J. (2002) *Britain: Labour Government Targets Single Parents and Disabled for Workfare*, World Socialist Website, www.wsws.org.

Jones, L. (1997) *Our America (Illinois)*. New York: Simon & Schuster.

Leyshon, A. and Thrift, N. (1995) 'Geographies of Financial Exclusion: Financial Abandonment in Britain and the US', *Transactions, Institute of British Geographers*, 20 (3), 312–314.

London Europe (2003) *Supplement to LONDONLINE*, newsletter of the Mayor of London, 4, Summer, 7 pages.

London Poverty Working Group (2003) *Tackling Poverty in London: Consultation Paper*, Technical report prepared by London Poverty Working Group, London, available from author.

Lott, T.L. (1999) *The Invention of Race: Black Culture and the Politics of Representation*, Oxford: Blackwell.

MacLeod, G. (2002) 'From Urban Entrepreneurialism to a Revanchist City? On the Spatial Injustices of Glasgow's Renaissance', in N. Brenner and N. Theodore (eds), *Spaces of Neoliberalism*, Oxford: Blackwell.

Mascuse, P. (2000) 'Cities in Quarters', in G. Bridge and S. Watson (eds), *A Companion to the City*, Oxford: Blackwell.

Mercer, K. (1997) *Welcome to the Jungle*. London: Routledge.

Naymik, M. (2003) 'Debt Service: Financial Disaster Threatens a Major Eastside Community Center', website www.clevescene.com/issues/1990-0.7-08/news.html.

Norton, A. (1993) *Republic of Signs*. Chicago: University of Chicago Press.

O'Malley, M. (2004) 'Baltimore Today', video prepared for distribution by the City of Baltimore, website www.ci.baltimore.med.us/news/digitalharbor/.

Peck, J. (2001) *Workfare States*. New York: Guilford Press.

Peck, J. and Ward, K. (eds) (2002) *City of Revolution: Restructuring Manchester*, Manchester: Manchester University Press.

Piven, F.F. (2005) *The War at Home: The Domestic Costs of Bush's Militarism*. New York: Guilford Press.

Raco, M. (2003) 'Remaking Place and Securitizing Space: Urban Regeneration and the Strategies, Tactics, and Practices of Policing in the UK', *Urban Studies*, 10 (5/6), 1869–1887.

Redd, C. (2004) Discussion with Planner, City of Chicago, 17 July.

Rufford, N. (1997) 'Prisons to Offer Telesales Jobs', *LaborNews*, 11 August, website www-pluto.informatik.uni-oldenburg.de/ ~ also/welar150.html.

Short, J. and Yeong-Hyun, K. (1999) *Globalization and the City*, New York: Addison-Wesley Longman.

Sowell, T. (1985) *The Economics and Politics of Race*, New York: William Morrow and Company.

Thompson, T., Parris, P. and Bleuford, A. (2002) 'They're Lethal, Unfeeling – And No One Can Touch Them', *Guardian Unlimited*, website http://observer.guardian.co.uk/focus/story/0,69003,706463,00.html

Wacquant, L. (2002) 'From Slavery to Mass Incarceration: Rethinking the Race Question in the US', *New Left Review*, 13, 41–104.

Wagner-Pacifici, R. (1994) *Discourse and Destruction*, Chicago: University of Chicago Press.

Weir, N. (2003) Discussion with Planner, City of Chicago, 14 May.

White, J. (2004) 'Portrait of Black Chicago'. Technical report for the National Archives, Washington, DC, available from National Archives, Washington, DC.

Williams, W. (1987) 'Why Do We Pay the Price?', *Chicago Tribune*, 12 January, p. 12.

Wilson, D. (2005) *Inventing Black-on-Black Violence: Discourse, Space, Representation*. Syracuse, NY: Syracuse University Press.

Wilson, D. (2007) *Cities and Race: America's New Black Ghetlo*, London: Routledge.

Zukin, S. (1995) *The Cultures of Cities*, Oxford: Blackwell.

13 CIVILITY AND ETIQUETTE

Mick Smith and Joyce Davidson

This chapter

○ Identifies the city as a site where ideas of what is 'civil' and what is 'incivil' are defined, worked through and refined

○ Considers contemporary debates about the social use of public space in the light of historically-inherited ideas of public comportment and etiquette

○ Argues that current concerns about the decline of the public realm may be overblown given city life has always generated moral panics about the city's capacity to encourage the licentious and the anti-social

Introduction

Why does the city epitomize the height of sophisticated, fashionable and civilized behaviour and yet is simultaneously regarded as fostering everything 'polite society' might deem depraved, rude, uncivil and immoral? For example, a visit to an art gallery, theatre, or classy restaurant may be imbued with 'refinement', 'good taste' and 'propriety', but could hardly be further from, say, the 'loutish' drunken night club revellers brawling and vomiting currently regarded as endemic to city-centre streets in the UK (see Figure 13.1). Yet both are regarded as part of urban life, with an apparent rise in anti-social behaviour often leading to news headlines about a decline in manners, a loss of moral authority, a need to instil 'higher' standards of behaviour, especially in young people. For instance, as part of his campaign to 'clean up' New York, Mayor Rudolph Giuliani (1998) argued the need to 'teach civility as a subject in the classroom'. 'Civility' was also a key theme in

Figure 13.1 Anti-social Britain (after Hogarth)

George W. Bush's inaugural presidential address and one now widely adopted by conservative politicians.

Perhaps though, the civil and uncivil are not as straightforwardly oppositional as these current campaigners try to suggest. Certainly, a return to 'traditional' values and manners hardly seems the answer since these same concerns and conflicts have always been a feature of city-life. As Sweet (2002) argues, politeness in eighteenth-century England was a 'quintessentially urban concept' associated in particular with London. Yet Hogarth's cartoons of this same period, like 'Gin Lane' (1751) (see Chapter 10 and Figure 10.1), depicted a complete breakdown in social standards due to the over-consumption of cheap alcohol. However, if so many of the urban poor sought refuge from the conditions of their existence through gin drinking and riotous behaviour, this wasn't simply because they were somehow less 'civilized' than those who regarded themselves as their social betters. There is also a kind of Catch-22 situation here, where the very existence of certain forms of manners, etiquette, and civility often operated to ensure such groups continued social exclusion from classes with better social and economic prospects.

In any case, the moral superiority claimed by the socially powerful has often been both superficial and hypocritical. The 'family values' famously espoused by the middle-class of Victorian London frequently served as a façade that hid this same class's social dependence upon and economic ties to the 'Dickensian' existence led by a much larger urban underclass, including thousands of prostitutes whose purportedly 'indecent' activities were funded by the visits of these very same 'gentlemen'. So, then as now, manners were at least as much a matter of public performance and social distinction (for example, in drawing class and gender divisions) as of a person's actual moral character. Consequently, we should be aware that the call for civility, in whatever form, always has hidden socio-political overtones.

Of course, those experiencing social exclusion will sometimes actively develop forms of fashionable incivility to mark their oppositional status to established norms, as for example in punk rock or the 'hippie' culture of the 1960s. On other occasions certain sections of society, for example, the homeless, beggars, or even street vendors, are singled out by authorities as socially disruptive 'problems' to be removed from 'public' spaces. Here we might think of decisions by many local authorities to ban the drinking of alcohol on streets, something arguably intended to remove the hopeless and homeless from public view rather than either addressing their real problems or tackling genuine dangers to public safety (Dixon et al., 2006). The very notion of a public space thereby becomes redefined to exclude certain kinds of people and/or forms of behaviour. Such labelling of given groups as 'anti-social' can also fuel *moral panics* (Goode and Ben-Yehuda, 1996), where a perceived threat to social norms of 'decent' behaviour elicits disproportionate responses. For example, newspaper stories in Britain recently generated panic about groups of teenagers wearing hooded tops congregating in town centres and shopping malls (Wintour, 2006). Some large malls attempted to ban 'hoodies' even though some businesses in the very same centres actually sold them! Perhaps the teenagers' real crime lay in their failure to conform to the consumerist norms expected of mall visitors – they were guilty of 'hanging out' rather than shopping (Jackson, 1998).

The role of civility, manners, and etiquette in city life is then both socially complex and also historically variable – after all, drinking a gin and tonic now has rather different cultural connotations from those suggested in Hogarth's famous etching. The theatre, too, provides another example of changing cultural associations. It was long regarded as a haven of loose morals among both performers and spectators. As late as the nineteenth century, audience 'participation' would normally include joining in with speeches and familiar songs, booing, heckling, stamping feet and shouting to demand an immediate encore of a particular scene, throwing objects at the actors and even full-scale riots. Twenty-two people died in a riot in 1849 at the Astor Place Opera House in New York. Gradually, a much clearer split developed along class lines between 'genteel' arts like opera and classical music, where audiences became increasingly emotionally constrained by codes of etiquette, and such 'boorish' behaviour was relegated to popular entertainment like boxing or wrestling matches (Kasson, 1990).

To trace some of these social and historical complexities we will first examine how civility is publicly performed according to different 'rules' in specific times and city locations. We then examine the ways in which 'civility' has been associated with discourses of progress (civilization) epitomized by the city, and the role of public opinion and social pressures in maintaining certain patterns of behaviour. Following this we introduce several key social theorists in order to develop a view of civility as a primary indicator of that which social circumstances permit to be expressed in public, and that which is socially repressed (which we refer to as an 'urban unconscious'). The politically contentious nature of exactly what should and should not be expressed, what is and isn't regarded as polite, politic, or socially appropriate, is illustrated through debates over public art works and the gentrification of relatively poor neighbourhoods as they change into fashionable residential locations for a wealthier clientele who possess rather different cultural sensibilities. Finally, we bring these themes together via a critical appropriation of Norbert Elias's (1982) notion of the 'civilizing process.'

'Performing' etiquette on Broadway

> New York. More sirens here, day and night. The cars are faster, the advertisements more aggressive. This is wall-to-wall prostitution. And total electric light too. And the game – all games – gets more intense. It's always like this when you're getting near the centre of the world. But the people smile. Actually they smile more and more, though never to other people, always to themselves. (Baudrillard, 1988: 14)

Jean Baudrillard's verbal snap-shots of his travels in America capture some of the conflicting images frequently associated with modern urban life: its neon commercialization, the predominance of impersonal economic exchanges over collective life, the city's pace, playfulness, and aggression, the ambiguous and strangely alienating pleasures that, like mythic sirens' calls, lure the unwary into this intense (but apparently also intensely superficial) city. New York is not just the commercial but also the cultural 'centre of the world', as Paris billed itself in the late nineteenth century. But to say this raises an obvious question: what now counts as *cultural capital* (Bourdieu, 1984), what gives New York its contemporary civic cachet? Is it simply a matter of staying 'ahead of the game', in Baudrillard's sense, of an increasingly rapid but ultimately meaningless exchange of fashions and styles? If so, this would indeed seem superficial and almost entirely inconsequential. Alternatively, though, perhaps such surface changes might themselves reflect that city's ability to produce avant-garde but nonetheless almost universally recognizable expressions of something more profound, something that might, however differently construed, be termed an 'authentic' urban experience. If so, then New York's tastes, styles and behaviours could be thought of as paradigmatically urban, as embodying somewhat idealized models of the particular and often problematic social relations associated with modern cities. Styles might thus contain more social substance than is casually apparent since manners and mannerisms, tastes, ornament and body language might all be regarded as ways of comporting citified selves in relation to others in such a way as to ensure a degree of civility in an otherwise potentially very anti-social landscape, where even smiling at someone else might be asking for trouble.

From this perspective, the vibrant theatricality of city life is not just for show, but is typical of the kind of *performative* norms (of forms of 'acting out') that regulate urban co-existence. In the terms used by Judith Butler (1990), these performances are actually constitutive of self-identities and of culture as such. As she argues in relation to gender roles, the individually embodied habits that make up the patterns of behaviour expected from, and the demeanour associated with, people playing certain social roles require 'a performance that is *repeated*. This repetition is at once a re-enactment and re-experiencing of a set of meanings already socially established; and it is the mundane and ritualized form of their legitimation. Although there are individual bodies that enact these significations by becoming stylized into gendered modes, this "action" is a public action' (Butler, 1990: 140).

Butler helps us understand how publicly reiterated performance defines the individual's persona: it marks them out as belonging or not belonging in certain social milieux and provides the necessary cues to allow others to understand how (and how not) to interact and communicate with them. Take, for example, the postures and rituals associated with smoking a cigarette, a practice itself regarded as uncivil in many (but by no means all) social circles, and one now *explicitly* regulated in many Western urban public spaces. There are clearly socially significant differences in the ways a cigarette is held, inhaled and exhaled, placed in or taken out of the mouth both within and between, say, the 'pub-culture' of the Glasgow 'hard-man', the 'party culture' of the Hollywood starlet, or the various 'street cultures' of majority black or Latino neighbourhoods in North American inner cities, like parts of New York's South Bronx. Such performances might all be regarded as forms of urban 'etiquette' in the sense that they help constitute individuals within certain socially pre-defined, but more or less contextually flexible, limits and expectations. They are forms of behaviour that embody the unwritten rules and subliminal conventions necessary to facilitate social understandings, reiterated behaviours that thereby define in-groups and out-groups, strangers and friends, amid the fragmented and constantly shifting patterns of urban existence. As Butler suggests, they also help legitimize and impose these same, often quite restrictive, social boundaries on newly recruited individuals.

Writing about nineteenth-century New York, Domosh (1998) discusses the complex patterns of behaviour that circumscribed acceptable behaviour in public places. She explicitly refers to the streets of New York as a 'public stage' where etiquette is not only defined in terms of dress codes and mannerisms, for example, doffing the hat and slightly tilting the head as a mark of recognition and respect, but in terms of spatial and temporal arrangements. It was, for example, regarded as improper, 'a breach of respectability', for a 'lady' to be seen on Broadway before 11 in the morning or after 3 in the afternoon. This was partly because, in the evening, Broadway became the primary site of street prostitutes. Yet, ironically, on Sunday mornings the very same place was the accepted and fashionable location for the white urban bourgeoisie to promenade in all their finery after attending church. Performing within the expectations defined by current rules of etiquette and conduct actually served to maintain social distinctions along lines of class, gender, and ethnicity in what was, and remains, a culturally heterogeneous cityscape. (Indeed, sociologist Pierre Bourdieu (1984) has argued that maintaining and exemplifying social distinctions and stratifications is the key feature of all fashions, tastes and forms of etiquette.) Such expectations were not, however, entirely fixed or uncontested. They could be gradually transgressed in terms of a 'polite politics' that had cumulative effects on established norms of public opinion, altering them over time. New York's streets were, Domosh (1998: 223) suggests, a 'theatre where scripts could be manipulated'.

One question that seems to arise here is whether such gradual changes contribute to a larger story of socio-historical progress or whether they merely reflect constantly shifting fluctuations and fashions – are they merely a series of scenes with no unifying structure? A second question concerns how these performances are elicited from and maintained in diverse individuals.

Civic 'improvement', public opinion and civil inattention

There seems to be something of a paradox that, at least since the Enlightenment, dominant Western discourses have tended to associate civility with what are claimed to be *universally* applicable notions of human moral and aesthetic development. These idea(l)s were associated with the city in so far as it both came to represent and was supposed to bring to fulfilment a generalizable model of social progress and self-improvement, one embodied in the adoption of 'civilized', 'cultured' or 'refined' behaviour and the acquisition of 'good' taste. The city was not just the hub of industry, the centre of new technical inventions, but also came to epitomize advanced, educated and sophisticated social relations (Williams, 1985). Despite actual social differences 'culture' and civility were, in this 'humanistic' and 'democratic' sense, something all should aspire to and not merely relativistic descriptions of the contrasting modes of existence (culture*s*) that, to this day, characterize urban existence.

The paradox then is that civility has been promoted as a unifying urban theme even though it establishes and maintains social and cultural divisions. This is well illustrated in the accompanying passage from an account of the socially 'dangerous' classes of New York from 1872 (see Extract 13.1). Here we have an explicit discourse of gradual and supposedly universal moral improvement that still clearly accepts the existence of an extremely hierarchical society of higher and lower classes, masters and servants, good and bad breeding. Interestingly, the author also portrays New York's industrial mobility as a moral improvement on old world and rural communities since its lack of temporal and spatial stability and its competitive 'climate' brings about an urban form of 'natural' selection that means the bad habits (which he associates entirely with the poor) are less likely to be passed on to the next generation.

Extract 13.1: From Brace, C. L. (1872) *The Dangerous Classes of New York and Twenty Years Work among Them*, New York: Wynkoop and Hallenbreck, pp. 46–7.

In the US a boundless hope pervades all classes; it reaches down to the outcast and vagrant. There is no fixity, as is often the case in Europe, from the sense of despair. Every individual, at least till he is old, hopes and expects to rise out of his condition.

The daughter of the rag-picker or vagrant sees the children she knows, continually dressing better or associating with more decent people; she beholds them attending public schools and improving education and manners; she comes in contact with the greatest force the poor know – public opinion, which requires a certain decency and respectability among themselves. She becomes ashamed of her squalid, ragged, or drunken mother. She enters an industrial school, or creeps into a Ward School, or 'goes out' as a servant. In every place, she feels the profound forces of American life: the desire for equality, ambition to rise, the sense of self-respect and the passion for education.

These new desires overcome the low appetites in her blood, and she continually rises and improves. If Religion in any form reach her, she attains a still greater height over the filthy and sensual ways of her parents. She is in no danger of sexual degradation, or of any extreme vice. The poison in her blood has found an antidote. When she marries it will inevitably be to a class above her own. This process goes on continually throughout the country and breaks up criminal intercourse.

Moreover, the incessant change of our people, especially in the cities, the separation of children from parents, of brothers from sisters, and of all from their former localities, destroy that

(Continued)

continuity of influence that bad parents and grandparents exert, and do away with those neighbourhoods of crime and pauperism where vice concentrates and transmits itself with ever increasing power. The fact that tenants must forever be 'moving' in New York, is a preventative of some of the worst evils among the lower poor. The mill of American life, which grinds up so many delicate and fragile things, has its uses, when it is turned on the vicious fragments of the lower strata of society.

Extract 13.1 illustrates how civility paradoxically extends into the modern city's associations with cosmopolitan notions of individual liberty and equality in terms of escaping from the parochial traditions and moral restraints deemed characteristic of earlier and more insular communities. For Brace, the fact the city breaks emotional attachments to family and community seems entirely positive, not because the poor are thereby free to do as they please, but because their behaviour becomes subject to the 'improving' pressures and authority of public opinion. Sociological analyses of city-life have, however, tended to regard public opinion with more suspicion and explicitly recognize the ambiguous nature of the individualistic freedoms associated with urban existence. For example, Simmel's classic essay, 'The Metropolis and Mental Life' (1903), argues that the press of crowded bodies, constant change, and lack of personal ties necessarily fosters mutual indifference, blasé attitudes and social reserve. Metropolitan individuals must maintain a lonely intellectualized distance from others in order to protect their mental life from the constant impingement of excessive social stimulations.

Erving Goffman (1963) was later to develop such insights through the notion of *civil inattention*, the reciprocal ways in which each person exhibits respectfulness towards the 'personal space' of the strangers that surround them (especially in urban situations). For example, people avoid looking directly at other passengers on public transport like the London Underground or New York's subways. He refers to such behaviours as 'situational proprieties', forms of social etiquette that allow people to maintain some control over their self-identities and immediate environs through a pretence of indifference to the potentially invasive gaze, presence and judgement of others. Following these usually implicit rules of conduct helps maintain social distance, by *performing* on cue we become 'part of the crowd' and avoid becoming subject to public scrutiny. This, however, entails a delicate balancing act since we also risk losing ourselves, our individuality, in the crowd and where such proprieties fail to give adequate protection, or individuals are especially susceptible, the social intensity of public spaces can become overwhelming. One way of thinking about agoraphobia – the intense fear of social spaces (long-defined as an urban phenomenon) – is in terms of the way sufferers experience dreadful feelings of depersonalization in busy areas. They subsequently avoid such areas in ways that might be thought akin to a form of 'stage-fright' or performance anxiety (Davidson, 2003: 82).

The urban unconscious

How, then, are we to think about these apparent contradictions and how are we to understand the city as simultaneously an iconic cultural centre and the actual home of various cohabiting or competing cultures distinguished by their (conscious and subliminal) adoption of very different styles, morals, tastes and behaviours? As suggested above, these contradictions are not simply surface phenomena. They are also, as Simmel and Goffman suggest, more than matters of individual psychology. They are woven into the social fabric of the city itself and played out in its myriad cultural fields: political, musical, emotional, architectural, and so on. Perhaps, then, we need to begin to think of civility, etiquette and culture itself in terms of their composition within, and contributions to, the historical and geographic rhythms of urban life, its ideals and realities. And in doing so we need to retain a critical awareness of the constant presence of hierarchical power relations and how these play out in the expression and repression of different cultural forms and ideals, how what is deemed permissible and 'proper' also defines what is culturally 'improper'. What passes as an acceptable part of the urban 'scene' also defines its 'ob-scene', the repressed and illegitimate aspects of that culture.

What if, as Henri Lefebvre (1994: 36) suggests, 'it turned out ... that every society, and particularly (for our purposes) the city, had an underground and repressed life, and hence an unconscious of its own'? If we take Lefebvre's proposition seriously, then we might begin to think about the 'lived experience' of the city in a rather different way, as something that often escapes the frameworks of urban analysis commonly applied to it. Urban studies would need to include more than representations and discussions of plans, architecture, and the spatial arrangements of people and places, more than just accounts of complex social interactions and practices. It would also have to encompass understandings of what we might term the spatial imagery and imaginary of city life with all their contradictions, all their covert as well as overt effects.

Lefebvre (1901–91) was certainly not the first to recognize the importance of an 'underground' aspect of urban existence that has fascinated many writers on city life. Lefebvre's originality, though, lies in emphasizing the breadth of the repressed's creative – though certainly not always welcome – contributions to the production of (urban) social space. This 'concrete unconscious' is not just associated with politically, morally, and legally repressed realms like criminality, prostitution or political intrigue, but also with the expressive sources of artistic endeavours, dreams, and the creative processes by which we develop as different human individuals. It might also be said to provide the symbolism and images through which the everyday life of inhabitants can conflict with the social expectations of those wielding economic and political power, for example, in the graffiti inscribed on city walls (de Certeau, 1988; Gardiner, 2000). And, just as graffiti can sometimes exemplify the artistic self-expression of its producers, it can also be considered a sign of disrespect for property and as disruptive social behaviour, a candidate for zero tolerance policing by those regulating the given social order.

Lefebvre's remark serves to bring out the constant tension between expressive and repressive aspects of social life, between creative differentiation from, and conformity to, current practices. What he refers to as the city's unconscious thus has little in common with popular psychoanalytic notions of the unconscious *per se*. Rather, this analogy serves to bring to our attention the presence of an obverse side, 'an obscure counterweight' to culture (Lefebvre, 1994: 208). The city's 'unconscious' is not something that is straightforwardly opposed to, or merely the opposite of, that conceptually conscious urbanism embodied in explicit plans, forms of organization, design of buildings, monuments and so on. It isn't, one might say, just its unknowable and hidden 'dark side'. Rather, it is part and parcel of how the city reproduces itself in ways that fail to conform to its externally transmitted image as a harmonious whole, a bastion of progress and civilization, a centre of knowledge and governance, and so on.

This might begin to explain how the culture of the city, caught up as it is within an ideology of social and material progress, is inherently contradictory. While striving to represent itself as on the cusp of a historical wave of cultural advancement and moral improvement, the city is just as easily depicted as a locus of iniquity, disease, dirt, squalor and depravity, as, for example, in the London of Charles Dickens, the Paris of Victor Hugo or, more recently, though less didactically, the heroin culture of Edinburgh in Irvine Welsh's *Trainspotting* (1993). But these seemingly irreconcilable views of urban life are not just indicative of different attitudes and analyses; they are also extreme examples of the uneasy cohabitation of cultural imaginaries within the shifting grounds of social expression and repression that constitute the production of city spaces. To be part of the urban fabric is, it seems, to be regarded as culturally 'sophisticated' in the dual sense of being 'discriminating in taste' or 'highly advanced' and, simultaneously, of being 'deprived of natural innocence', 'not plain, honest or straightforward' (*Oxford English Dictionary*).

The interdependency of the expressive and repressive aspects of urban culture is sometimes made particularly visible in certain locations. This explains the fascination that the shopping arcades (*Passagen*) of the mid to late nineteenth century – once the home of fashionable purveyors of elegance to the urban aristocracy and bourgeoisie – had to early twentieth-century theorists interested in the imagery of everyday urban life, such as Siegfried Kracauer (1889–1966) and Walter Benjamin (1892–1940). These now (mainly) dingy, déclassé and decaying thoroughfares, half-empty except for tacky souvenir and tawdry clothes shops and popularly regarded as the haunt of a criminal underclass, seemed to offer architectural as well as ideational passages to the repressed urban unconscious. Kracauer's description of the decline of the Linden Arcade in Berlin (see Extract 13.2) is a case in point. Kracauer was fascinated by the ways in which critical analysis could re-evaluate the surface appearances of everyday life revealing the hidden social substance of cultural styles and fashions. Similarly, in his massive but uncompleted Arcades Project (1999), Benjamin called upon a *mélange* of dream-like imagery, obscure documents and *objet trouvé* (neglected items found by chance) to try to recall and redeem these suppressed, repressed and, in his own word, 'labyrinthine' aspects of Parisian city life. In this way he hoped to show how the rational organization of urban existence is actually shot through with the strange rituals, patterns and hidden influences that momentarily domi-nate civil society and how 'fashion (the uncritical acceptance of transient social norms) functions as camouflage for quite specific interests of the ruling class' (Benjamin, 1999: 71). For Benjamin, Kracauer and Lefebvre alike, the return of the repressed always offers socially and politically disruptive possibilities: the 'dangerous classes' indeed challenge the city's façade of civility.

Extract 13.2: From Kracauer, s. (1995 [1930]) 'Farewell to the Linden Arcade', in *The Mass Ornament: Weimar Essays*, Cambridge, MA: Harvard University Press, pp. 338–42.

The time of the arcades has run out … The peculiar feature of the arcades was that they were passageways, ways that passed through the bourgeois life that resided in front of and on top of their entrances. Everything excluded from this bourgeois life because it was not presentable or even because it ran counter to the official world view settled in the arcades. They housed the cast off and the disavowed, the sum total of everything unfit for the adornment of the façade. Here in the arcades, these transient objects attained a kind of right of residence, like the gypsies who are allowed to camp only along the highway and not in town. One passed them by as if one were underground, between this street and the next. Even now the Linden Arcade is still filled with shops whose displayed wares are just such passages [Passagen] in the composition of bourgeois life. That is, they satisfy primarily bodily needs and the craving for images of the sort that appear in daydreams. Both of these, the very near and the very far, elude the bourgeois public sphere – which does not tolerate them – and like to withdraw into the furtive half-light of the passageway, in which they flourish as in a swamp. It is precisely as a passage that the passageway is also the place where, more than anywhere else, the voyage which is the journey from the near to the far and the linkage of body and image can manifest itself. …

What united the objects in the Linden Arcade and gave them all the same function was their withdrawal from the bourgeois façade. Desires, geographic debaucheries, and many images that caused sleepless nights were not allowed to be seen among the high goings-on in the cathedrals and universities, in ceremonial speeches and parades. Wherever possible, they were executed, and if they could not be completely destroyed they were driven out and banished to

(Continued)

the inner Siberia of the arcade. Here, however, they took revenge on the bourgeois idealism that oppressed them by playing off their own defiled existence against the arrogated existence of the bourgeoisie. Degraded as they were able to congregate in the half-light of the passageway and to organize an effective protest against the façade culture outside. They exposed idealism for what it was and revealed its products to be kitsch. The arched windows, cornices, and balustrades – the Renaissance splendour that deemed itself so superior – was examined and rejected in the arcade. Even while traversing it – that is, effecting the movement appropriate to us alone – we could already see through this splendour, and its pretentiousness was unveiled in the light of the arcade. The reputations of the higher and highest ladies and gentlemen, whose portraits with their guarantees of fidelity stood and hung in the display windows of the court painter Fischer, fared no better. The ladies of the kaiser's court smiled so graciously that this grace tasted as rancid as its oil portraits.

Whether espousing revolutionary, reformist, reactionary, or just 'romantic' alternatives, these disruptive counter-currents challenge the manner in which, for dominant modern discourses, being 'cultured' becomes almost synonymous with contemporary city life, with a movement from nature and the rural to the urban and the *urbane*. This movement is, as Raymond Williams (1985) showed in his analysis of literary themes, *The Country and the City*, always subject to conflicting moral judgements. The cultural evaluations of city sophisticates and detractors alike are in part, Williams argues, historically acquired tastes dependent upon the emergence of different *structures of feeling*. These socially produced patterns of emotionally mediated responses play crucial – but often unrecognized – roles in determining, for example, how city life is envisaged and whether the surrounding countryside is viewed through the lens of nostalgic romanticism or as merely a haven for 'primitive' rural idiocy.

Contesting urban imagery

Such different responses can surface in debates about the iconography and imagery of city life, especially when dominant discourses try to expunge repressed, but still resistant, social histories. For example, Dolores Hayden (Extract 13.3) recounts debates concerning a proposal to commemorate Cincinnati's past through a controversial new sculpture crowned by 'winged pigs' intended to highlight its role as 'Porkopolis', the first North American centre for industrial-scale pig slaughtering. Hayden is primarily interested in the public repression of histories of everyday life, all too often hidden behind trappings of civic pride and civility, for example, the all-pervasive monuments to bearded 'city fathers' (see Chapter 5). Hayden's account clearly illustrates how a discourse of civility plays a key role in public discussions about the moral, aesthetic and political significance of contested developments and how this language is informed by differing structures of feeling. Although the debate Hayden presents is portrayed as one between city 'sophisticates', for whom pigs symbolize an uncivilized rural animality that needs to be transcended, and those 'politely' welcoming the 'witty' porcine statuary, one can certainly envisage yet other responses. Some, for example, might think it the epitome of tastelessness to either repress *or* make a joke out of a history of mass-slaughter, even if those concerned are pigs and not humans. Such responses would presumably be tied to those marginalized structures of feeling dubbed romantic structures that nonetheless resist urban culture's constant, excessive, and not always necessary, repression of natural as well as social history (Philo, 1998).

Extract 13.3: From Hayden, D. (1997) *The Power of Place: Urban Landscapes as Public History,* 1995 Cambridge, MA: Harvard University Press, pp. 72–3.

In the nineteenth century, Cincinnati was 'Porkopolis.' Between 1830 and 1870, the first assembly line was developed there for turning pigs into ham. The landscape suffered while the local economy prospered. The canal ran red with blood from the slaughterhouses; German American immigrant workers came home spattered with blood and smelling of offal. For the celebration of the city's bicentennial in 1989, artist Andrew Leicester proposed a new gateway for Sawyer Point Park that climaxed with a tribute to Porkopolis in the form of four winged pigs atop a suspension bridge. (A collation of images chosen by the artist from local landscape history were also part of the piece ... .)

The citizens erupted in debate about the pigs. Were they a symbol of a smelly industry the city had outgrown? Were they insufficiently genteel? Or did they capture the city's spirit in a witty and wonderful way? The city council heard testimony from both camps. One indignant opponent argues, 'We are now poised on entering the twenty-first century as a technologically sophisticated city and pigs are a rural image.' He suggested that the pigs represented greed and sloth. Another complained, 'I don't think we need the burden of being called the pig city of the world.' Proponents claimed, 'We've got enough old statues with bearded men on horses in this town and its time to have something else.' One advocate noted, politely, 'I would be pleased to see their porcine presence.'

Amid wild applause and loud laughter from partisans, the debate flourished. At least three city council members donned pig snouts before the vote to show their support for the artist, while schoolchildren had their say as well. ... Leicester's winged pigs won, and the piece was constructed. They stand atop the gateway today because Cincinnati's residents chose to remember Porkopolis as an essential part of the city's historic landscape and make it part of a compelling new public history.

This interweaving of civility's expressive and repressive functions is, of course, further complicated by those extensive and pervasive social divisions within the urban landscape we have already alluded to along lines of gender, ethnicity, class, language, status, nationality, and so on. Hayden's work does much to highlight the way such divisions have themselves been linked to forms of social and historical repression in contradictory urban landscapes. These distinctions too depend on more than 'objective' categories and *conscious* and *explicit* cultural presuppositions. They too have their own 'unconscious' ramifications in Lefebvre's sense. They too are associated, whether positively or negatively, with differing kinds of acquired tastes, patterns of behaviour, and structures of feeling that, on the one hand, provide counterweights to idealized impositions of a unitary cultural ideal, a single supposedly 'civilized' state of being, but simultaneously reveal potential grounds of social estrangement. Racism, for example, is almost always justified by its proponents in terms of the supposed cultural inferiority, the lack of civility, manners and taste, shown by those categorized as both different and inferior. The challenge then, as Hayden (1997: 246) remarks, is how 'to respect and nurture a diverse urban public'.

Taste, gentility and gentrification

City spaces are landscapes within which dominant and subaltern discourses and identities are played out, for example, in processes of *gentrification* (see Chapter 6) in inner-city areas. A moment's reflection shows this notion, even if only in the minds of geographers, involves both conscious cultural ideals and aspects of Lefebvre's repressed urban unconscious. The very term 'gentrification' is linked by numerous associations to morally loaded expressions of gentility, to the appearance of supposedly more 'refined' (sophisticated) and less 'coarse' tastes and behaviour on the part of those aspiring gentlemen and women, the want-to-be members of 'polite society', who increasingly come to inhabit such spaces. It is also, as Neil Smith has argued, linked to dominant discourses about the advance of culture through 'taming' a very un-natural environment misrepresented as the equivalent of an urban wilderness or even 'jungle'. It thereby expresses an arrogant (and also presumably predominantly white and masculine) 'frontier mentality', a language of urban pioneerism that 'conveys the impression of a city that is not yet socially inhabited' (Smith, 1996: 340). This urban myth informs every aspect of the social and economic processes by which 'run-down' areas are deigned 'improved' as they go 'up-market'.

Interestingly, though, having made these insightful comments, Smith too pushes these cultural processes to the back of conscious geographical analysis. The imagery and imaginary of gentrification is itself repressed by a much more typical and predominant form of urban analysis focusing on tangible, explicit and quantifiable economic processes. The urban frontier, he argues, has a 'quintessentially economic definition' (Smith, 1996: 345). But the appearance of a delicatessen instead of a Spar or 7–11 isn't just a mark of new residents' increasing economic wealth, it also signifies their different sensibilities, their mimicry of the 'high' culture of an idealized upper-class, a *gentry*. Of course, depending upon one's social perspective, gentrification also links to other competing structures of feeling that might evoke resentment at both the loss of, say, a supposedly more 'genuine' or 'vital' working-class community, and what is taken to be the insensitive intrusions of newcomers 'putting on' airs and graces. But the fact that these cultural ideals involve a social imaginary, that, for example, they employ utopian or dystopian images of particular urban communities, does not mean they are 'unreal'. These competing imaginaries are tangibly 'acted out' in changing patterns of social performance that, for example, serve to reconstitute gender roles through what Liz Bondi (1991: 195) refers to as 'gentrification as cultural production'.

The issue of gentrification reveals the presence of a dominant and dominating cultural ethos at work in these urban transformations, an ethos that links in complex ways privileged aesthetic and moral ideals with economic and political power. It might initially be tempting to think of this as a form of cultural '*hegemony*' in the sense theorized by Antonio Gramsci (1982), whereby the ideas and ideals of the ruling class, including their claim to social leadership, become surreptitiously instilled and embodied in the common-sense understandings and habitual practices of the general population. From this perspective, what is generally accepted as 'good taste', 'good behaviour', 'good manners', and so on, reflect the interests of those in power, and actually serve to support the current socio-political hierarchy.

But this notion of hegemony is perhaps too limiting, since it is often taken (rightly or wrongly) to imply the existence of an over-arching, uniform, consensual, largely non-coercive and, from the perspective of everyday existence, almost irresistible, cultural totality rather than the dynamic, fractured, dissenting, anarchic and potentially disruptive urban unconscious that, for Lefebvre, is culture's constant dialectical companion. To return to our example, gentrification is certainly not – as a rather static understanding of hegemony might suggest – universally accepted as social 'improvement' but is actually resisted on many levels (Bondi, 1999; Newman and Wyly, 2005). Nor is it entirely consensual since it obviously operates through mechanisms of economic exclusion as, for example, house prices rise in an area, effectively forcing residents out, and through overt policing, for example, the forcible relocation of those street vendors and streetwalkers now regarded as morally and aesthetically contaminating presences.

Civility and the civilizing process

Through concepts like 'urban unconscious', 'performance', 'civil inattention', and so on, we begin to understand something of the complex cultural roles of social mores and moral codes, the manners, etiquette, practices, tastes and sensibilities that together compose much more than just the surface appearances of the individual inhabitants, communities and architectures of urban life. Codes of civility and etiquette should not be considered just as socially imposed *prohibitions* but, as Lefebvre suggests, aspects of the (social) *production* of space. This 'cultural' aspect of urban space might be thought under three inter-linked and recurrent themes:

- civility (*politesse* and the polite society),
- civil (political and policed) society, that is, the public realm of the citizen (city dweller) and of public opinion,
- civilization (envisaged as a process of cultural evolution and expansion stemming from that first idealized city-state, the *polis* of ancient Athens).

Each of these terms unfolds within a different temporal and geographical field and at different scales, running, for example, from the micro-politics of how to conduct oneself when dining out to the purportedly progressive conduct of history in its entirety. Nevertheless we might by now suspect all will harbour repressed elements of an urban unconscious and each will well up into and inform the others. Understanding, then, requires that we take account of the dynamic and constantly shifting relations between what is expressed and repressed in each field and their interrelations.

We also need to try to understand to what extent these relations and their material ramifications in city life are made explicit or left implicit. By this we mean not only what is left unsaid – for example, if a certain person or behaviour is described as 'civilized' – but also whether these cultural attributes operate at the level of conscious rule-following or habituated non-conceptualized behaviour, such as knowing when and where shaking hands will be appropriate. (Bourdieu, 1984, uses the term *habitus* to refer to the individual's incorporation, through ongoing processes of socialization, of flexible, open-ended strategies that eventually constitute a subliminal 'feeling' for what behaviour is appropriate in particular contexts. This is clearly the mode of operation of many interpersonal forms of etiquette and a vital ingredient in understanding how far one can push social boundaries.)

Gaining even a rudimentary understanding of these complex possibilities would be an impossible task without theoretical generalizations and it is worth concluding with one of the most influential attempts to combine an account of civility, civil society and civilization, namely Norbert Elias's *The Civilizing Process*. Elias (1982: xii) charts what he calls 'the psychical process of civilization' in terms of both a history of evolving manners and the relationship between this process and centralized state formation in Western Europe. As with Simmel's 'mental life', Williams' 'structures of feeling', and Lefebvre's 'urban unconscious', this 'psychical' process is not reducible to individual psychology but is intended to reveal underlying patterns, movements, and trends of social expression and repression as they develop in and through the material processes of everyday life. Elias refers to the outcomes of these different processes by which individuals and societies are historically co-constituted as *figurations*.

Elias does not naïvely accept a unitary ideal of culture as given, nor present a celebratory account of the 'rise' of civilization. Not only does he begin *The Civilizing Process* with a detailed exposition of the historical sociogenesis of 'civilization' and 'culture' as guiding, but sometimes contrasted, concepts in Western societies, but he pays special attention to their potentially repressive as well as liberating influences. For Elias, the civilizing process is marked by an increasing delicacy about, and self-regulation of, bodily processes in public places. Drawing on etiquette manuals and other historical sources, he traces the ways in which, for example, it gradually becomes regarded as good manners that individuals should use a handkerchief, rather than a hand or sleeve, to blow their nose and courteous to refrain from scratching or spitting in the street, rather than just over the dinner table! In the words of one book of manners from 1859, 'Spitting is at all times a disgusting habit. ... Besides being coarse and atrocious, *it is very bad for the health*' (quoted in Elias, 1982: 128, original emphasis). Using Butler's terminology, civility becomes a matter of the self-regulation and social repression of bodily *performance*, of what might otherwise, without civilization's influence, be thought to come 'naturally'.

While Elias attempts to provide a non-judgemental sociological account of the figurational dynamics of the civilizing process, it is nonetheless envisaged as primarily a unidirectional trend. Bizarrely, he also suggests that the civilized individual's development from (pre-social) child to mature and socialized (civilized) adult recapitulates this social history in miniature. Elias might thus be said to surreptitiously buy into certain aspects of the view that modern Western society exemplifies a more 'advanced', or at least more 'highly' evolved, social order. Useful as his emphasis on the cultural control and repression of bodily expression is, it is actually very difficult to regard recent, and constantly shifting, fashions, sensibilities and tastes, for example in terms of tattoos and body piercing, as straightforwardly indicative of this purportedly general trend. Examining changing evaluations of eating practices in public streets in the face of reduced lunch hours and the widespread availability of fast-food, Valentine (1998: 202) actually suggests 'eating, or "grazing" on the street appears to be breaking down some of the social codes that helped to maintain boundaries. ... The city street no longer appears to be a space where 'private' bodily propriety is quite so rigidly regulated by the public gaze.'

Summary

In this chapter we have seen 'civility' is both an ideal and a form of publicly regulated sociability that finds expression in the city's architecture and artefacts as well as its people. As such, it still contributes to the continued existence of urban civil society, despite (though some might argue because of) its repressive functions. But the challenge for contemporary urban life, perhaps again exemplified in New York, is to avoid sliding straight from that situation which Domosh described as a 'polite politics' governed by the restrictive gaze of public opinion, into a post-9/11 state of authoritarian surveillance of a public not allowed to hold any dissenting opinions or behave differently, where everyone is always deemed a potential member of the 'dangerous classes'. The threat here is that this creates a situation where civil society is simply, and literally, impossible. Those concerned to develop a liveable urban existence might look to the imaginaries of the city's own concrete unconscious, what has been repressed, in order to provide alternative figurations of the kinds of sociability and sensibilities that allow the city's culturally diverse inhabitants to become the creative producers of their own urban spaces. The challenge is, as it always has been, to rescue civility from conformity.

References

Baudrillard, J. (1988) *America,* London: Verso.

Benjamin, W. (1999) *The Arcades Project,* Cambridge, MA: Harvard University Press.

Bondi, L. (1991) 'Gender Divisions and Gentrification: a critique', *Transactions of the Institute of British Geographers* new series, 16, 190–98.

Bondi, L. (1999) 'Gender, Class and Gentrification: enriching the debate', *Environment and Planning D: Society and Space* 17, 261–82.

Bourdieu, P. (1984) *Distinction: a social critique of the judgement of taste,* London: Routledge.

Brace, C.L. (1872) *The Dangerous Classes of New York and Twenty Years Work among Them,* New York: Wynkoop and Hallenbreck.

Butler, J. (1990) *Gender Trouble: feminism and the subversion of identity,* London: Routledge.

de Certeau, M. (1988) *The Practice of Everyday Life,* Berkeley: University of California Press.

Davidson, J. (2003) *Phobic Geographies: the phenomenology and spatiality of identity,* Aldershot: Ashgate.

Dixon, J., Levine, M. and McAulay, R. (2006) 'Locating Impropriety: Street Drinking, Moral Order, and the Ideological Dilemma of Public Space', *Political Psychology* 27 (2), 187–206.

Domosh, M. (1998) 'Those "Gorgeous Incongruities": Polite Politics and Public Space on the Streets of Nineteenth-Century New York City', *Annals of the Association of American Geographers* 88 (2), 209–26.

Elias, N. (1982) *The Civilizing Process,* Oxford: Blackwell.

Gardiner, M.E. (2000) *Critiques of Everyday Life,* London: Routledge.

Giuliani, R.W. (1998) 'The Next Phase of Quality of Life: Creating a More Civil Society', available at http://www.nyc.gov/html/rwg/html/98a/quality.html (accessed 4 August 2006).

Goffman, E. (1963) *Behaviour in Public Places: notes on the social organisation of gatherings,* London: Collier Macmillan.

Goode, E. and Ben-Yehuda, N. (1996) *Moral Panics: the social construction of deviance,* Oxford: Blackwell.

Gramsci, A. (1982) *Selections from the Prison Notebooks,* London: Lawrence and Wishart.

Hayden, D. (1997) *The Power of Place: urban landscapes as public history,* Cambridge, MA: MIT Press.

Jackson, P. (1998) 'Domesticating the Street: The Contested Spaces of the High Street and the Mall', in N. Fyfe (ed.), *Images of the Street: planning, identity and control in public space,* London: Routledge.

Kasson, J.F. (1990) *Rudeness and Civility: manners in nineteenth-century urban America,* New York: Hill and Wang.

Kracauer, S. (1995) *The Mass Ornament: Weimar essays,* Cambridge, MA: Harvard University Press.

Lefebvre, H. (1994) *The Production of Space,* Oxford: Blackwell.

Newman, K. and Wyly, E. (2005) 'Gentrification and Resistance in New York City', *Shelterforce Online* 142. Available at: http://www.nhi.org/online/issues/142/gentrification.html.

Philo, C. (1998) 'Animals, Geography, and the City: Notes on Inclusions and Exclusions', in J. Wolch and J. Emel (eds), *Animal Geographies: place, politics, and identity in the nature–culture borderlands,* London: Verso.

Simmel, G. (1903) 'The Metropolis and Mental Life', in D.L. Devine (ed.), *Georg Simmel on Individuality and Social Forms,* Chicago: University of Chicago Press.

Smith, N. (1996) 'Gentrification, the Frontier, and the Restructuring of Urban Space', in S. Fainstein and S. Campbell (eds), *Readings in Urban Theory,* Oxford: Blackwell.

Sweet, R.H. (2002) 'Topographies of Politeness', *Transactions of the Royal Historical Society* 6 (12), 355–74.

Valentine, G. (1998) 'Food and the Production of the Civilized Street', in N. Fyfe (ed.), *Images of the Street: planning, identity and control in public space,* London: Routledge.

Welsh, I. (1993) *Trainspotting,* London: Secker and Warburg.

Williams, R. (1985) *The Country and the City,* London: Hogarth Press.

Wintour, P. (2006) 'No More Misbehaving', *The Guardian,* 26 July 2006.

14 HOUSE AND HOME

Sarah L. Holloway

This chapter

○ Argues that urban scholars have often downplayed the significance of the domestic sphere

○ Focuses on the importance of the home as a material (physical) site, a space of social reporduction and a place where dominant cultural values can be reworked

○ Concludes that the conceptual division of public and private space is perhaps unhelpful, and insists that the home needs to be considered as an 'unbounded' space

---

Introduction

The themes of house and home are the poor relations of urban studies, generating less sustained interest over time than, say, the changing geography of the urban hierarchy, world cities, place marketing, segregation in the city and so on. In this respect, urban studies is but one element of a wider academic picture that privileges the supposedly public over the notionally private. Marston (2000), for example, makes the point that recent theorizations of scale, driven by an interest in capitalist accumulation, have focused on the global, the national and the urban, overlooking the home's role as a site of production, consumption and social reproduction. Notwithstanding this outsider status, progress has been made in recent decades and house and home are now gaining a place on the urban agenda.

This chapter reviews these developments and considers the ways in which urban theorists – including both those who might self-define in this way and others who might not but nevertheless write about urban areas – have challenged the orthodoxy that our interest in housing stops at the front door and have opened up the 'black boxes' of the house and the home. The chapter is hence divided into three sections, each of which examines a particular take on the notion of house or home. The chapter turns first to questions about the design and consumption of houses; the second section introduces debates about social reproduction in the home, focusing on both domestic labour and parenting cultures; and the final section considers notions of homeplace, and the possibility that the home might not only be a site of oppression but also a site of resistance.

Houses

Urban studies has had a longstanding interest in housing and its links with broader concerns about urban land markets and the development of urban form. The purpose of this chapter is not to review the healthy diversity or research on housing (which has been considered extensively elsewhere – see Knox and Pinch, 2000) but to focus instead on other questions about the ways people design and consume houses, questions which have gained increasing visibility over the past couple of decades.

The first scale at which an interest in housing design emerged in urban studies is at the neighbourhood level. Dolores Hayden's (1981, 1984) research on the design of American suburbs demonstrates different assumptions about family life and how they shape the production of different built environments. For example, in the introduction to *Redesigning the American Dream* (1984) she contrasts the development of Vanport City, Oregon, and Levittown, New York. Vanport City was designed to meet the needs of a wartime labour force, which included women as well as men, and people from diverse economic and ethnic backgrounds. It therefore sought to provide affordable rented housing that was designed to be energy efficient and low maintenance. It had very good support services, including maintenance crews for DIY jobs in the home, extensive childcare services at little cost to workers, and cooked food services. In contrast, Levittown, which was developed just after the war, was designed to house white working-class families: the scheme was not racially integrated, but houses were relatively inexpensive. The design of houses, each with their individual plot, emphasized privacy, and the size of the lot relative to the house meant there was room for extensions when the family grew, as well as for a private garden (see Figure 14.1). Men were expected to be breadwinners and maintain traditional gender roles at home. 'Levitt [the developer] liked to think of the husband as a weekend do-it-yourself builder and gardener: "No man who owns his own house and lot can be a Communist. He has too much to do," asserted Levitt in 1948' (Hayden, 1984: 7). Women were expected to be too busy raising children to want to go out to work and therefore childcare, support services and suitable transportation were not provided.

Hayden's (1981, 1984) research starts to demonstrate how ideas about how different types of people should live are written into the urban landscape by

Figure 14.1   Levittown, c. 1955

planners and developers. Research in a variety of different contexts shows that while these ideas make some styles of living more likely than others, they do not straightforwardly determine behaviour because different groups can choose to consume or use the city in unexpected ways. Indeed, historical research documents the ways in which the nascent women's movement mobilized positive understandings of women's role in the home in setting out to challenge and change wider urban and national processes and open hitherto closed spheres up to (middle-class) women (Walker, 1998; Marston, 2000; Mackintosh, 2005). However, a classic study by Duncan and Duncan (1997) demonstrates that 'people power', in which residents act together to shape the future of their urban environment, is not always progressive. In a study focusing on an affluent American landscape, Duncan and Duncan examined how ideas about the need to conserve aesthetically beautiful natural and historical landscapes came together with the ideology of possessive individualism (a belief in private ownership, democracy among equals, and local control) to legitimate planning policies which enabled rich residents to exclude less well-off groups from their neighbourhood (see Extract 14.1).

Extract 14.1: From Duncan, N.G. and Duncan, J.S. (1997) 'Deep suburban irony: the perils of democracy in Westchester County, New York', in R. Silverstone (ed.), *Visions of Suburbia*, London: Routledge, pp. 161–2, and 176.

We will focus primary attention on Bedford Village, which is among the most affluent towns in the county and whose landscape is maintained by some of the most exclusionary zoning practices anywhere in the US. Approximately 80 per cent of the town is zoned for single-family houses on a minimum of four acres, approximately 95 per cent for houses on one or more acres, and less than 1 per cent for two-family dwellings or apartments. ...

Bedford's residents are extraordinarily vigilant and at times aggressive in retaining and maintaining the dirt roads lined with dry stone walls. Conservationists protect its brooks, ponds and wetlands with great zeal and at great cost to the town; and historical preservation committees buy, maintain, restore, and/or closely regulate the most minute architectural details of the white wooden shops in the village centre.

Bedford's zoning (greatly compounded by the fact that much of the surrounding country is nearly as exclusive) causes hardship in the form of overcrowded, over-priced housing, much of which is located far from suburban and New York City jobs. It produced a segregation of rich and poor and as a result an inequality of public services, especially education. As one commentator put it: 'The exclusionary laws are not completely explicit: there are no zoning maps divided into racially or economically restricted areas, so labelled. But there are thousands of zoning maps which say in effect: "Upper-Income Here", "Middle to Upper Income Here", "No Lower-Income Permitted Except as Household Employees", "No Blacks Permitted"' (Reeves, 1974: 304).

Although the causes of inequities wrought by exclusionary zoning are largely structural, we found that they were reinforced by hegemonic ideologies subscribed to not only by those whose material interests are served by structural arrangements, but also by those whose interests are not.

...We suggest that there is an aestheticization of the suburban politics of exclusion focused on landscape. This aetheticization operates through appeals to both historical preservation and environmental conservation as well as the ideals of the pastoral and the picturesque. This aestheticization is effective, we argue, because the American ideology of individualism influences people to seek concrete, visual evidence of individual successes and failures; such evidence is often sought in the residential landscape. Residents of the most sought-after residential areas attempt to maintain and enhance the beauty (and hence prestige) of those areas by excluding new development, especially higher-density, less expensive housing. Appeals to the virtually universal values of history and environment allow the exclusion to proceed with support not only from the excluders, but from a significant proportion of the excluded as well. We refer to this as the aestheticization of politics.

This interest in the gender, class and racial assumptions which are written into the design and maintenance of the suburbs can also been seen at the level of individual houses. Hayden (1981, 1984) was herself interested in suburban architecture, and her work was complimented by a feminist design collective which examined the assumptions written into the designs of post-war British homes (Matrix, 1984). This field has seen a flurry of renewed interest in the early twenty-first century in 'the ways in which domestic architecture and design are inscribed with the meanings, values and belief that both reflect and reproduce ideas about gender, class, sexuality, family and nation' (Blunt, 2005: 507). Integral to this work has been an analysis of the links but also the breakages between design ideals and the ways people actually live their lives in these spaces (Blunt, 2005). Thus, for example, Bryden (2004) examines the gendered, religious and caste assumptions which were built into the traditional space of the courtyard house, or *haveli*, in Jaipur, India. Nevertheless, while she notes the principles of Vastu are written into the design of the building and appreciated by its residents, these families also design and use space within the building in their own ways.

Concern with the design and use of domestic architecture has been paralleled by a growth of interest in the ways in which people design their home interiors. Leslie and Reimer (2003) examine the rise of modernist furniture retailing and marketing in the 1990s, and examine its gendered incorporation into the home. By contrast, Clarke (2001) examines the importance of interior design in working-class homes in London and considers how presentation of the home acts as a projection of one's Self. Kelly, for example, is a 40-year-old, single mother with two school-aged children. She wants to dissociate herself from her traditional Jamaican working-class roots, and identifies herself as middle-class despite her low income and lack of qualifications. Kelly counters her unease at living in social housing by decorating her house according to formal design principles: her lounge, for example in monochrome, featuring a mirrored wall along with white paintwork, carpet and sofa. The room is now beginning to show wear and tear, and Kelly wants to revamp the room: she is currently seeking a new partner, and wants the room to reflect the image she projects of herself. Lola and Philipe moved to Britain from Chile 10 years ago: Lola is a full-time mother, Philipe an electrician. Their home decoration is centred on their children. The main living area has not been decorated since they moved in and is furnished with garden chairs, a makeshift table and an old television. By contrast, the children's bedrooms are like show homes: they blend European tastes with Chilean touches, such as laminate flooring, home carpentry and the use of bright colours. In both these families there is an enormous disparity between the great amount of thought and time given to how a place should look (as if it were constantly on show as part of the public sphere), and the minimal number of times it will actually be seen by an outsider. This leads Clarke (2001: 42) to conclude: '[t]he house objectifies the vision the occupants have of themselves in the eyes of others and as such it becomes an entity and process to live up to, give time to, show off to' (see also Tolia-Kelly, 2004).

A second development which has caused interest in the domestication of artefacts within the home has been the rise in new technologies. In the wider field of urban studies the rise of the information society has prompted interest in twin processes of centralization and decentralization in the urban system, as well as changes in spatial relations within cities (Castells, 1989; Kitchen, 1998). In relation to the home, mirror-imaged discourses abound in public and policy debate as these technologies are seen to be ushering in a great new era of home-centredness in the smart-house of the future, or conversely the end of contemporary urban life as Internet-connected homes will become disconnected from their local worlds (see Graham and Marvin, 1996; Gumpert and Drucker, 1998). Literature in the social studies of technology (Callon, 1987; Latour, 1993; Law, 1994; Star, 1995) suggests that both utopian and dystopian visions of the place of home in the wired world are likely to be overly simplistic. In this approach computers are not viewed as invariant objects with a predictable set of effects, but rather as 'things' that materialize for people in diverse social situations and which may, therefore, vary as much as the contexts in which they are used (Law, 1994; Bingham et al., 1999). Indeed, technology may play very different roles within different 'communities of practice' (Wenger, 1987) and so may emerge as a very different tool within them (Bingham et al., 1999). The implication is that the outcomes for different types of home can only be ascertained through empirical research. Extract 14.2 details one such empirical research project which explores the implications of children's ICT use in the urban West (see Adams, 1999, and McGrail, 1999, for examples focusing on homeworking and surveillance technologies).

Extract 14.2: From Holloway, S.L. and Valentine, G. (2001) 'Children at home in the wired world: reshaping and rethinking home in urban geography' *Urban Geography*, 22 (3), 562–83, pp. 578–9.

We have highlighted two intersecting scales at which we might assess how homes are being (re)shaped through children' domestic use of ICT. Firstly, our focus on the micro-geographies of computer location within the home highlights the different ways in which technology is domesticated in different households, and thus the varied consequences it has for social practices within and the meaning of home in different families. In some homes the availability of material resources, alongside parental beliefs that their children are capable of managing their own ICT usage, means children are able to use ICT in their bedrooms. This practice is positively endorsed by some parents who feel that that computer usage in the child's bedroom protects them from the dangers of, or dangerous use of, adult public space (cf. Livingstone et al., 1997). Equally, however, other families domesticate technology in other ways, defining the computer as a social machine and keeping it in a family room. Rather than seeing the child's bedroom with an on-line PC as a safe space, they see it as inherently risky, giving children access to unsuitable materials on-line and, more importantly, strangers on-line access to their children. Moreover, many families who keep a computer in a family room do so not only to protect children and the

(Continued)

privacy of individual family members' bedrooms, but also to ensure a social atmosphere surrounding computer use. The implications ICT has for the meaning of home – whether the home continues to be conceived of as a haven from outside dangers, or as a space which now needs to be protected from penetration by these outside dangers – is thus dependent on the processes of domestication. ...

The second scale at which we have considered the home builds on earlier work by 'critical geographers' (Hanson and Pratt, 1988; Moss, 1997) who suggest that our conception of home needs to be broadened outwards to include the neighbourhood and wider urban resources. In examining the importance of children's use of ICT for this more broadly conceived notion of home we show that, contrary to popular discourses ... these homes are not becoming dislocated from the locale, or more home-centred in a narrow sense. Rather, children mediate their use of ICT through their locally embedded social networks. For some children this means forming friends with other computer users, but for most it means integrating some use of ICT into their everyday activities. This is not to say that the home is not mediated in new ways. The use of on-line computers by some children does link their homes beyond the neighbourhood to the wider world, thus allowing them a broader diversity of experiences than they would previously have been exposed to at home. Interestingly, these on-line experiences can lead to the re-embedding of children's lives in the local, as the information gathered is used to inform local social practices. As such these 'cultures of computing' ... are neither simply global nor local, but reproduced through social relations which are not 'organised into scales so much *as constellations of temporary coherences*' (Massey, 1998, pp. 124–5, original emphasis).

Homemaking

The second 'take' on home this chapter examines is the conception of the home as a place of homemaking or, in academic terms, social reproduction. Researchers in urban studies have developed important insights into questions about work and production, ranging from studies of the importance of global economic change, through inter-city competition and place marketing, to the possibility of concerted action by the labour movement (see Chapter 6). Since the late 1980s feminist writers have insisted this focus on production needs to be mirrored by an interest in social reproduction, because work and home are interrelated rather than separate spheres (Hanson and Pratt, 1988). This relationship works at a number of interlocking levels. We can, for example, see the ways in which a home neighbourhood acts as a source of socialization and affects the job search strategy of individuals (Hanson and Pratt, 1995), as well as the ways some companies perceive suburban environments as low-cost locations, assuming there will be a plentiful supply of women workers who will accept lower wages as domestic responsibilities make it

Table 14.1   Couples' breadwinner type by women's occupation

Breadwinner type	Women's contribution(%)	All	Professional women	Clerical women	Manual women
Strong male breadwinner	<25	32	14	30	51
Moderate male breadwinner	25–<35	20	16	20	22
Weak male breadwinner	35–<45	22	22	29	16
Dual breadwinner	45–<55	17	30	16	6
Female breadwinner	55+	9	17	6	5

Source: Adapted from T. Warren (2003: 744)

difficult to travel further afield to find better paid jobs (England, 1993, 1995). Taking a step further into the family home, other research has suggested that women's attitudes to mothering and paid employment are in part shaped by, and shape, local moral geographies of mothering – locally-embedded ideas about the ways good mothers should care for their children (Dyck, 1996; Holloway, 1999). Jarvis et al. (2001) draw a range of diverse influences together as they talk about the work–life balance in cities, and critique models of sustainable city development which do not take seriously the ways in which real people in households combine home and work.

This interest in the links between home and work was an important step in opening up the home to further academic scrutiny. The changes witnessed in the labour market at the end of the twentieth and the beginning of the twenty-first century – including a broad-scale feminization of the labour force as well as a shift in job types away from manufacturing towards service sector employment (McDowell, 2001) – have further increased interest in the ways in which reproductive work is managed within the home. Warren (2003) examines patterns of productive and reproductive work in professional, clerical and manual dual-waged households. Table 14.1 shows that 73% of working-age, dual-waged couples could be classified as having some form of male-breadwinner arrangement, but that this falls to 52% when women are in professional employment compared to 89% where women are in manual work. For middle-class women in particular, these figures show a shift away from traditional dependence on male breadwinners.

Table 14.2 shows how couples in the same survey share caring work. The traditional pattern of female responsibility for caring work is still seen in 72% of households, but it is being shared or done by men in 28% of households. What is particularly interesting is the class divide in these patterns. Professional women share or are the major breadwinner in 47% of cases; however, only 36% share caring work or live in households where men are responsible for caring work. In

Table 14.2   Couples' caring work by women's occupation

Caring work: couple carers types	All	Professional women	Clerical women	Manual women
Female carers	72	65	72	79
Dual carers	26	34	26	19
Male carers	2	2	2	1

Source: Adapted from Warren (2003: 745)

contrast, while only 11% of women in manual work are the breadwinners or share breadwinning with their partners, 20% share caring or live in households where men are responsible for caring work. Warren argues that this reflects a greater use of split scheduling in these households: men and women maximized income and minimized childcare costs by women working, often part-time, in the hours their partners were not at work, resulting in men's greater involvement in childcare. By contrast, middle-class women have taken on greater financial responsibility within households without a similarly significant decline in their caring work.

One response among professional and managerial dual-earner couples to women's greater involvement in the labour market has been an expansion in the use of paid domestic workers such as cleaners, nannies and au pairs. In effect, work previously done by middle-class women is not shared with their partners but undertaken by a paid employee, usually working-class women. Gregson and Lowe (1994) trace the rise of these forms of work and the employment relations upon which they depend in the home. For example, looking at cleaning work in the home they found that cleaners have a considerable amount of autonomy in both their day-to-day practices (in which they were generally free to construct their own work routines, deciding what needs to be done and which jobs they would not do) and their employment terms and conditions (where they were free to specify when they would work and how dirty a house they would agree to clean). This autonomy, Gregson and Lowe concluded, was central to their main-tenance of self-respect while undertaking stigmatized employment. On some occasions these cleaners were treated as valued members of the family, perhaps because their work enabled other women to have more free time to themselves or to spend with their children; on other occasions they were seen as more dis-tant employees, not least because some cleaners worked while their employers were out and face-to-face encounters were thus rare. Notwithstanding that the growing employment of cleaners represents a major shift in the ways reproduc-tive work is being undertaken in the urban West, it provides only a limited chal-lenge to traditional gender ideology. Both the cleaners and the households employing them saw women as the employers, because cleaners were seen to be relieving middle-class women, not households, of work they would otherwise have had to do themselves (Gregson and Lowe, 1994).

Although Gregson and Lowe's study of domestic workers in the UK was largely underlain by a class analysis, Pratt's research in North America also points to the racialization of labour markets. Extract 14.3, for example, shows some of the stereotypes associated with Filipino and British nannies in Canada (see Pratt, 1997, for further details). The material consequences of such stereotyping is that British nannies command higher wages, while their Filipino counterparts are required to undertake more cleaning alongside their childcare work.

Extract 14.3: From Pratt, G. (1997) 'Stereotypes and ambivalence: the construction of domestic workers in Vancouver, British Columbia' *Gender, Place and Culture* 4 (2), 159–77, pp. 162–3.

The distinction between European and Filipina was made by all agents and they were remarkably obliging in clarifying the profiles, and the strengths and weaknesses of each (even if there was some disagreement on what counted as a strength or weakness). European nannies are constructed as professionals; Filipinas as servants. One agent expressed this quite graphically; the same point was made by almost all agents:

AGENT G: Depend on what you are looking for, what you want. My personal view, if you have a baby and you want someone to lick your house clean: Filipino girl. Go for that.

PRATT: [Paraphrasing a common stereotype] Because they'll love the baby?

AGENT G: That's right. If you have kids 3, 4 years of age and up, and you want interaction, you want them to go to the park, you want them to play arts and crafts, do thing, you're better off with a European.

PRATT: Because the traditions are …

AGENT G: Intellectual level, communication level, openness, a lot of things. Your average Filipino girl is a quiet, shy personality. She does her job and that's the most important. The house has to be clean, spotless when God's coming home, sorry, the parents are coming home. Every … the kids, they come second. No! I want the kids to come first. The house could be second.

In general, the stereotypes of Filipinas expressed by agents in Vancouver in 1994 are remarkably similar to those deployed by Americans, as colonisers in the Philippines from 1898 to 1946. … A common stereotype in the early twentieth century was of the Filipino as passive mimic, incapable of independent thought (Rafael, 1993; Doty, 1996). As Doty observes, although stereotypes are instantiated in particular contexts, they draw upon globalised representations, and histories of previous stereotypes are sedimented within contemporary ones. In contrast to Filipinas, much was made of British nannies' credentials, especially the 2-year nanny 'degrees':

(Continued)

AGENT D: I mean they're top of the line nannies. Norland nannies are actually better than NNEB[1]. When they come out of Norland, they have a uniform with a number on it. That's their official Norland kind of...

PRATT: But are there jobs for them in England?

AGENT C: And they're good jobs for that class of nanny, that educated. So they're not thrilled to come here. And when they do come, they're quite particular about what they want to do.

AGENT D: They're actually quite nice. Provided you put them in the right job. But the majority of Canadians, whether you've got money or you haven't, are really down to earth. And if you're replacing mommy in the home, you had better be able to stir fry the vegetables and water flowers. You know what I mean? That, that sort is what we're looking for more and more. Not slaves. They don't want slaves.

Aside from drawing attention to the way that agents acknowledge the educational credentials of British nannies, the dynamics of class and colonialism (in this case between Britain and Canada) are also at work in this conversation, with Canadians, regardless of income, portrayed as down-to-earth.

This debate about responsibilities for domestic work, including cleaning and the care of children, are highly Western in nature. Much less research is available concerning caring work in the home in the global South. Where it is available it provides a stark reminder of global inequalities across urban environments. Robson (2000), for example, focuses on the caring work undertaken by some young people in urban Zimbabwe. In the context of an AIDs pandemic, and structural adjustment policies which have forced reductions in health spending, the issue facing some young people is not whether their mother or some other woman will care for them and clean the home, but rather their own role as carers. Lack of access to healthcare services means some young people – most often those from low-income backgrounds, those in lone-parent households, and female rather than male siblings – will find themselves providing personal healthcare and undertaking significant levels of domestic work, activities which often force their withdrawal from school.

In the West, the interest in the organization of domestic labour in the home has been paralleled by research focusing on the cultures surrounding the care of children. Valentine (1997), for example, examines how ideas about good mothering and fathering shape the way parents allow their children to use public space. She argues that since the nineteenth century motherhood has been glorified as women's chief vocation; and this idealized picture was played out in practice as the women in her study retained primary responsibility for childcare. In the context of fears over stranger danger and road traffic accidents, it was mothers who took the central role in keeping their children safe during the day. They were, for example, chiefly responsible for ferrying children to and from school and particularly organized play activities which are seen as a safe and constructive alternative to outdoor play. This role means they are on the frontline when determining children's spatial range: they are not only best placed to judge their children's growing competence, as the ones who enforce the rules they are also the first to come under pressure from children to allow them more freedom. Valentine demonstrates that mothers' decisions are not solely made within the context of the family, but are also shaped by local parenting cultures. Although an individual mother may ultimately decide to go her own way, she must nevertheless negotiate powerful local social norms about what children should and should not be able to do at different ages. Fathers in Valentine's study, by contrast, usually claimed a more authoritarian role in policing rather than setting boundaries, but in some households this role also remained the mother's responsibility. These debates about local moral geographies of mothering or parenting cultures have been extended in other research which examines attitudes to preschool education and how mothers access care that allows them leisure time away from home and children (Holloway, 1998a, 1998b).

Homeplace

The third 'take' on the home which this chapter examines is the notion of home as homeplace. Researchers in urban studies have expended considerable effort on researching social difference within the city, in particular looking at minority group's experiences of migration, settlement, segregation and division (see Chapters 11 and 12 for an overview of this work). This section takes this interest in minority groups and considers their position within the space of the home.

The earliest 'minority' group to attract attention in the home were women. By the 1960s and 1970s feminists were beginning to point out women's responsibility for domestic labour and childrearing (discussed above) was leaving many feeling isolated in the home with feelings of guilt that they were not wholly fulfilled by their role as housewives (Friedan, 1963; Gavron, 1966). This second-wave feminism recast the home not as a haven but as a site of oppression. This naming of the problem was an important move. However, it was not a problem shared equally by all women, but rather was a concern of white, middle-class housewives. Full-time mothering in the home was not an option many working-class women could afford (McDowell, 1999), and, as bell hooks pointed out, the home for black women was as much a site of resistance as a site of oppression (see Extract 14.4). She argues that black women have been important in transforming homeplaces into spaces of care for black children living in white racist societies, a space which could affirm their identity. She thus sees black women's role in the home as potentially politically subversive rather than simply a reflection of their oppression as women.

Extract 14.4: From hooks, bell (1991) *Yearning: Race, Gender and Cultural Politics*, Boston: South End Press, pp. 42, 46.

Their lives were not easy. Their lives were hard. They were black women who for the most part worked outside the home serving white folks, cleaning their houses, washing their clothes, tending their children – black women who worked in the fields or in the streets, whatever they could to make ends meet, whatever was necessary. Then they returned to their homes to make life happen there, This tensions between service outside one's home, family, and kin network, service to white folks which took time and energy, and the effort of black women to conserve enough of themselves to provide service (care and nurturance) within their own families and communities is one of the many factors that has historically distinguished the lot of black women in patriarchal white supremacists society from that of black men. Contemporary black struggle must honor this history of service just as it must critique the sexist definition of service as women's "natural" role.

[...She continues later] In our family, I remember the immense anxiety we felt as children when mama would leave our house, our segregated community, to work as a maid in the homes of white folks. I believe that she sensed our fear, our concern that she might not return safe to use, that we could not find her (even though she always left phone numbers, they did not ease our worry). When she returned home after working long hours, she did not complain. She made an effort to rejoice with us that her work was done, that she was home, making it seem as though there was nothing about the experience of working as a maid in a white household, in that space of Otherness, which stripped her of dignity and personal power.

Looking back as an adult woman, I think of the effort it must have taken for her to transcend her own tiredness (and who knows what assaults or wounds to her spirit had to be put aside so that she could give something to her own). Given the contemporary notions of 'good parenting' this may seem like a small gesture, yet in many post-slavery black families, it was a gesture parents were often too weary, too beaten down to make. Those of us who were fortunate enough to receive such care understood its value. Politically, our young mother Rosa Bell, did not allow the white supremacist culture of domination to completely shape and control her psyche and her familial relationships. Working to create a homeplace that affirmed our beings, our blackness, our love for one another was necessary resistance. We learned degrees of critical consciousness from her. Our lives were not without contradictions, so it is not my intent to create a romaticized portrait. Yet any attempts to critically assess the role of black women in liberation struggle must examine the way political concern about the impact of racism shaped black women's thinking, their sense of home, and their modes of parenting.

Since hooks's intervention the notion that the home might be a site of resistance as well as oppression has been considered more broadly in urban studies. Johnston and Valentine (1995), for example, explore the role of the home in same-sex families and Hyams (2003) examines mid-teenager Latinas in financially deprived neighbourhoods of Los Angeles. Home, in terms of neighbourhood, does give these young women self-esteem as they think the wild reputation of their run-down neighbourhood makes them look cool in the eyes of others. However, their ability to negotiate this neighbourhood independently is limited by fairly strict parental control and their own concerns about age-inappropriate male gaze. The home itself is a place of labour for many of these young women, who are heavily involved in domestic labour in a way that is not the case for male siblings. In terms of homeplace, Hyams concludes that a minority of these young women find the family home an affirming space (as hooks suggests), and the majority have to try to carve a special place out within it away from the rest of their families, and in overcrowded homes, some of the teenagers are only able to find this freedom in their minds.

These studies are exploring notions of home among those with access to permanent housing. However, as Daly makes clear in Chapter 15, homelessness is also a considerable problem in the city. Beazley (1999) explores the ways in which homeless children in Java, many of whom are rural-to-urban migrants, create a sense of home for themselves while living on the city streets. She demonstrates that understandings of children's home villages remain important to them, but they also create an alternative family of other children to provide feelings of membership and social support in the face of harassment by the authorities. Tikyan subculture, which involves some visual signs such as earrings and tattoos and cultural practices such as drug taking and glue sniffing, are an important aspect of this, giving them a feeling of belonging. Once living and working within the city, children identify with different territories (a road, a bus terminal, etc.) and regard these places as safe spaces compared with the rest of the city, as safe at least as their original homes. As Beazley (1999: 198) argues, 'by creating these feelings of security and belonging, such spaces have, in a sense, become their alternative homes'.

Summary

This chapter has reviewed the ways in which ideas about houses, homemaking and homeplace have been opened up for academic enquiry within urban studies. Focusing first on houses, the chapter examined how older interest in housing design has been mirrored in more recent years by a growing interest in home consumption. Feminist interests in social reproduction have challenged urban theorists to take seriously the ways in which daily and generational reproduction is achieved in the space of the home. Finally, critiques of the home as a site of women's oppression have raised a broad-ranging interest in the home as a potential site for the politics of resistance.

These foci are deliberately diverse, as the chapter has sought to illustrate some of the ways urban theorists might engage with the home (there are undoubtedly other avenues which could also be pursued). However, there is one common thread running through these accounts and this is the conception of home as open rather than bounded, being influenced by and influencing a wider variety of processes urban theorists might more commonly study at broader spatial scales. In advocating that urban studies embraces the home, then, the chapter is not advocating small-scale isolated studies, but rather a broader engagement with issues in which we already have a long track history of study. It is only by including houses and home we can produce a more fully urban, urban studies.

Note

National Nursery Examination Board. This qualification indicates that the person successfully completed a two-year full-time course covering ages 0–7.

References

Adams, P. (1999) 'Bringing globalization home: a homeworker in the information age', *Urban Geography* 20 (3), 356–376.

Beazley, H. (1999) 'Home sweet home? Street children's sites of belonging', in S.L. Holloway and G. Valentine (eds), *Children's Geographies: Playing, Living, Learning*, London: Routledge.

Bingham, N., Holloway, S.L. and Valentine, G. (1999) 'Where do you want to go tomorrow? Connecting children and the Internet', *Environment and Planning D: Society and Space* 17, 655–672.

Blunt, A. (2005) 'Cultural geography: cultural geographies of home', *Progress in Human Geography* 29 (4), 505–515.

Bryden, I. (2004) 'There is no outer without inner space: constructing the *haveli* as home', *Cultural Geographies* 11 (1), 26–41.

Callon, M. (1987) 'Society in the making: the study of technology as a tool for social analysis', in W. Bikker, P. Hughes and T. Pinch (eds), *The Social Construction of Technological Systems*, Cambridge, MA: MIT Press.

Castells, M. (1989) *The Informational City: Information Technology, Economic Restructuring, an the Urban-Regional Process*, Oxford: Blackwell.

Clarke, A.J. (2001) 'The aesthetics of social aspiration', in D. Miller (ed.), *Home Possessions: Material Culture behind Closed Doors*, Oxford: Berg.

Doty, R.L. (1996) *Imperial Encounters*, Minneapolis: University of Minnesota Press.

Duncan, N.G. and Duncan, J.S. (1997) 'Deep suburban irony: the perils of democracy in Westchester County, New York', in R. Silverstone (ed.), *Visions of Suburbia*, London: Routledge.

Dyck, I. (1996) 'Mother or worker? Women's support networks, local knowledge and informal childcare strategies', in K. England (ed.), *Who Will Mind the Baby? Geographies of Child Care and Working Mothers*, London: Routledge.

England, K. (1993) 'Suburban pink collar ghettos: the spatial entrapment of women?', *Annals of the Association of American Geographers* 83 (2), 225–242.

England, K. (1995) ' "Girls in the office": recruiting and job search in a local clerical market', *Environment and Planning A* 27 (10), 1995–2018.

Friedan, B. (1963) *The Feminine Mystique*, London: Gollancz.

Gavron, H. (1966) *The Captive Wife: Conflicts of Housebound Mothers*, London: Penguin.

Graham, S. and Marvin, S. (1996) *Telecommunications and the City: Electronic Spaces, Urban Places*, London: Routledge.

Gregson, N. and Lowe, M. (1994) *Servicing the Middle Classes: Class, Gender and Waged Domestic Labour in Contemporary Britain*, London: Routledge.

Gumpert, G. and Drucker, S.J. (1998) 'The mediated home in the global village', *Communications Research* 25 (3), 422–438.

Hanson, S. and Pratt, G. (1988) 'Reconceptualising the links between home and work in urban geography', *Economic Geography* 64 (4), 299–318.

Hanson, S. and Pratt, G. (1995) *Gender, Work and Space*, London: Routledge.

Hayden, D. (1981) *The Grand Domestic Revolution: A History of Feminist Designs for American Homes, Neighborhoods and Cities,* Cambridge, MA: MIT Press.

Hayden, D. (1984) *Redesigning the American Dream: The Future of Housing, Work and Family Life*, New York and London: W.W. Norton and Co.

Holloway, S.L. (1998a) ' "She lets me go out once a week": mothers' strategies for obtaining "personal" time and space', *Area* 30 (3), 321–330.

Holloway, S.L. (1998b) 'Local childcare cultures: moral geographies of mothering and the social organisation of pre-school education', *Gender, Place and Culture* 5 (1), 29–53.

Holloway, S.L. (1999) 'Mother and worker?: the negotiation of motherhood and paid employment in two urban neighbourhoods', *Urban Geography* 20, 438–460.

Holloway, S.L. and Valentine, G. (2001) 'Children at home in the wired world: reshaping and rethinking the home in urban geography', *Urban Geography* 22 (3), 562–583.

hooks, bell (1991) *Yearning: Race, Gender and Cultural Politics.* Boston: South End Press.

Hyams, M. (2003) 'Adolescent Latina bodyspaces: making homegirls, homebodies and homeplaces', *Antipode* 35 (3), 536–558.

Jarvis, H., Pratt, A.C. and Wu, P.C.-C. (2001) *The Secret Life of Cities: The Social Reproduction of Everyday Life*, Harlow: Prentice Hall.

Johnston, L. and Valentine, G. (1995) 'Wherever I lay my girlfriends, that's my home: the performance and surveillance of lesbian identities in domestic environments', in D. Bell and G. Valentine (eds), *Mapping Desire*, London: Routledge.

Kitchin, R. (1998) *Cyberspace: The World in the Wires*, Chichester: John Wiley.

Knox, P. and Pinch, S. (2000) *Urban Social Geography*, Harlow: Prentice Hall.

Latour, B. (1993) *We Have Never Been Modern*, London: Harvester Wheatsheaf.

Law, J. (1994) *Organising Modernity*, Oxford: Blackwell.

Leslie, D.A. and Reimer, S. (2003) 'Gender, modern design and home consumption', *Environment and Planning D: Society and Space* 21 (2), 293–316.

Livingstone, S., Gaskell, G. and Bovill, M. (1997) 'Europäische Fernseh-Kinder in veränderten Medienwelten', *Television* 10 (1), 4–12.

Mackintosh, P.G. (2005) 'Scrutiny in the modern city: the domestic public and the Toronto Local Council of Women at the turn of the twentieth century', *Gender, Place and Culture* 12 (1), 29–48.

Marston, S.A. (2000) 'The social construction of scale', *Progress in Human Geography* 24 (1), 219–242.

Massey, D. (1998) 'The spatial construction of youth cultures', in T. Skeleton and G. Valentine (eds), *Cool Places: Geographies of Youth Cultures*, London: Routledge.

Matrix (1984) *Making space: Women and the Man-made Environment*, London: Pluto Press.

McDowell, L. (1999) *Gender, Identity and Place: Understanding Feminist Geographies*, Cambridge: Polity Press.

McDowell, L. (2001) 'Father and Ford revisited: gender, class and employment change in the new millennium', *Transactions of the Institute of British Geographers* 26, 448–464.

McGrail, B.A. (1999) 'Communication technology and local knowledges: the case of "peripheralized" high-rise housing estates', *Urban Geography* 20 (2), 303–333.

Moss, P. (1997) 'Negotiating spaces in home environments: older women living with arthritis', *Social Science and Medicine*, 45, 23–33.

Pratt, G. (1997) 'Stereotypes and ambivalence: the constructions of domestic workers in Vancouver, British Columbia', *Gender, Place and Culture* 4 (2), 159–177.

Reeves, R. (1974) 'The battle over land', in L. Masotti and J. Hadden (eds), *Suburbia in Transition*, New York: The New York Times Company.

Rafael, V.L. (1993) 'White love: surveillance and nationalist resistance in the US colonization of the Philippines', in A. Kaplan and D.E. Pease (eds), *Cultures of United States Imperialism*, Durham, NC, and London: Duke University Press.

Robson, E. (2000) 'Invisible carers: young people in Zimbabwe's home-based healthcare', *Area* 32 (1), 59–69.

Star, S. (ed.) (1995) *Cultures of Computing*, Oxford: Blackwell.

Tolia-Kelly, D. (2004) 'Locating processes of identifications: studying the precipitates of re-memory through artefacts in the British Asian home', *Transactions of the Institute of British Geographers* 29 (3), 314–329.

Valentine, G. (1997) ' "My son's a bit dizzy." "My wife's a bit soft": gender, children and cultures of parenting', *Gender, Place and Culture* 4 (1), 37–62.

Walker, L. (1998) 'Home and away: the feminist remapping of public and private space in Victorian London', in R. Ainsley (ed.), *New Frontiers of Space, Bodies and Gender*, London: Routledge.

Warren, T. (2003) 'Class- and gender-based working time?: time poverty and the division of domestic labour', *Sociology* 37 (4), 733–752.

Wenger, E. (1987) *Communities of Practice*, Cambridge: Cambridge University Press.

15   HOUSING AND HOMELESSNESS

Gerald Daly

This chapter

O   Shows that homelessness is a social problem that is particularly pronounced in cities

O   Reviews literature that demonstrates the urban causes and consequences of homelessness in different national contexts

O   Demonstrates that urban scholarship and activism can help reduce homelessness and mitigate against its most negative consequences

---

Introduction

While not exclusively an urban phenomenon, homelessness is unquestionably a more visible and pressing concern in cities than in rural areas. Accordingly, it has been the focus of urban research from a variety of perspectives for many years. One purpose of this chapter is to explore these different approaches in order to demonstrate the validity of multi-disciplinary analyses of complex urban problems.

Homelessness is a fluid concept. People move from shelter to the streets frequently. Some spend much of their time in transition from one phase to another, while others shift physically among several locations. As a result, it is essential to find a suitable definition to fit a highly diverse population and a phenomenon with multiple causes. Related to this is the cause–effect dilemma. With respect to alcohol and drugs, for example, there is an ongoing debate about whether they are a cause of homelessness or whether they come into play once people are forced to live on the street? This difficulty in differentiating between cause and effect is inherent in much of the research. Some observers explain homelessness by emphasizing individual responsibility over structural factors. Others point to the pervasive influence of such systemic factors as poverty, unemployment and globalization (Burrows et al., 1997) One's stance on this problem will affect, in turn, the question of appropriate responses. How much should be done by government and social service agencies to assist homeless people? To what extent should reliance be placed on self-help initiatives? While it is important to intervene positively and in timely fashion, many homeless individuals stress the need to recognize they are not merely passive or helpless and that assistance emanating from a charity model or a sense of moral duty often fosters dependency.

This quandary highlights another key issue. How we respond to homelessness depends on how we perceive those without shelter. The Self/Other dichotomy which informs understandings of the homed/homeless affects our notion of what constitutes suitable policies or programmes. Increasingly, urban researchers take the position that, to fully illuminate these problem areas, one must examine the situated knowledge possessed by homeless people. In other words, it is essential to observe the experience of homelessness through the eyes of people without shelter in order to understand the complexity of homelessness, their lives on the street, and how they are affected by the policy decisions of others (Liebow, 1993; Bridgman, 2003).

Because homelessness is such a difficult issue, built on shifting policy ground, it is essential to begin by discussing the key issue of how homelessness has been defined, and how changing definitions affect policy responses. This is followed by a brief analysis of causal factors and the roles of government and the social service industry. Within this context it is then possible to describe the experiences of homeless people, focusing on *survival strategies* and the important role of *situated knowledge*. The final section of the chapter examines research approaches taken over the past few decades by academics and practitioners seeking to arrive at a fuller understanding of homelessness.

Defining homelessness

Definitions of homelessness reflect different purposes, values, ideologies, and political agendas. Narrow demarcations ignore people who are not on the streets or in emergency shelters. When responses concentrate on the vulnerability and helplessness of the client population, their potential political power is devalued while their dependence on the social service system is encouraged. The nature of homeless populations, the causes of their condition, and their capacity for self-help have been widely discussed by urban researchers. These are important issues because, in many instances, descriptive labels unfairly categorize individuals and simplify across a varied range of experiences.

Language, then, is instrumental in the social construction of reality. Language is political. It is revealing of how we look at issues, and the way we use language serves to convey society's messages of authority. The life stories of homeless individuals may be silenced as a result. These processes serve not just to define social order and to set the political agenda, but also to marginalize, to devalue, and to hold 'the other' at arm's length.

Because so many people who are inadequately sheltered will lose their housing for at least a short time, most urban researchers believe that homelessness must be broadly construed to include those at risk. As interpreted in policy and programmes, definitions determine the amount and type of aid provided and to whom. Hutson and Clapham (1999) note that narrow constructions of homelessness attribute causes to individual weaknesses or pathologies; hence, the response is to see these as private troubles most appropriately dealt with as welfare cases. Homelessness, however, also may be viewed as a socially constructed phenomenon, a result of systemic influences or structural defects that must be dealt with by the state, and assistance provided as of right.

People are generally acknowledged to be homeless if they lack adequate shelter in which they are entitled to live and can do so safely. At the extreme, they are sleeping rough. Others live under a roof but their accommodation is deficient in terms of security or basic amenities (i.e. heat, lighting, water, bathroom). In other instances, the homeless are forced to share space with family, friends or relatives, and lack basic privacy and living space. Homelessness may be experienced by individuals from a variety of social backgrounds, and deprivation depends on the extent to which the absence of shelter is combined with social exclusion, isolation and economic poverty (Sibley, 1995; Pleace, 1998).

Homeless people may accordingly be described in a variety of ways (e.g. demographic and familial characteristics, economic status, housing history, health problems, degree of disability, work record). Often homelessness is characterized in terms of presumed causes or precipitating problems, accidental homelessness (resulting from natural disasters or exogenous events), structural (relating to poverty, for example), economic (unemployment resulting from deindustrialization), political (refugees), or social (discrimination).

People also may be described in terms of the duration of homelessness or their degree of vulnerability, ranging from those who are absolutely or chronically without shelter, individuals who are episodically homeless (e.g. migrant workers, street youth, or women fleeing domestic abuse), those who are temporarily homeless as a result of an extraordinary event (sudden unemployment, severe health problem, death of household head, eviction), through to those individuals who are vulnerable or at risk of homelessness because of poverty or indebtedness.

Causes of homelessness

The inability to separate results from causes leads to the mystification of homelessness and diverts attention from underlying structural issues. By creating a form of analysis that hides the cause of problems, this process entrenches the status quo. As soon as the problem is mystified, it can be denied or dismissed as unwieldy or even intractable. How else can we explain the persistence of homelessness, how a social service industry has evolved and grown around this issue, how, despite widespread media coverage, there has been no social transformation and very little long-term success in eradicating this social ill (Hopper, 2003)?

While it is difficult to distinguish between root causes and precipitating events, or between individual factors and global or structural trends, urban researchers have explored the causal factors of homelessness. These include globalization and urbanization (Sassen, 2001), structural economic changes, poverty and deinstitutionalization. In many cities these structural shifts have ushered in a new era marked by increased immigration, high unemployment, movement of industrial jobs to the developing world, the rise of a large low-paid service sector, and privatization of social services (Hutson and Liddiard, 1995; Burrows et al., 1997; Fitzpatrick, 2005).

The concentration of global capital in large urban centres has altered the relationship between capital and labour. The nature of work has been changed fundamentally, polarizing the workforce – relatively prosperous knowledge workers at one end of the spectrum, poorly paid temporary or part-time contract workers at the other. One can expect to find a high concentration of women, immigrants and visible minorities in these poorly paid positions, which are usually minimum wage, dead-end jobs without benefits. Urbanization, then, can be held accountable in large part for the decline of full-time, secure jobs, replaced by a patchwork of precarious employment for those who lack choice (Vosko, 2000).

This pattern of residualization is mirrored in the housing market. As a result of global economic change and uneven development, local housing markets are characterized by reduced shelter options. An acute shortage of affordable rental units has ensued, along with declining social benefits relative to the cost of living, an inability to pay for available housing, and a rapid increase in evictions. Facilitating this process are public allocation policies driven by politicians determined to cut social expenditures. In recent years, government housing policies in many Western countries have emphasized privatization and tax subsidies for middle- and upper-income home-owners at the expense of social housing. The problem of homelessness, then, is directly linked to social policy and political decisions regarding public spending for shelter and social programmes (Wallace, 1990; Metraux and Culhane, 1999). These decisions, in turn, are tied to social values and choices.

Globalization's pernicious effects are evident as well in the form of spatial and class disparities (Takahashi et al., 2002). NIMBYism and gentrification in downtown districts (Lyon-Callo, 2001) led to the loss of low-income housing, rooming houses, and single-room occupancy (SRO) units and the exclusion of homeless people from parts of the city (Franck, 1991); these were accompanied by residential displacement and the conversion of rentals to condominiums. Not coincidentally, these trends have occurred in tandem with evictions and a rise in homelessness. Neil Smith (1996) hence refers to the process of 'conquest' inherent in urban policies which displace low-income residents, who are deemed disposable (see Extract 15.1).

Extract 15.1: From MacLeod, G. (2002) 'From urban entrepreneurialism to a "revanchist city"? On the spatial injustices of Glasgow's renaissance', *Antipode* 34 (3), 602–24, pp. 605–6.

Spurred on by the unrelenting pace of globalization and the entrenched political hegemony of a neoliberal ideology, throughout the last two decades a host of urban governments in North America and Western Europe have sought to recapitalize the economic landscapes of their cities. While these 'entrepreneurial' strategies might have refueled the profitability of many city spaces across the two continents, the price of such speculative endeavor has been a sharpening of socioeconomic inequalities alongside the institutional displacement and 'social exclusion' of certain marginalized groups. One political response to these social geographies of 'actually existing neoliberalism' sees the continuous renaissance of the entrepreneurial city being tightly 'disciplined' through a range of architectural forms and institutional practices so that the enhancement of a city's image is not compromised by the visible presence of those very marginalized groups. For some scholars, these tactics are further spiked with a powerful antiwelfare ideology, a criminalization of poverty, rising levels of incarceration, and a punitive or 'revanchist' political response. ... The fragile maintenance of value inscribed into this recommodification of space is ever more intricately dependent on a costly system of surveillance – performed through a blend of architectural design, CCTV, private 'acceptable' patterns of behavior commensurate with the free flow of commerce and the new urban aesthethics.

...The new urban glamour zones conceal a brutalizing demarcation of winners and losers, included and excluded. Indeed, in some senses we might speculate that the lived spaces of the neoliberal city symbolize an astonishingly powerful geographical expression of the erosion of Keynesian ideals of full employment, integrated welfare entitlement, and 'social citizenship' – not least in that when compared to many earlier rounds of municipal investment, which sought to engineer projects aimed at a 'mass public,' the new initiatives appear to be 'reclaiming' public spaces for those groups who possess economic value as producers or consumers to the virtual exclusion of the less well-heeled.

A parallel trend to gentrification is deinstitutionalization, predicated on a conviction that large psychiatric hospitals failed to provide tangible benefits to patients with mental illness, to the discovery of psychotropic drugs which allowed for treatment outside institutions, and to an optimistic belief in the compassionate capacity of the community to care for those released from institutions (Krieg, 2001). This concept took hold in part because it appealed to governments intent on containing healthcare costs. However, the savings realized by the closure of institutions were not redirected to special-needs housing or to community care. Moreover, deinstitutionalization is grounded on misleading notions that patients would, on release, become self-sufficient. In practice, patients released (or potential patients who are refused admittance to institutional care) are cast adrift without adequate supports. Many find themselves homeless or relegated to emergency shelters or hostels.

Care in the community depends on family and friends as well as a well-resourced social service sector. However, the social service industry – including public officials, consultants and representatives of non-profit agencies – often does not include the presumed beneficiaries. Because they lack a collective voice and are not organized, individuals on the street are represented by proxies whose interests may be self-serving. These relationships often constitute a control system based on a charity model and on naïve assumptions about the need to dictate terms to the recipient population. A self-perpetuating network, characterized by common interests, mutual dependencies and benefits, it has fashioned a web of interdependent communities based on self-interest. While most social service providers are well intentioned, they are, nevertheless, motivated by a desire to exert control over the purse strings, policy-setting and the allocation of resources. Ultimately, they decide who will be helped and who will not (Daly, 1997). In his research on women in shelters, Eliot Liebow commented: 'It is difficult to appreciate the intensity of feeling, the bone-deep resentment that many of the women felt at always having to answer questions, often very personal, and often the same ones, over and over again. But having to answer questions was part of the price they paid for being powerless' (Liebow, 1993: 137).

The experiences of homeless people

A number of urban researchers have examined the ways society categorizes and responds to homeless individuals. Rather than describing them as disaffiliated, some suggest it is more appropriate to examine the notion of social exclusion as it is applied to individuals on the street. These people are consigned to the periphery of public consciousness because, by failing to conform, they are seen to violate social norms and offend public sensibilities. As a consequence, homed populations may deal with them by strategies of distancing designed to minimize or displace feelings of resentment, fear, contempt, guilt, shame or conflict. In doing so, a cycle of disinterest is generated, allowing mainstream society to shun collective responsibility (Daly, 1998: 124–125).

The principal behaviours of street people are designed to ensure survival. Their strategies may appear spontaneous, random, contradictory, illogical, or even self-destructive. But judgements of this sort fail to comprehend the imperative of adaptation in a world that does not conform to dominant conceptions of normality. The environment of homeless people, while apparently characterized by disorder and chaos, has an order of its own. Frequently, individuals sleeping rough lack the power to control events that affect their lives. Chance plays a major role. The system within which they operate is not in equilibrium. They are powerfully influenced by external forces, and seemingly small events can become amplified, often with devastating consequences. When urban researchers describe this world they need to define reality in situational terms. An open, contextual and dynamic conceptualization of street reality – situated knowledge – is needed; one that is grounded in an attempt to understand the coping strategies created by homeless persons; one that reveals how they manage to live with isolation, anxiety, frustration and rejection (Koegel, 1992; Sibley, 1995).

For people without housing, days may be marked by endless walking and waiting. They keep moving, often at the command of police, shelter operators or store-owners. They wait: for shelter, food, welfare benefits, healthcare, and for short-term jobs. Homeless individuals are victims of harassment and violence because they are vulnerable, often alone, and lack political clout or even civil rights (Smith, 1996). Public indifference ensures that much of this crime goes unreported. Chronically homeless individuals, many of them elderly, disabled or malnourished, are at risk of crime. Both heterosexual and homosexual rape are common. Residents of emergency shelters or hostels also may be the perpetrators of crime, preying on other shelter occupants who are defenceless.

Those who are able seek alternatives to the shelters. Parks are used, if available, as are beaches, woods, ravines, alleys, vehicles, abandoned buildings, train tunnels, storage containers and spaces under highways or bridges. Single homeless individuals in particular depend on social networks. Because of their single status, however, many are marginalized and suffer from social exclusion (Pleace, 1998). They move frequently, but remain within a relatively circumscribed area of the city, defined by the location of such institutions as hostels, drop-ins, food banks, soup kitchens, rooming houses, and casual labour pools (Wolch and Dear, 1993). These areas typically have a concentration of retail outlets used by street people (coffee shops, beer and liquor stores, pawn shops, and pubs or taverns), public facilities (libraries, transit, public baths) and support services (hospitals, street health clinics, counselling and detoxification centres, methadone clinics, welfare and employment offices). People on the street supplement their incomes by piecework, by panhandling, selling newspapers, recycling cans or bottles and, in some cases, by prostitution.

Women and men differ significantly in the nature of their networks, activity patterns, and the ways in which they use the city. Men spend more time alone and on the streets, while at least half of homeless women are accompanied by children. Typically, shelters or hostels for men require they leave early each day and they are not allowed to return until late afternoon. Women's refuges generally permit them to remain within the sheltered surroundings for longer stays; they provide a number of services under one roof, while traditional men's hostels offer only space for sleeping – although some also provide meals and bathing facilities (Bridgman, 2003). Men range more freely throughout their particular part of the city, hanging out at drop-ins or on street corners, and staking out turf at specific locations where they can congregate or sleep undisturbed. Women are more restricted in their movements because they have children in tow; they are more vulnerable and constantly must be alert to security threats. As a consequence, some women acquire a male friend or protector to ensure their safety and to keep aggressive men at bay. One characteristic shared by men and women on the street is their high level of distrust, which increases the longer they are sleeping rough. A common observation by homeless individuals is that it is hard to find someone they can trust (Rich and Clark, 2005).

Men and women often are homeless for different reasons. Men usually succumb to homelessness because of loss of employment, leading to eviction and, often, family connections, all of which may be aggravated by illness, injury or substance abuse. Women without shelter sometimes are referred to as situationally homeless as a result of immediate economic or domestic problems, because they have been abused, or because of mental illness. Once on the streets, it is highly likely that physical and sexual abuse will continue, and perhaps worsen, unless they can gain entry to a woman's refuge. Battered women usually leave home abruptly after being beaten repeatedly. Consequently, they do not anticipate being homeless. Most see their situation as temporary and are unable directly to address their problems. Paralyzed by fear, they use denial as a coping mechanism. Wherever they go, they feel unsafe, afraid their abuser will stalk them. Many homes for women, as a result, keep their locations confidential (Bassuk, 1997) (see Extract 15.2).

Extract 15.2: From Liebow, E. (1993) *Tell Them Who I Am: The Lives of Homeless Women*, New York: The Free Press, pp. 41–2.

Set against … intermittent, catch-as-catch-can programs for people on the street is the fact that homelessness in general, and living on the street in particular, are times of great stress, instability, and insecurity. Such conditions are known to aggravate alcohol, drug, and mental health problems and to render their carriers maximally resistant to treatments that are of uncertain effectiveness even under the best of conditions … [A]lmost every shelter had a policy of refusing entry to persons with serious and active substance abuse or mental health problems. The rationale for these exclusionary policies – the need to maintain peace and good order and to protect the health and safety of the staff and shelter residents – was unassailable, but the effect was to force those with the greatest need for shelter and health care onto the street where the possibilities for either were remote indeed.

Homeless women are typically younger, and many are visible minorities, particularly in North America, but in many European cities as well. Despite being jobless for long spells, women (especially those accompanied by children) are on the streets for much shorter periods than are men, in part because women typically have priority for housing. They are less likely to have institutional contacts with mental health or criminal justice systems, but are in greater need than men of most social services (Hagen, 1990; Takahashi et al., 2002).

The children of homeless women represent a tragic failure of urban society. Children without permanent homes are in jeopardy; in addition to illnesses, they have a high incidence of learning disorders, emotional difficulties, and problems coping with school. As they reach their teens young people on the streets fall prey to pimps and drug dealers. With little education and no skills they have virtually no hope for a better future (Buckner et al., 2001; Dachner and Tarasuk, 2002).

Other high-risk groups include people with HIV/AIDS, those with disabilities, both mental and physical, frail older people on low fixed incomes, refugees and new immigrants, and individuals who have been deinstitutionalized. These groups are not always well served by existing social service arrangements. There is not so much a lack of programmes as inappropriate or inflexible delivery mechanisms. For traditional service providers it has been difficult to adapt services to the needs of these individuals (Freund and Hawkins, 2004).

Virtually all people without adequate shelter confront problems with health and insecurity. Ironically, within a stone's throw of the most advanced medical establishments in major cities, thousands of individuals on the street suffer from an array of health issues, some of which contribute to homelessness; others are the result of sleeping rough. More than one-third are in poor health and their mortality rates are three to four times greater than those of the population at large. In addition to economic deprivation, itself one of the most serious health hazards, they suffer from skin diseases, diabetes, hunger/nutritional deficiencies, sleep deprivation, hypertension, respiratory problems, a number of infectious diseases and numerous mental health issues. Among street youth there is an extraordinarily high rate of physical and sexual abuse, unplanned pregnancy, and HIV infection (Stoner, 1995).

Several of the health problems noted above are aggravated by chronic alcohol and drug abuse. The severity of alcoholism among homeless adults is frequently a consequence of their destitute situations. It is less severe among the newly homeless. People on the street may drink to cope with cold weather, isolation and depression. Because it dulls pain, induces euphoria, and fills idle time, alcohol is accepted as one of the drugs of choice and as a means of fostering sociability among men. There is no doubt, however, that alcohol and other drugs contribute to health problems, accidents and fatalities, and hasten the death of some homeless individuals. Drug use varies widely between men and women and among different racial/ethnic groups (Dear and Takahashi, 1992). The drug of choice also varies depending on cost and availability: most common in the US are heroin, methadone, crystal meth and crack, followed by cocaine and marijuana.

Health problems frequently are precipitated by loss of shelter, particularly as a result of eviction (Crane and Warnes, 2000). Tenants may become homeless after being intimidated by landlords who evict them in order to raise rents or to convert their buildings to condominiums. Some landlords shut off heat and water, illegally lock residents out, and arrange arson or assaults to convince them to leave. Because they are alone, are unaware of their rights as tenants, and lack political or economic power, they may become victims of harassment, violence and illegal eviction.

Children, in particular, suffer ill effects from lack of safe, secure shelter. Wright concluded that poor health and chronic physical illness among children in shelters 'contribute to the cycle of poverty ... interfere with, if not preclude, normal labor force participation, and with it, the ability to lead an independent adult existence. ... Poor health may be one mechanism by which homelessness reproduces itself in subsequent generations' (Wright, 1989: 72–73).

Research approaches

In the past, descriptions of homeless people ranged from typologies of the homeless and street dwellers in the nineteenth century (see Extract 15.3), pictorial depictions of street children (Riis, 1890 [1971]) to sociological surveys of hobos (Anderson, 1923) to sympathetic portraits of destitute families during the Depression, to reports by advocates in the 1980s designed to dramatize the plight of street people. In the mid-1980s the magnitude of homelessness and the nature of emergency shelter dominated discussion. Research concentrated on quantitative surveys and debates raged over the accuracy of population counts (Rossi, 1989). Gradually, the notions of vulnerability and shelter uncertainty gained acceptance and researchers began to take account of hidden or invisible homelessness and the characteristics of people affected.

Extract 15.3: From Mayhew, H. (1861) *London Life and the London Poor, Vol. 1*, London: Griffin, Bohn and Company, Stationer's Hall Court, p. 3.

The nomadic races of England are of many distinct kinds – from the habitual vagrant – half-beggar, half-thief – sleeping in barns, tents, and casual wards – to the mechanic or tramp, obtaining his bed and supper from the trade societies in the different towns, on his way to seek work. Between these two extremes there are several mediate varieties – consisting of pedlars, showmen, harvest-men, and all that large class who live by either selling, showing, or doing something through the country. These are, so to speak, the rural nomads – not confining their wanderings to any one particular locality, but ranging often from one end of the land to the other. Besides these, there are the urban and suburban wanderers, or those who follow some itinerant occupation in and round about the large towns. Such are, in the metropolis more particularly, the pick-pockets – the beggars – the prostitutes – the street-sellers – the street-performers – the cab-men – the coachmen – the watermen – the sailors and such like. In each of these classes – according as they partake more or less of the purely vagabond, doing nothing whatsoever for their living, but moving from place to place preying upon the earnings of the more industrious portion of the community, so will the attributes of the nomade tribes be found to be more or less marked in them. Whether it be that, in the mere act of wandering, there is a greater determination of blood to the surface of the body, and consequently a less quantity sent to the brain, the muscles being thus nourished at the expense of the mind, I leave physiologists to say. But certainly be the physical cause what it may, we must all allow that in each of the classes above-mentioned, there is a greater development of the animal than of the intellectual or moral nature of man, and that they are all more or less distinguished for their high cheek-bones and protruding jaws – for their use of a slang language – for their lax ideas of property – for their general improvidence – their repugnance to continuous labour – their disregard of female honour – their love of cruelty – their pugnacity – and their utter want of religion.

A more subtle and comprehensive schema emerged with the use of the ecological model which views homelessness as a consequence of the interplay between personal factors and structural causes. This model is modified somewhat by an approach known as social ecology, employed by researchers exploring the spatial, racial and class disparities of the city and the marginalization of people and space (Hutson and Liddiard, 1995; Sibley, 1995; Smith, 1996; Takahashi et al., 2002).

The study of a different culture or way of life – *ethnography* – examines the experience of homeless people through participant observation, assuming that these lives can only be understood contextually. Researchers immerse themselves in the lives of people on the streets, participating in their daily routines, conversing with them, earning their trust, while recording and interpreting what they see and hear. The ethnographic portraits emerging from this process, though based on a much smaller sample size than quantitative surveys, have enlarged our understanding of this world. They provide rich, contextual descriptions of the social and physical environments, the people, and their coping strategies. Ethnographic studies also serve as correctives to survey research that excludes certain populations or discounts particular characteristics. Koegel's (1992) work, for instance, presents an insightful glimpse into the world of mentally ill homeless people, and Dordick's (1997) research revealed the importance of supportive social networks – 'the principal currency of survival' – to homeless people. Ethnographic research also has been incorporated in evaluations of programmes and service delivery, providing useful contrasts between the objectives of providers and homeless individuals (Bridgman, 2003).

To date, the next step, longitudinal studies following people over time, have not been given much attention. Urban research in this area is potentially valuable in determining the nature and characteristics of people who are episodically homeless and in ascertaining why, when and how they determine when to seek shelter and when to return to sleeping rough. Longitudinal research also could be invaluable in following individuals who leave shelters to determine how they are coping, what supports or interventions appear to be most helpful in assisting the transition from shelter to housing, and what other variables play a role in this movement towards increased housing security. Urban researchers have, however, successfully followed individuals over time through the use of shelter databases. Most large shelters now have systems in place to track unduplicated counts of individuals over time. Studies in major US cities used such databases to determine that about 3% of the population in New York City and Philadelphia used shelters over a 3–5 year period (Metraux and Culhane, 1999).

It also can be quite revealing when people have the opportunity to sketch or map out their view of the places they inhabit or use on a daily basis. Cognitive mapping has been employed by researchers to determine what homeless people do on the street, routes they follow, people they interact with, locations where they sleep, eat or drop in for coffee, as well as places they avoid out of fear for their safety or because they feel unwanted (Wolch and Dear, 1993).

Because homelessness is costly to society and to the individuals affected, considerable attention has been directed at determining its costs. Assessment is difficult because homeless people typically have a range of issues which are dealt with by an array of services provided by a host of public and non-profit agencies. Isolating the effectiveness of only one of these strategies poses formidable statistical and methodological hurdles. One way to begin to address these issues is to examine the situation from a number of perspectives: from the point of view of the individual, family members, service providers, public agencies, and society at large. After arriving at methods for effective evaluation, it is necessary to determine if the costs incurred today are justified by long-term savings in the social and economic burden of continued homelessness.

In addition to analyses of cost effectiveness, services and service providers also may be evaluated in terms of their successes and failures in assisting homeless individuals and families. Researchers examine service-utilization data to ascertain which programmes are most appropriate and offer the best return on resources. In addition to the basics – emergency shelters, food banks and soup kitchens – the range of services offered include legal aid and advocacy, drop-ins, substance abuse programmes, mediation services, life skills training, mobile health clinics, education projects, supportive housing for particular need groups, such as those with HIV/AIDS, SRO units for adult singles, youth programmes, transitional and permanent housing, and self-help housing planned and managed by residents (Daly, 1996: 210–237).

Another way of researching homelessness is to undertake comparative studies. These are difficult in part because the availability and comparability of data vary substantially from one country to another. In making comparisons and viewing issues at the national level researchers may overlook or obscure important differences, or even contradictory trends, at the regional or local levels. Nevertheless, comparative research permits us to more carefully analyze our own system with the benefit of different perspectives and insightful reflection. A variant of this approach, the cross-cultural study, has been employed to explain the peripheral status of outsider communities in relation to mainstream society (Sibley, 1995).

This type of analysis helps researchers to appreciate the Self–Other dilemma and associated ethical concerns. It is difficult for researchers to place themselves in another's shoes. Studies of homeless people may be intrusive. Inevitably, they are value-laden. Relationships between individuals are affected by preconceptions, power and status differences that stand in the way of communication and understanding. A great deal of sensitivity and grounded research is required to bridge this gap. However, as Sibley observed, 'by attempting to understand the peripheral culture, we can develop a critical view of ourselves. If such introspection is to be profitable, however, we must be prepared to discard the conceptual equipment with which we are burdened by our education' (Sibley, 1981: 199).

Summary

Governments in a number of Western countries have reacted to global economic change by deregulating and privatizing public sector activities. Social services, in particular, have been targeted for reduction and contracting out. When the public sector opts out of these areas the void often is filled by for-profit private enterprise, by firms paying minimal wages and offering little security to a marginalized workforce of ethnic minorities and recent arrivals, many of them women. Poverty and homelessness intersect, with their effects concentrated in identifiable groups: single mothers, children and visible minorities. The presence of these groups in urban centres has accentuated class and racial polarization and new forms of spatial and class disparities as those with financial means flee to gated communities or insular condominiums. These processes, associated with globalization and privatization, worsen disparities and are so narrowly construed that they yield a relatively impoverished array of perceived solutions to the issues of poverty and homelessness.

An alternative exists. It is possible for public officials and service providers, along with members of the public, to view people without shelter as integral to solutions and, therefore, politically and socially capable of participating in programmes they help to design and manage. This is the most likely path out of homelessness for them and out of complicity in the construction of homelessness for those who provide services. Homeless people are a dynamic, heterogeneous array of individuals with capacities that are usually overlooked by professional care-givers. Many of these people have considerable coping skills and could serve as a resource rather than a drain on society. The desire to get off the streets is strong. If allowed to exercise power and control over the programmes which shape their future, many can become advocates for themselves. A number already have. To realize this ambition, however, professionals must relinquish some control. Case management approaches must give way to self-help initiatives. Homeless individuals themselves must be allowed to regain their individuality, dignity and self-esteem, and to assume personal responsibility for following one of the paths out of homelessness.

References

Anderson, N. (1923) *The hobo: the sociology of a homeless man*, Chicago: University of Chicago Press.

Bassuk, E.L. (1997) 'Homelessness in female-headed families: childhood and adult risk and protective factors', *American Journal of Public Health* 87 (2), 241–248.

Bridgman, R. (2003) *Safe haven: the story of a shelter for homeless women*, Toronto: University of Toronto Press.

Buckner, J.C. et al. (2001) 'Predictors of academic achievement among homeless and low-income housed children', *Journal of School Psychology* 39 (1), 45–69.

Burrows, R., Pleace, N. and Quilgars, D. (eds) (1997) *Homelessness and social policy*, London: Routledge.

Crane, M. and Warnes, A. (2000) 'Evictions and prolonged homelessness', *Housing Studies* 15 (5), 757–773.

Dachner, N. and Tarasuk, V. (2002) 'Homeless squeegee kids: food insecurity and daily survival', *Social Science and Medicine* 54 (7), 1039–1049.

Daly, G. (1996) *Homeless: policies, strategies, and lives on the street*, London and New York: Routledge.

Daly, G. (1997) 'Charity begins at home: a cross-national view of the voluntary sector in Britain, Canada and the US', in M. Huth and T. Wright (eds), *International critical perspectives on homelessness*, London: Praeger.

Daly, G. (1998) 'Homelessness and the street', in N. Fyfe (ed.), *Images of the street: planning, identity and control in public space*, London and New York: Routledge.

Dear, M. and Takahashi, L. (1992) 'Health and homelessness', *Western Geographical Series* 27 (1), 185–212.

Dordick, G. (1997) *Something left to lose: personal relations and survival among New York's homeless*, Philadelphia: Temple University Press.

Fitzpatrick, S. (2005) 'Explaining homelessness: a critical realist perspective', *Housing Theory and Society* 22 (11), 1–17.

Franck, K. (1991) 'Overview of single room occupancy housing', in K. Franck and Ahrentzen (eds), *New households, new housing*, New York: Van Nostrand Reinhold.

Freund, P. and Hawkins, D. (2004) 'What street people reported about service access and drug treatment', *Journal of Health and Social Policy* 18 (3), 87–93.

Hagen, J. (1990) 'Designing services for homeless women', *Journal of Health and Social Policy* 1 (3), 1–16.

Hopper, K. (2003) *Reckoning with homelessness*, Ithaca, NY: Cornell University Press.

Hutson, S. and Clapham, D. (eds) (1999) *Homelessness: public policies and private troubles*, London: Cassell.

Hutson, S. and Liddiard, M. (1995) *Youth homelessness: the construction of a social issue*, London: Carfax.

Koegel, P. (1992) 'Through a different lens: an anthropological perspective on the homeless mentally ill', *Culture, Medicine and Psychiatry* 16 (1), 1–22.

Krieg, R.G. (2001) 'An interdisciplinary look at the deinstitutionalization of the mentally ill', *The Social Science Journal* 38 (13), 367–382.

Liebow, E. (1993) *Tell them who I am: the lives of homeless women,* New York: The Free Press.

Lyon-Callo, V. (2001) 'Making sense of NIMBY: poverty, power and community opposition to homeless shelters', *City and Society* 13 (2), 183–209.

MacLeod, G. (2002) 'From urban entrepreneuralism to a "revanchist city"? On the spacial injustices of Glasgow's renaissance', *Antipode* 34 (3), 602–624.

Mayhew, H. (1861) *London life and the London poor,* London: Griffin, Bohn and Company.

Metraux, S. and Culhane, D. (1999) 'One-year rates of public shelter utilization by race/ethnicity, age, sex and poverty status for New York City (1990 and 1995) and Philadelphia (1995)', *Population Research and Policy Review* 18 (3), 219–236.

Pleace, N. (1998) 'Single homelessness as social exclusion: the unique and the extreme', *Social Policy and Administration* 32 (1), 46–59.

Rich, A. and Clark, C. (2005) 'Gender differences in response to homelessness services', *Evaluation and Program Planning* 28 (1), 69–81.

Riis, J. (1890 [1971]) *How the other half lives,* New York: Dover.

Rossi, P. (1989) *Down and Out in America,* Chicago: University of Chicago Press.

Sassen, S. (2001) *The global city: New York, London, Tokyo,* Princeton, NJ: Princeton University Press.

Sibley, D. (1981) *Outsiders in urban societies,* Oxford: Basil Blackwell.

Sibley, D. (1995) *Geographies of exclusion: society and difference in the West,* London: Routledge.

Smith, N. (1996) *The new urban frontier: gentrification and the revanchist city,* London and New York: Routledge.

Stoner, M. (1995) 'Interventions and policies to serve homeless people infected by HIV and AIDS', *Journal of Health and Social Policy* 7 (1), 53–68.

Takahashi, L., McElroy, J. and Rowe, S. (2002) 'The sociospatial stigmatization of homeless women with children', *Urban Geography* 23 (4), 301–322.

Vosko, L. (2000) *Temporary work: the gendered rise of a precarious employment relationship,* Toronto: University of Toronto Press.

Wallace, R. (1990) 'Homelessness, contagious destruction of housing, and municipal service cuts in New York City', *Environment and Planning A* 22 (1), 5–15.

Wolch, J. and Dear, M. (1993) *Malign neglect: homelessness in an American city,* San Francisco: Jossey-Bass.

Wright, J.D. (1989) *Address unknown: the homeless in America,* New York: Aldine de Gruyter.

SECTION FOUR
ORDER AND DISORDER

Although only the final chapter in this section is explicitly labelled as such, most of the authors in this section are concerned, in their different ways, with exploring a path between those two pervasive urban clichés, the city as dream and the city as nightmare. Spurred on by numerous popular imaginations of the city, cities have been seen historically, and arguably increasingly so today, within this binary. Although having intuitive (and 'box office') appeal, these ways of thinking of the city probably hide more than they reveal. Despite this, they have had a worrying grip on the imaginations and the practices of many involved in the actual shaping of the city. The authors here are hence highly critical of seeing the city as either dream or nightmare, and all argue that the reality lies somewhere between these poles.

The dreams and nightmares the authors in this section explore take many forms. The section opens with Donald McNeill's account of the world of power, politics and urban policy and those ultimate urban dreamers, city mayors. In this chapter McNeill reviews various attempts to understand and to locate power within cities and is critical of the notion of an overarching urban theory, instead arguing for the importance of geographical specificity in accounts of power in the city. The nightmares that Alan Latham and Derek McCormack explore are concerned with the supposed acceleration of urban life, and the consequences for those who cannot keep up. Like McNeill, Latham and McCormack are critical of some of the more hyperbolic accounts of the speeding up of city life, recognizing instead that if cities are speeding up in some ways, they must be slowing down in others. The authors explore this both historically and in the context of the post-industrial city, considering the ways in which speed and slowness shape the texture of urban life – and thus effect a politics of mobility that confines as much as it enables.

As we saw in Section One, town planning has long been about dreaming, imagining and creating ideal societies. When things go wrong – as John Gold (Chapter 4) suggested was the case in the instance of some post-war high-rise development – we are in the realm of urban nightmare. In his chapter, Malcolm Miles considers the planners and their dream of creating conflict-free cities. Again, the complexity of the city brings the achievement of such a dream into doubt – not least when groups of citizens do not behave in the way planners had anticipated they would. The arrogance of planners to assume that a 'one-size fits all' solution will provide more commodious urban living is questioned by Miles, who concludes that urban conflict undermines many attempts to create a city, but suggests that conflict is not inevitable.

Yet the ultimate urban nightmare for many is crime, and the fear of crime. Poll after poll suggests that this is the urbanite's greatest fear, and politicians routinely argue that this poses the greatest threat to civic society. Stephen Herbert explores two contrasting paradigms of the causes, and responses, to crime – the instrumental, 'broken windows' thesis and the situational crime prevention that follows this, and the constitutive paradigm. The latter does not focus on individual criminals but instead sees crime as the result of a broad range of urban social phenomena. Jon Coaffee and David Wood further explore the taming of the supposed nightmare of urban crime through situational crime prevention in their chapter on surveillance and terrorism. They consider the role of reinforcing the city against threat, both internal and external, in the history of cities and focus on recent terrorist threats to particular cities and the responses by governments. In their account, scale is crucial to understanding the ways in which response to threat is shaping cities and the lives of citizens.

The section is rounded off by Stuart Aitken's chapter, which pulls together the various strands that constitute urban dreams and nightmares. He does this through an analysis of various urban representations and their affects. Aitken explores the world of urban dreams and nightmares and focuses on some pathologies of the urban – agoraphobia, vertigo and schizophrenia – and the ways in which social theorists have explored the relationship between the city and mental life. Working through a variety of texts – film, music, fiction – Aitken identifies the way that dreams and nightmares are deeply embedded within our collective psyche, and hence shows that urban studies will remain poised between diagnosing the city of violence and a proscribing a sublime city of escape.

16   POLITICS AND POLICY

Don McNeill

This chapter

○ Argues that city politics is not a simple process where an elite takes control, but involves complex processes of coalition-making, brokering and electioneering

○ Underlines the importance of space in political processes by considering the way that political territories are reconfigured to serve particular interests

○ Suggests that urban theory is often too monolithic and unbending to provide a nuanced understanding of city politics

---

Introduction

Urban areas and cities are often described as engines – sites of economic development and resource consumption and production that drive their wider regional or national economies. To calibrate these engines requires a series of policy interventions, investment decisions, and ordering mechanisms empowering or disempowering various groups, classes, elites or individuals. However, it is not always clear where the power lies, and the 'political landscape' can often be cloudy. While we may be aware that the likes of Los Angeles, Sydney, Shanghai or Rome are central to our urban imaginary, the question of whether these entities are *powerful* is not an easy one to answer. And cities – whether understood as territories, institutions, or ideas – are constituted by power relations, which in turn should lead us to question their ontological status. The field of urban politics and policy is huge, spreading across disciplines as diverse as human geography, urban sociology, political science, urban planning, real estate finance, and economic development.

This chapter is organized into three sections. First, it discusses the concept of the city mayor, and outlines issues of 'city voice', machine politics, electoral power, and introduces the notion of scale and autonomy Second, it reviews the idea of urban governance, and suggests that electoral politics is only one aspect of the exercise of power in cities. It highlights Mike Davis's study of the power elites of Los Angeles, an example of the construction of growth coalitions and civic boosterism. Third, it focuses on issues of territory, and examines the political machinations surrounding boundary construction, metropolitan politics, and the disjuncture often arising between local government boundaries and *de facto* city activities. Drawing on the reconfiguration of Kuala Lumpur within the Malaysian government's globalization strategies, it highlights the importance of geographical specificity, and suggests the development of an overarching, singular 'urban theory' is insufficient in capturing the spatial specificity of urban politics.

The mayor and the city

To begin, it might be worth considering the role of the mayor in city politics. Why? Because mayors have historically been considered the 'first citizen' of a city, and as such can be seen to be a personification of the complex stories around which cities are structured. They are also very power-conscious individuals, who are engaged not only in the awarding of contracts, the mediation of resource claims, and the pleasing of electorates, but they are also acutely aware of their own weakness: they have a temporary grip on the levers of power endowed by elections, and the budget at their control is usually far below what they would like.

The study of mayors and city politics is often associated with the example of a party machine (and machine politics). Large American cities were long-dominated by machine politics. In New York, the 'Tammany Hall' machine led by 'Boss' Tweed, allowed the Democrats to organize New York politics so tightly that they were consistently elected between 1854 and 1934; in Chicago, the mayoralty was held by Richard J. Daley between 1953 and 1976, who held the ability to appoint (via ward supervisors) public posts from street cleaners to department heads (Judd and Swanstrom, 2006). Such machines often have a close system of political rewards, where favourable votes are paid back with job appointments, a process often known as clientelism. While such machines were often seen as vehicles of social mobility for new immigrants (such as the Irish community), they were arguably one of the few modes by which working-class minorities could gain a purchase on city politics. The machines 'prospered in rapidly growing industrial cities that required massive expenditures on roads, bridges, street-car systems, schools, and parks' (Judd and Swanstrom, 2006: 65–6), and as such have been of declining significance in US cities in recent years.

Nonetheless, speaking 'for' cities is central to the political arts of urban leaders around the world, and it may be suggested the mayor provides the linkage between cities and external agents, be they national governments or major public and private investors. This relationship is cemented by a performative relationship to mayoral governance, encapsulated in three interrelated roles. First, mayors 'embody' cities. Many are born in the city they represent (and here representation has a double meaning), and will have some sort of relationship to the essentialized characteristics of its inhabitants (often accent, sense of humour, or 'inheritance' of attributes of an idealized predecessor). Here, they communicate with their voters/ fellow citizens. Second, they will act as animator of city space. Rather than pursuing an abstract notion of territory, mayors often strive for visibility in the everyday life of the city, especially at times of crisis (consider Rudy Giuliani's presence around the site of the World Trade Center on and after 9/11). From presiding over ceremonies to high-profile public walkabouts, they will seek to become 'everyman', in experiencing the city's urban life (and we should consider the gendered nature of such terms, including the more general idea of the 'city fathers'). Finally, they are likely to provide some sort of narrative in their press conferences and public appearances about the immediate past, present and future of the territory they represent, shaping and responding to a public discourse concerning crime, fear of terrorism, the economic climate, and so on (McNeill, 2001).

Surprisingly, given that city politics is often surrounded by webs of intrigue, political scientists have often limited themselves to the study of voting behaviour and electoral geography. Interestingly, the conduct of such analysis is often done at an abstract level, where quantitative indicators (the amount and distribution of votes cast) are explored in greater depth than the specific conduct of the elections themselves. ÓTuathail (1998: 84) has drawn attention to how:

> election returns become raw material for scientific data manipulations and hypothesis testing. The discursive politics of elections is often reduced to a battle between models of electoral cleavage. ... It is indeed a pity that many of those who work on electoral geography are hostile to questions of discourse and signification.

However, as Brown et al. (2005) demonstrate, local elections can be analyzed both in terms of qualitative method (such as discourse analysis), or – more conventionally – through quantitative analysis, where the pure representation of the electorate is expressed through countable votes. Elections provide condensations of more general political narratives, but their inherent spatiality is often revealed in voter question times, party slogans, political 'stumping', and sensitive planning issues. Electoral campaigning often uses the city as a kind of political theatre, and as the mediatization of politics has become more sophisticated, so symbolic acts, gestures and personality 'performances' become more important. Elections can be seen as 'events' punctuating four-or five-year cycles of governance, and bringing into the open contested issues of taxation and service provision, voter insecurity and political (dis)satisfaction (McNeill, 2001).

This idea of being the voice of the city thus puts the mayor at the centre of the articulation of political discourse and performance. Extract 16.1 provides an interpretation of the institution of mayor of London, and argues that the *personality* of Ken Livingstone will be a very significant factor in that institution's power and influence.

Extract 16.1: From Pimlott, B. and Rao, N. (2002) *Governing London*, Oxford: Oxford Universtiy Press, pp. 19–20.

If London needed to learn from foreign experience, it was not at all obvious what the main lesson should be. Indeed, international examples provided a variety of different approaches, with little indication that any was particularly successful in practice or self-evidently applicable to British conditions. What problems should be addressed? In the late 1980s and early 1990s a key issue in urban leadership appeared to be the ability of metropolitan government to promote and project a city's image and maximize its attractiveness as a location for global business. City-wide governments and mayors were judged internationally on their ability to provide such a focus. London seemed to lack one after 1986. Its absence was seen as giving other cities a competitive edge. 'Whether or not this perception is correct', according to one study, 'London is certainly taking a risk by not having some way to promote itself and also of providing outsiders with a view of the City's future development' (Greater London Group, 1992: 4).

This was seen as a primary challenge: if London was to remain in the same league as the major cities of North America, Europe, and the Far East, it needed the political arrangements to facilitate it. Such a logic led to what was seen as the nearest or most compatible model at hand. However, transposing a US-style mayor into the British local government context was bound to be a radical move. … Whereas the mayor system was both long-standing and widespread in US cities, reflecting the separation of powers at the federal level, the British tradition had been one of collective responsibility through a council-based system in which mayors, where they existed, were generally figureheads (Rao, 2000).

[…]

Whether the new London will turn out to be a success or failure depends, critically, on the first few mayoral incumbents – and, most of all, on the first. Ken Livingstone's effectiveness, in turn, will depend on his ability to work within the institutional and political constraints. The formal political resources at his disposal are limited, while his own political style is interventionist, flamboyant, even charismatic. Time will show whether Livingstone turns out to be a 'crusader' or is forced to retreat to a 'broker' role.

Far from his days as radical left-wing leader of the Greater London Council (GLC), a major thorn in the side of Thatcherism during the early 1980s, Ken Livingstone's current mayoralty of London has been characterized by a more pragmatic approach. While his enthusiasm for affordable housing and desire to revive the publicly subsidised cultural scene of the 1980s are characteristic of progressive Left politics, his other key urban policy planks – based upon a pro-developer attitude, a land-use strategy dubbed the 'zone 1 plan' due to its centralizing logic, and a flat-rate congestion charge – reveal the complexities within which this erstwhile socialist must now operate (McNeill, 2002a, 2002b).

The example of Livingstone can be used as illustration of the significance of scale and the degree of political autonomy enjoyed by mayors. At a formal level, cities can be seen as at the foot of a hierarchy of nested scales. In reality, the ability to govern a city of the size and complexity of London is comparable to running a large region or even a small nation-state. However, political actors all have varying capacities to govern, and city councils are notorious for their inability to conduct anything other than a reactive politics, circumscribed by central government budget restraints, regulatory watchdogs and legislative controls. By 2005, Livingstone's rehabilitation within the Labour Party was such that central government announced the strengthening of the mayor's powers.

Some of these complexities relate to the centrality of London within Thatcherism, at once a hugely important accumulation centre through global financial reregulation, and a graphic illustration of Peck and Tickell's (2002: 386) argument that the Thatcher/Major regimes were more concerned with 'de(con)struction of "anticompetitive institutions" ... than with the purposeful *construction* of alternative regulatory structures'. After 15 years without a London government, the Greater London Authority has replaced the GLC, an institution designed to construct the alternative strategic body so lacking under Thatcher and Major. Under New Labour, the presence of Livingstone as an executive mayor (albeit a hamstrung one) has produced some very unexpected policy directions, yet his hostility to a totem of Labour economic policy – the Private Finance Initiative – reveals the complexity of the London political scene. Furthermore, while Livingstone has a close dialogue with major property developers, he has remained a firm supporter of trades unions, such as those of the transport workers.

The study of the degree of autonomy which councils have was a major academic industry for much of the 1980s and 1990s. This can be fundamentally important when it comes to lobbying: the direct or indirect representation to central or regional government for more public spending for infrastructure, housing and other public services. However, there are clearly defined limits to local autonomy, which calls for 'local' politics to be analyzed as part of a wider spatial politics. Books such as *City Limits* by Paul Peterson (1981) argue that any attempt by city governments to follow genuinely redistributive politics is limited by the liquidity of investment capital. High local taxation levels would be met by firm or investor relocations, either to competing cities or, more likely, to suburban councils. Stone (1989: 239) notes that: 'Voting power is certainly not insignificant, but policies are decided mainly by those who control concentrations of resources.' To follow any meaningful redevelopment or growth strategy, politicians had to learn to construct complicated coalitions between varying social and business groups, in a far looser way than the party machines discussed above. The next section explores some of these issues.

Urban governance, property and power elites

Early work on growth coalitions emerged in studies such as Logan and Molotch (1987) and stressed the creation of political fora (either formal, as in chambers of commerce, or informal, as in personal networks) among actors with a shared interest in maximizing economic growth in a locality. They identified competing groups which organized around the 'politics of growth': on the one hand, firms and institutions with a vested interest in job growth, new factories and shopping malls, and so on; on the other, 'quality of life' groups, particularly environmentalists and homeowners, who sought to preserve amenity (and, often, the values of their homes). Business elites seek growth for the generalized increase in profitability gained from resale, development and rent, and municipal councils seek tax revenues which are derived from these profits. Similarly, regime theorists such as Clarence Stone (1989) have argued for an understanding of business elites in terms of their influence upon organized politics within a city, in this case Atlanta. Stone begins his account like this:

> What makes governance in Atlanta effective is not the formal machinery of government, but rather the informal partnership between city hall and the downtown business elite. This informal partnership and the way it operates constitute the city's regime; it is the means through which major policy decisions are made. (Stone, 1989: 3)

A key issue in this case was the way racial politics were significant within decision-making processes. For Stone, Atlanta of the 1970s and 1980s was driven by a public–private partnership based upon a reconciliation between African-American electoral power and a white corporate business elite (see also Rutheiser, 1996). In other cases, as in the shift from municipal socialism to urban entrepreneurialism in Manchester (Quilley, 1999, 2000), specific political traditions of party organization and cultural identity structure the local state's response to urban restructuring.

Subsequent work has criticized such approaches as being excessively localistic (blind to external pressures) or voluntaristic (e.g. Jessop et al., 1999; Jonas and Wilson, 1999), endowing local actors with excessive power or capacity to act. Such a backlash, which has seen theorists focusing more on the structural constraints placed on actors due to macroeconomic restraints, is often taken as an indication in the shift in the governing principles of national economies. Here, the macroeconomic theory of Keynesianism, which stressed strongly centralized government direction to stimulate employment and redistribute public spending, has largely been replaced. Both within the leading industrialized states, as well as countries subject to swingeing fiscal straitjackets imposed through institutions such as the World Bank, a growing number of scholars have drawn attention to neo-liberalism as a defining moment in urban development (e.g. Kipfer and Keil, 2002; Swyngedouw et al., 2002; Weber, 2002).

In part, this is a reflection of what has come to be known as urban entrepreneurialism (Hall and Hubbard, 1998; Harvey, 1989). Here, territorial governments are conceptualized as firms operating within a competitive marketplace, having to pitch a series of place products – health services, labour markets, consumption spaces, policing – to discerning consumers. The limits to the analogy should be obvious – few people can switch and choose their governments as frequently as they can their favourite television show or toothpaste – but the underlying message of inter-urban competition for limited consumer spending, apparently footloose firms, touched a nerve with many urbanists.

The key issue here is how local states fund their services. One aspect of neo-liberalism is an increasing dependence on property tax revenues, which makes local treasuries seek to stimulate the private real estate market. However, as Weber (2002: 537) argues, 'reliance on the erratic capital markets to reinvigorate devalued properties often jeopardizes the fiscal health of cities'. This has led to the rise of fiscal 'enclaves', patches of urban territory with separate financial rules, such as districts redeveloped under tax increment financing (TIFs), business improvement districts, and development corporations (Swyngeduow et al., 2002; Weber, 2002). TIFs are interesting in their focus on blighted areas. Whereas under Keynesianism, such areas were often recipients of federal subsidy, they are now driven by market-led criteria. Measured by the presence of unused buildings and low relative property values within the municipality, TIFs allow for local states to issue bonds to finance large-scale redevelopment of an area, based upon the future tax revenues. However, this reduces the financial autonomy of local states:

> Dependence on financial markets contributes to the public sector's loss of time sovereignty, as investors have much shorter time horizons than states. A small cadre of highly specialized underwriting shops for TIF deals have become important agents in the networks of urban fiscal governance. These boutiques are able to charge higher spreads for TIF debt because of the speculative risks involved. Meanwhile, cities woo bond-rating agencies with exaggerated claims of performance in hopes of securing better grade – and therefore less expensive – debt (Weber, 2002: 536).

As Jessop et al. (1999: 147) note, unlike regulation theory, which explores changes in the labour process, 'the microeconomic foundations of growth machine theory are grounded in the transformation of the built environment'.

However, despite the growing sophistication of such analysis, there can be a tendency for theorists to explore case studies of individual cities in order to substantiate urban theory as an end in itself, rather than to use theory as a tool to explain real cities and power elites. As a corrective to this, consider Extract 16.2 from *City of Quartz*, by Mike Davis (1991).

Extract 16.2: From Davis, M. (1991) *City of Quartz: Excavating the Future in Los Angeles,* New York: Verso, pp. 100–102.

Popular images of power in Los Angeles are curiously contradictory. On the one hand is the common belief, almost folk legend, that LA is ruled by an omnipotent Downtown establishment, headed by the *Times* and some big banks, oil companies and department stores. On the other hand is Chandler's lofty avowal, echoed by journalists of the 'there is no "there, there" school, that power in Southern California is fragmented and dispersed, without a hegemonic centre.

Both images exploit partial truths. For the half century between the Spanish-American and Korean wars, the Otis–Chandler dynasty of the *Times* did preside over one of the most centralized – indeed, militarized – municipal power-structures in the US. They erected the open shop on the bones of labor, expelled pioneer Jews from the social register, and looted the region through one great real-estate syndication after another. Important residues of their power – and looting – remain inscribed in Downtown, influencing the present Bradley regime, even if the Old Guard is being supplanted by more powerful players from Tokyo, Toronto and New York.

[...]

Obviously the poly-centered complexity of the contemporary system of elites is no longer susceptible to the diktat of any single dynasty of Mr Big. But if Los Angeles has long ceased being a hick town with a single 'executive committee of the ruling class', it is still far from being a mere gridwork of diffused wealth and power. Political power in Southern California remains organized by great constellations of private capital, which, as elsewhere, act as the permanent government in local affairs. What is exceptional about Los Angeles is the extreme development of what remains merely tendential in the evolution of other American cities.

In this extract, Davis draws attention to the genealogy of elites, the fact that wealth is often concentrated in families and passed on through inheritance, even if it is often an 'old guard' under threat by external business groups. Peck (1995) calls these people the 'movers and shakers' of political affairs, individuals who are likely to merge successful business careers with some kind of political activity, whether this be through standing for office, or through business lobby groups, or through stealthier modes of operation such as bribes and kick-backs. However, it is clear one of the dominant trends in urban studies has been the response of politicians to the growing mobility of capital. This raises a question about whether power is 'centred' or whether it is fragmented or dispersed. This is a very important idea.

Here, we can see that one of the key issues of urban governance is the fact that external actors have major influence on structures of city governance. As can be seen in the extract from *City of Quartz*, 'the Old Guard is being supplanted by more powerful players from Tokyo, Toronto and New York'. The choice of actors is interesting: some come from a different part of the USA, some from across a nearby territorial border, and some from relatively far away (though Japan's proximity to California is not accidental). So, we have to understand city politics as being structured both by actors who are locally dependent and those who are relatively remote. Kevin Cox (1993, 1998) has demonstrated the significance of localities for the success of firms, and he is careful to delineate the variable scales at which these institutions operate. However, there is a major challenge in the whole nature of how cities, and urban economies, are conceptualized:

> For a long time now, work on urban economies has been framed in terms of points, lines and boundaries. Thus cities have been seen as entities that can be cut up into centres and peripheries, positioned against other cities in an urban hierarchy, sites of production or consumption, but rarely both. ... Yet what seems clear about modern economies is that they can no longer be contained by these kinds of habits of thought because they are always both global and local, here and there, in between. They are increasingly structured around flows of people, images, information and money moving within and across national borders (Amin and Thrift, 2002: 50).

We can no longer – if we ever could – discuss cities as islands of wealth. In reality, all urban areas by necessity grew as a means of exploiting agricultural hinterlands, giving rise to an early distanciated economy (Amin and Thrift, 2002). The growing complexity of global capitalism has, however, changed how we conceptualize urban politics. For example, Roberts and Kynaston (2001: 182) conceptualize London as a 'global portal', a restless site of scheduled air routes, wired up via a deregulated telecommunications market, operated by a cosmopolitan workforce and thriving on the utility of English as a global language, 'an unreplicatable communications asset'. The material reality of this situation is often expressed anecdotally, and it has been left to ethnographers, geographers, sociologists and cultural theorists, rather than political scientists, to locate the unique power relations of such financial sites.

One such theorist is Kris Olds (2001), who provides one of the most sophisticated existing accounts of transnational cultures and the remaking of cities, through case studies of Shanghai and Vancouver. Olds aims to identify the spatiality of globalization, through case studies of Li Ka-Shing and Richard Rogers, the former property developer, the latter an architect. These people have:

> the capacity to shape, in a very influential manner, the ongoing constitution of networks that reach across space via the circulation of intermediaries. In other words, these powerful actors are made powerful through their relations, their associations – their power is contained in relationships that are bound up in ever-shifting networks. Even then though ... their power to achieve goals has to be performed and implemented with considerable effort, and their successes are by no means guaranteed (Olds, 2001: 13).

The significance of this work lies in its ability to interweave the stories of agents (ranging from personal ambition to portfolio diversification) with structural analysis. In his Shanghai case study, Olds explores the limited ability of foreign architects to shape local built environments, despite their apparent autonomy. Similarly, Fainstein (2001) provides a comparative framework for understanding the role of property elites and state policy in the redevelopment of major financial centres, using case studies of London and New York, which displays a tension between mobile capital and embedded firms and public governments. So, cities are increasingly dominated by discourses and distanciated practices of capital movement. However, there can be a tendency in some discussions to ignore the constitutive power of territory and spatial politics in shaping urban policy, as the next section explores.

Urban policy and territorial strategies

> The crucial point ... is that territorialization, on any spatial scale, must be viewed as a historically specific, incomplete, and conflictual *process* rather than as a pregiven, natural or permanent condition (Brenner, 2004: 43, original emphasis).

Often, cities or city-regions have been described using the metaphor of the 'engine' or 'motor' of their national economies, whether this be Silicon Valley, Tokyo, Lagos or Sydney. However, while the 'nation' itself is an intangible, imagined, performed construct, urban policy is one of the mechanisms by which central governments can calibrate and lubricate their economies, and hence provide for their electorates and populace. To this end, central governments will usually take a great interest in urban areas that contain promising locational, resource and human attributes (and those that fail to do so may find themselves on a political downslide). These areas may be the target of specialized funding streams (sometimes known as pork-barrel politics when seen to be 'bribes' to voters), or may be given exceptional political status *vis-à-vis* other areas within the nation-state. The Special Economic Zones in China are one of the most striking examples of this, and highlight the importance of uneven development as a concept, where differing degrees of political resource allocation, or unevenly applied tax or land-use regulations, can 'splinter' territory (Graham and Marvin, 2001). This is particularly important when we consider that definitions of the 'city', as opposed to the urban, or the city-region, have become more and more complex, and where economic geographies of growth and decline become spatialized in ways disrespectful to the ordering logic of administrative, or formal government, institutional structures. Political geographers are providing some interesting interpretations of these disjunctures. To illustrate this, consider the following recent case studies drawn from Italy, Malaysia and the USA.

First, Bialasiewicz (2006) explores the wealthy Italian administrative region of the Veneto, which is home to 4.5 million people and which has emerged as one of the key contributors to up-scale global consumption niches. However, the shape of this space is difficult to define, and has become known as the *città diffusa*, which 'has no centre, no piazza. It has developed along a series of straight lines: the *statali*, the state roads that provide the only visible skeleton giving form to this new socioterritorial "organism" (Bialasiewicz, 2006: 47–8). Here, a reactionary, anti-foreigner politics has emerged in the form of the regionalist Lega Nord, where some of the family-oriented businesses that have emerged in the post-war period have reacted against migrant labour. This raises an issue of how to conceptualize politics and policy within a context of place dynamism: 'the shifting ways in which places are articulated within wider spatial divisions of labour and global flows of goods, people and information. ... It might be intriguing, then, to ... ask: what sort of politics is emerging today in the Veneto's "dislocated" spaces of diffuse industrialization?' (Bialasiewicz, 2006: 64).

Second, as the extract from Tim Bunnell's account of the Malaysian Multimedia Super Corridor demonstrates (Extract 16.3), large infrastructural projects such as airports, skyscraper office buildings, light rail systems, motorways, technology parks and information technology networking are fused with a series of economic strategies promoting national economic growth.

Extract 16.3: From Bunnell, T. (2004) *Malaysia, Modernity and the Multimedia Super Corridor: A Critical Geography of Intelligent Landscapes*, London: Routledge, p. 65.

The spectacular rise of the Kuala Lumpur skyline in the 1990s (Figure 16.1) was the most visible sign of the city's increasingly global orientation. I have alluded already to the growing role of the city as *the* national centre for advanced producer services connecting it up to regional and global financial markets. Even in the previous decade, architectural observers were reading 'global' influences from the proliferation of internationalist high-rise buildings in the commercial centre (Yeang, 1987). While such signs are clearly significant, however, there is a danger here of underplaying the active role of *in situ* authorities and everyday inhabitants in fostering *global* urban landscape change. To consider merely how Kuala Lumpur was increasingly subjected to transnational flows or marked by the architectural icons of global capital is to posit globalisation as an external force impacting upon an essentially passive city. In the 1990s, Kuala Lumpur and the larger urban region of which it forms part (the Klang Valley), were characterised by unprecedented attempts by federal authorities to discursively and materially reconstruct urban space and subjectivities in 'global' ways.

(Continued)

Figure 16.1   Kuala Lumpur skyline

A key aspect of this reconstruction in greater Kuala Lumpur, as elsewhere, was large-scale infrastructural projects, particularly for transport connectivity and the provision of premium commercial office space. The increasingly 'mega'-scale of such projects in East and Southeast Asia in the 1990s meant that they were frequently located on sites beyond existing urban boundaries (Olds, 1995). The 10,000 hectare Kuala Lumpur International Airport, for example, was located at Sepang, some 60 km south of Kuala Lumpur, making known a new growth corridor from the national capital. In the 'high-tech' times of mid-1990s Malaysia, it was this area which became the Multimedia Super Corridor (MSC) and so the locus of further globally-oriented infrastructure development. The global reorientation of Kuala Lumpur–Klang Valley was thus bound up with the emergence and official recognition of an extended Kuala Lumpur Metropolitan Area (KLMA).

Here, Bunnell argues that the MSC was part of a wider project of identity reconstruction, of altering the subjectivity of what it means to be Malaysian, spearheaded by the controversial Malaysian Prime Minister, Mohammed Mahathir.

Third, consider the recent vote over the future political shape of Los Angeles, as described by Purcell (2001). Here, the emergence of a powerful lobby advocating the secession of the San Fernando Valley (of around 1.8 million people) from the City of Los Angeles is illustrative of the growing number of suburban political movements in the USA that are lobbying for smaller units of government, with lower tax bills. Interestingly, however, the Valley VOTE movement spearheading this movement (which was ultimately unsuccessful in a 2002 ballot) contained two usually counterposed movements which coalesced to support secession. Both saw the City of Los Angeles as being too bureaucratic, aloof and inefficient, drawing from the Valley more than they gave back in terms of services. Yet both had very divergent interests that they would be prepared to argue out should secession be granted. Business elites aim to promote growth and investment, reduce business taxes and ease land-use planning conditions; homeowner groups seek to tighten planning laws, which under the status quo are too permissive (Purcell, 2001: 623). However, it was clear that there are powerful regionally-organized business elites active in opposing secession, which gives a clue as to the limits of claims to local autonomy, which led to the failure of the secession movement.

Taken together, these examples illustrate the significance of spatiality and territory for any understanding of political organization. It also highlights the difficulty of building generalizable urban theories from single case studies. Indeed, the Western (or rather US) dominance of urban theory is one that is being increasingly challenged. As Graham and Marvin (2001: 35) argue, 'contemporary geographical divisions of power and labour on our rapidly urbanizing planet wrap cities and parts of cities into intensely interconnected, but extremely uneven, systems. These demand an international, and multiscalar, perspective.' Nonetheless, for commentators such as Jessop et al. (1999: 141), you do not have to be 'an unreformed structuralist to discern intriguing parallels and telling similarities in the responses of contemporary cities to wider forces, such as neoliberalism and economic globalization'. It is this challenging research agenda that should invigorate the study of urban politics and policy in coming years.

Summary

This chapter has sought to identify a series of issues concerning politics and policy in urban space, covering issues from mayoral identity and the power of civic 'voice' to sociological understandings of urban elites, and to the nature of cities and urban systems with broader territorial formations. This has involved a consideration of intertwining theoretical perspectives emananating from different sub-disciplines: the focus on discourse brought about by post-structuralism; the 'relational turn' in human geography, which avoids seeing cities as bounded containers for social action; the importance of fiscal and financial actions for the regulation of cities; the dramaturgy of elections; and issues of autonomy and spatial scale. It can be surmised from this is that the politics of cities, and of urban redevelopment, is not something that a single (sub)-discipline can explain.

Furthermore, it should be clear from the very limited range of examples and case studies used here that the terms we use to define the urban must be specified. This is particularly important because in major urban areas, the nature of territory is often constitutive of the political movements themselves. Whether it be the power struggles in and around Los Angeles (Davis, 1991; Purcell 2001), the 'unique socioterritorial fabric' of the Veneto (Bialasiewicz, 2006), the reconstituted, reimagined, replanned London of the Livingstone-led Greater London Authority, or the complex informational corridor space of Mahathir's Malaysia, spatial relations are a fundamental starting point for urban political analysis. Debates over scale and autonomy are bound into the specificities and contingencies of these spaces.

References

Amin, A. and Thrift, N. (2002) *Cities: Reimagining the Urban*, Cambridge: Polity.

Bialasiewicz, L. (2006) 'Geographies of production and the contexts of politics: dislocation and new ecologies of fear in the Veneto *città diffusa*', *Environment and Planning D: Society and Space*, 24 (1): 41–67.

Brenner, N. (2004) *New State Spaces: Urban Governance and the Rescaling of Statehood*, Oxford: Oxford University Press.

Brown, M., Knopp, L. and Morrill, R. (2005) 'The culture wars and urban electoral politics: sexuality, race and class in Tacoma, Washington', *Political Geography*, 24 (3): 267–91.

Bunnell, T. (2004) *Malaysia, Modernity and the Multimedia Super Corridor: A Critical Geography of Intelligent Landscapes*, London: Routledge.

Cox, K. (1993) 'The local and the global in the new urban politics: a critical review', *Environment and Planning D: Society and Space*, 11 (3): 433–48.

Cox, K. (1998) 'Spaces of dependence, spaces of engagement and the politics of scale, or: looking for local politics', *Political Geography*, 17 (1): 1–23.

Davis, M. (1991) *City of Quartz: Excavating the Future in Los Angeles*, London: Verso.

Fainstein, S. (2001) *The City Builders: Property Development in New York, and London, 1980–2000*, Lawrence, KS: University Press of Kansas.

Graham, S. and Marvin, S. (2001) *Splintering Urbanism: Networked Infrastructure, Technological Mobilities, and the Urban Condition*, London: Routledge.

Greater London Group (1992) 'Annex', in Coopers and Lybrand Deloitte, London, *World City: Report of Studies*, i. London: London Planning Advisory Committee.

Hall, T. and Hubbard, P. (eds) (1998) *The Entrepreneurial City: Geographies of Politics, Regime and Representation*, Chichester: John Wiley.

Harvey, D. (1989) 'From managerialism to entrepreneurialism: the transformation of governance in late capitalism', *Geografiska Annaler*, 71B (1): 3–17.

Jessop, B., Peck, J. and Tickell, A. (1999) 'Retooling the machine: economic crisis, state restructuring, and urban politics', in A.E.G. Jonas and D. Wilson (eds), *The Urban*

Growth Machine: Critical Perspectives, Two Decades Later, Albany, NY: SUNY Press, pp. 141–59.

Jonas, A.E.G. and Wilson, D. (eds) (1999) *The Urban Growth Machine: Critical Perspectives, Two Decades Later,* Albany, NY: SUNY Press.

Judd, D.R. and Swanstrom, T. (2006) *City Politics: The Political Economy of Urban America* (5th edn), New York: Pearson Longman.

Kipfer, S. and Keil, R. (2002) 'Toronto Inc.? Planning the competitive city in the New Toronto', *Antipode*, 34 (2): 227–64.

Logan, J.R. and Molotch, H.L. (1987) *Urban Fortunes: The Political Economy of Place,* Berkeley: University of California Press.

McNeill, D. (2001) 'Embodying a Europe of the Cities: the geographies of mayoral leadership', *Area* 33 (4): 353–9.

McNeill, D. (2002a) 'Livingstone's London: left politics and the world city', *Regional Studies*, 36 (1): 75–91.

McNeill, D. (2002b) 'Mayors and skyscrapers: Ken Livingstone, London and world city discourse', *International Planning Studies*, 7 (4): 325–34.

McNeill, D. (2003) 'Mapping the European Left: the Barcelona model', *Antipode*, 35 (1): 74–94.

ÓTuathail, G. (1998) 'Political geography III: dealing with deterritorialisation', *Progress in Human Geography*, 22 (1): 81–93.

Olds, K. (1995) 'Globalization and the production of new urban spaces: Pacific Rim mega-projects in the late 20th century', *Environment and Planning A*, 27 (11): 1713–43.

Olds, K. (2001) *Globalization and Urban Change: Capital, Culture and Pacific Rim Mega-Projects,* Oxford: Oxford University Press.

Peck, J. (1995) 'Moving and shaking: business elites, state localism, and urban privatism', *Progress in Human Geography*, 19 (1): 16–46.

Peck, J. and Tickell, A. (2002) 'Neoliberalising space', *Antipode*, 34 (3): 380–404.

Peterson, P. (1981) *City Limits,* Chicago: University of Chicago Press.

Pimlott, B. and Rao, N. (2002) *Governing London,* Oxford: Oxford University Press.

Purcell, M. (2001) 'Metropolitan political reorganization as a politics of urban growth: the case of San Fernando Valley secession', *Political Geography*, 20 (5): 613–33.

Rao, N. (2000) *Reviving Local Democracy: New Labour, New Politics?* Bristol: Policy Press.

Quilley, S. (1999) 'Entrepreneurial Manchester: the genesis of elite consensus', *Antipode*, 31 (2): 185–211.

Quilley, S. (2000) 'Manchester First: from municipal socialism to the entrepreneurial city', *International Journal of Urban and Regional Research*, 24 (3): 601–15.

Roberts, R. and Kynaston, D. (2001) *City State: A Contemporary History of the City of London and How Money Triumphed,* London: Profile.

Rutheiser, C. (1996) *Imagineering Atlanta: The Politics of Place in the City of Dreams,* London: Verso.

Stone, C.N. (1989) *Regime Politics: Governing Atlanta 1946–1988,* Lawrence, KS: University Press of Kansas.

Swyngeduow, E., Moulaert, F. and Rodriguez, A. (2002) 'Neoliberal urbanization in Europe: large-scale Urban Development Projects and the New Urban Policy', *Antipode*, 34 (3): 542–77.

Weber, R. (2002) 'Extracting value from the city: neoliberalism and urban redevelopment', *Antipode*, 34 (3): 519–40.

Yeang, K. (1987) *Tropical Urban Regionalism*, Singapore: MIMAR.

17 SPEED AND SLOWNESS

Alan Latham and Derek McCormack

This chapter

O Suggests that particular ideas about the pacing of city life have informed specific strategies of urban planning and ordering

O Critiques the literature which states that the inexorable speeding up of city life is emaciating urban social life

O Shows that speed and slowness are interrelated, with some aspects of city life slowing down as others speed up

---

Introduction

One of the most obvious features of contemporary cities is their speed. Modern cities appear to be fast. Indeed, in the popular imagination they appear to be endlessly accelerating. The experience of this speed and acceleration is something about which both academic and cultural commentators are profoundly ambivalent. Writing just a few years before the First World War the Italian futurist Marinetti (in Kern, 2003: 119) proclaimed 'the world's magnificence has been enriched by a new beauty; the beauty of speed. A racing car whose hood is adorned with great pipes that seems to ride on grapeshot is more beautiful than the *Victory of Samothrace*.' In a similar vein the modernist architect Le Corbusier in his *Vers une architecture* wrote of the need 'to use the motorcar to challenge our houses and great buildings' (in Wollen and Kerr, 2002: 22), while simultaneously dreaming of a vertical city where the tedious slowness of the pedestrian street would be replaced by expansive freeways and airplane traffic.

If both Le Corbusier and Marinetti deployed a rhetorical hyperbole characteristic of all *avant-garde* writing, their celebration of speed is by no means exceptional. At the same time as Le Corbusier was writing *Vers une architecture*,

American realtors were building a new kind of city, quite different from that imagined by continental European modernists, but premised nonetheless on the speed and spatial reach of technologies like the automobile and telephone. This new automobile city radically burst the boundaries of all previous urban forms – in 1931 the Californian real-estate commissioner Stephen Bornson (in Fishman, 1987: 155) wrote of how he saw 'California as a deluxe subdivision – a million acre subdivision'. And some forty years later the thoroughly sober and sensible urban geographer Peter Hall (1969: 272), in his prescient book *London 2000*, imagined a future London stretching out from Cambridge to Brighton and woven together by a vast network of high-speed 'expressways'. These motorways would be as much aesthetic as technological wonders: 'a brilliant sight ... now trenching over the side of railways, now flying over rooftops, now burrowing through the heart of reconstructed shopping and office centres.'

Paralleling these more celebratory hopeful accounts of the potentiality of speed to engineer new landscapes and capacities for transformation is an equally vigorous tradition of urban critique in which the acceleration of the contemporary city is seen as a process destroying all that makes space and place truly human. By the 1850s, when the poet and social critic Matthew Arnold ([1852] in Coleman, 1973: 130) published *The Future*, his famous meditation on the impact of mass urbanization on the English landscape, it had become *de riguer* among Europe's literary elites to lament the speed and chaos of the modern:

> This tract which the river of Time
> Now flows through with us, is the plain.
> Gone is the calm of its earlier shore.
> Border'd by cities and hoarse
> With a thousand cries is its stream.
> And we on its breast, our minds
> Are confused as the cries which we hear,
> Changing and shot as the sights which we see.

A few decades later the German philosopher Ferdinand Tönnies (1957 [1887]) gave sociological form to Arnold's critique in his classic *Gemeinschaft und Gesellschaft (Community and Society)*. Tönnies characterized the new industrial city as demanding a quickness of action and thought that ultimately divided urban dwellers from themselves. In a similar vein, but emerging from a quite different intellectual tradition, the left-wing social critic Walter Benjamin (1969 [1936]) strung an analytic bow connecting rapid industrialization, the acceleration of everyday life, and an apparent decline in the quality of collective life. 'Experience has fallen in value', he stated in a 1936 essay. And it was not hard to pinpoint the reason why. The vast expansion of technological capacity engendered by industrial capitalism had overwhelmed the individual's experiential abilities. Nothing demonstrated this severing of external reality and the individual's inner-life than the industrialized violence of the First World War's trench warfare. 'A generation that had gone to school on a horse-drawn streetcar now stood under the open sky in a countryside in which nothing remained unchanged but the clouds, and beneath these clouds, in a field of force of destructive torrents and explosions, was the tiny, fragile human body' (Benjamin, 1969 [1936]: 84–5). For many, Benjamin's observations continue to resonate up to the present. Indeed, it takes no great effort to find writers like the sociologist Kevin Robins (1997: 132) who writes 'when we think about cities now, we are likely to talk in terms of fragmentation, disintegration, disenchantment, disillusionment'. As a result of the explosion in quantity of images about it, 'the city,' he continues, 'is no longer imageable. It is becoming lost from view.'

Picking a path between these two critical traditions, this chapter will explore the ways in which speed – and with it countervailing eddies of slowness – has come to define the experience of the contemporary city. Our aim is not to offer a rapprochement between the twin narratives about speed and the urban experience outlined above. That is a project far beyond the scope of a mere book chapter. Rather, by presenting a series of ways of thinking about the relationships between speed and everyday urban experience, we wish to navigate between the extremes of an all-encompassing critical pessimism and a straight out celebration of the power of acceleration. With this aim in mind, the chapter follows a straightforward structure. Beginning with the emergence of the industrial city in the late eighteenth and nineteenth century, we will look at the main drivers of the modern city's speed-up. The everyday experience of the industrial city was transformed through the introduction of a range of new technologies and organizational forms. These transformations created both new ways of getting on with living in a city and new spatial-temporal imaginaries through which the altered speed and scale of the industrial city was given sense and structure. Following on from this historical overview, the chapter will then move on to how these historical trends have developed within the contemporary city. In particular, we will explore the ways in which the idea of speed and acceleration, fostered by a set of technological innovations that nineteenth- and early twentieth-century commentators could scarcely have imagined, has been read by a range of social critics as evidence not only of the death of the city and the urban, but indeed of sociality itself. These arguments – like those of earlier urban pessimists – are seductive and, as we will see, not without some substance. Yet, they are also prone to exaggeration, tending to amplify to excess the processes about which they write. What we need instead is an account of the city attentive to the ways in which urban life generates eddies of slowness alongside vortices of speed. As we discuss in the final section, this gives us cause to remain hopeful that rather than non-places of alienating acceleration, contemporary cities – whether they are faster than previous ones or just imagined to be so – remain centres for all sorts of solidarities and socialities.

Speed and the industrial city

Cities have not always been understood as sites of accelerating life. Medieval European cities, for instance, were as much places of rest, worship, and periodic festivity as they were of motion (Mumford, 1938). It is really only since the emergence of the twin motors of industrialism and capitalism – both profoundly *urban* in their provenance – that cities have become particularly closely associated with notions of speed and acceleration. To understand why we have come to make sense of cities in this way it is useful to look back at the origins of our contemporary obsession with speed. As the historian Stephen Kerns (2003: 130, 129) observes, our contemporary experience of speed is profoundly shaped by nostalgia for how much slower things used to be:

> [I]f a man travels to work on a horse for twenty years and then an automobile is invented and he travels in it, the effect is both an acceleration and a slowing. In an unmistakeable way the new journey is faster, and the man's sense of it is as such. But that very acceleration transforms his former means of travel into something it had never been – slow – whereas before it was the fastest way to go.

During the nineteenth century the most obvious generators of the accelerating transformation of urban life were the technological innovations underpinning the industrial revolution. Inventions like the steam engine, the blast furnace, the Spinning Jenny, and then later the railway and steam ship offered not only more efficient ways of undertaking existing tasks – generating power, producing iron, spinning cotton, travelling across land and sea. Through their efficiency they also transformed the ways in which production, consumption, and travel were experienced. And as the previous quote from Kerns suggests, this transformation of experience was neither linear nor predictable. It was not simply the case of having more of something – whether that meant larger factories, more commodities, or more speed. The texture and rhythm of everyday experience were also radically altered in all sorts of surprising ways.

Drawing on systems theory (Luhmann, 1995), we might say that transformations in the *scale* of the industrial system also transformed *how* society was organized and arranged. Or, put another way, certain key innovations during the industrial revolution managed to draw or organize around themselves a whole world of experiences, activities, routines, and other devices. They were not simply symptoms of transformation but dynamic elements of complex human and technical systems through which socio-spatial transformation, quite literally, took place. Perhaps nothing illustrates this more strikingly – certainly not in relationship to speed – than the birth of the railway and its rapid expansion across Europe and North America. As the urban historian Wolfgang Schivelbusch (1986 [1977]) shows in his book *The Railway Journey: The Industrialization of Time and Space in the 19th Century* (see Extract 17.1), the invention of the railway did not simply make getting from one place to another faster. It also generated a new spatial and temporal grammar, new ways of thinking about and articulating the relation between self, society, and space. And while initially experienced as a profoundly alien presence, over time – and with the rapid improvement of the quality of the technology itself – the railway quickly became part of the taken-for-granted reality of nineteenth-century society. Quite simply, while railroading one world out of existence, the railway simultaneously propelled another into being.

Extract 17.1 From Schivelbusch, W. (1986 [1977]) *The Railway Journey: The Industrialization of Time and Space in the 19th Century,* Berkeley: University of California Press, pp. 159–61.

The railroad related to the coach and horses as the modern mass army relates to the medieval army of knights. In the railroad journey, the traditional experience of time and space was demolished the way the individual experience of battle of the Middle Ages is abolished in the modern army. The early descriptions of the train journey as an experience of being 'shot' through space (with the train as the projectile) no longer seem merely accidentally associated with the military realm. Structurally, the train passenger is analogous to the solider in the mass army in being conditioned by the unit in which he functions as an integral part. The conditioning of the individual in a military context can now be seen as the earliest model of all subsequent and similar conditioning in the civilian economic world. In the modern army individuals are for the first time mechanized, or even subsumed, into an organizational scheme that is completely abstract and exterior to them. In the further history of the modern age, this condition becomes increasingly common in all spheres of life.

The nineteenth-century travellers gradually got accustomed to what at first seemed frightening: the demolition of traditional time-space relationships and the dissolution of reality. The travellers developed new modes of behaviour and perceptions, forms in which the new experiential context extended itself. 'Panoramic vision' was one of these innovations, as were the new general consciousness of time and space based on train schedules and the novel activity of reading while travelling.

These new modes of behaviour and perception enabled the traveller to lose the fear that he formerly felt towards the conveyance. ... In the early descriptions of the railroad, open fear or subliminal apprehension is evident as a fear of derailment, of velocity, of collisions. People used to the more leisurely technology of the previous era were still unable to comprehend that it became possible to travel safely in something that seemed like an enormous grenade. These fears had an actual technological base, as railroad technology in its first phase still suffered from 'gaps', or infantile maladies, which were a real enough source of danger. ... This interim situation was conducive to fear, but it did not last long. The conveyance was perfected, it worked ever more smoothly, and its disquieting idiosyncrasies

(Continued)

were, if not abolished totally, at least ameliorated, or 'upholstered'. The sinister aspect of the machinery that first was so evident and frightening gradually disappeared, and with this disappearance, fear waned and was replaced by a feeling of security based on familiarity. The traveller who sat reading his newspaper instead of worrying about the ever-present possibility of derailment or collusion no doubt felt secure. His attention was diverted from the technological situation in which he found himself and directed to an entirely independent object. One might say that he felt secure because he had forgotten how disquieting the technological conveyance still was, how tremendous and potentially destructive were the amounts of energy it contained.

As Schivelbusch is careful to emphasize, the transformative power of the railway was not simply a technical or technological matter but was connected intimately to the growth of capitalism as a system of economic, social, and spatio-temporal forces seeming to lead, inevitably, to the annihilation of space by time. As a set of forces working to bring everything into circulation and movement, capitalism is defined by the logic of speed and acceleration. If, to quote Benjamin Franklin, 'time is money', then of necessity it follows that the faster things get done the better. As necessity became a virtue, an emerging capitalist society increasingly valorized the imperative to accelerate (Harvey, 1989). This applied not only to the process of production. It applied equally to consumption, where the market economy sought to generate an ever more rapid turn over of products and fashions (Lehmann, 2000).

The development of the railway shared an affinity with these wider processes. Technical and technological innovations like the railroad were not simply productive of a certain experiential sense of speed. The way in which that speed came to be organized was intimately related to other transformations in the texture of everyday experience wrought by the new monetarized, capitalistic economy. Schivelbusch, for example, examines the parallels between the manner by which railway travel reduced the individual to 'a parcel', a simple object to be transported from point A to point B, and the more general tendency within capitalism to reduce the individual to a mere economic abstraction shorn of any individuality. Similarly, the panaromic gaze engendered through the speed of the rail travel, in which the landscape outside the train's windows is reduced to a flat panorama where individual details blur together, also came to be characteristic of much urban life. The organization of department stores' displays increasingly came to be structured around vast panoramas of commodities, and customers experienced the department store as if on a journey into a foreign land (Leach, 1994). And the sheer volume of people and traffic that had come to define the industrial city created a flattening of experience where, in the words of the German sociologist Georg Simmel (1950 [1903]: 415), 'the nerves [of the urban dweller] reveal their final possibility of adjusting themselves to the content and form of metropolitan life by renouncing their response to them'.

Simmel saw in the above mentioned 'blasé attitude' of the urban dweller the potential for the authentic realization of the individual personality. Most other contemporary commentators were much more sceptical. In 1881 the American psychologist George M. Beard published his influential *American Nervousness*, which catalogued the alleged dangers of an accelerating culture. In Germany, Willy Hellpach's *Nervosität und Kultur* spoke of how innovations in communication and transport technology were generating an 'overwhelming increase of normal metal processes', while the French social critic Gabriel Hanotaux wrote of a society in which 'we are burning our way during our stay in order to travel more rapidly' (Kern, 2003: 125–6). The German journalist and social commentator, Joseph Roth (1894–1939) offered perhaps the best response to this Greek chorus of criticism of the supposedly ever-accelerating industrial city. As he wrote, in 1924, of the railway junction:

> Whose dizzying velocity makes backwards sentimentalists fear the ruthless extermination of inner forces and healing balance but actually engenders life-creating warmth and the benediction of movement. In the triangles – polygons, rather – of tracks, the great, shining iron rails flow into one another, draw electricity and take on energy for their long journeys and into the world beyond: triangular tangles of veins, polygons, polyhedrons, made from the tracks of life: *Affirm them with me!* (Roth, 2004: 105, original emphasis)

The power of acceleration: the brave new world of the post-industrial city

While it was a veritable tangle, at least the complicated geometry affirmed by Roth was visible, tangible. The relationship between acceleration and urban life was marked out on the urban landscape – in the physical form of roads and railways. While the speed of both road and rail networks have undoubtedly continued to contribute to the acceleration of urban life, other factors have come into play since the time about which Roth writes. Most obviously, the relation between speed and the socio-spatial organization of cities has been complicated by the emergence of digital information and communication technologies. As with the case of transport technologies, the impacts of these more recent technologies have generated extensive criticism. One of the most significant commentators on these developments has been the Spanish urbanist Manuel Castells. In both *The Informational City* (1989) and in a magisterial three-volume set *The Information Age* (1996, 1997, 1998), Castells details the social, economic, political, and spatial transformations associated with the rise of information technologies. Castells' argument is about more than cities, but one of the most important points he makes is that the city needs to be understood in terms of its position in relation to the variable geometries of the *space of flows*. This space of flows is placeless, a space of 'timeless-time', yet it is thoroughly implicated in the organizational logics of physical space. This, then, is one of the key dynamics shaping the post-industrial city, the interaction between apparently abstract, virtual processes – a 'city of bits' (Mitchell, 1995) – and the lived experience of contemporary urban space (see also Harvey, 1989; Jameson, 1991; Bauman, 2000). It is also worth stressing that, perhaps not surprisingly, the experience of these dynamics is differentiated along socio-economic and spatio-temporal lines. There is a global geography to these processes – albeit one whose discussion is beyond the limited scope of this chapter. But even in Western cities we can discern this differentiation. Or put another way, not everyone has the capacity to manipulate time to productive ends, and as such, not everyone has the capacity to negotiate the tensions and inequalities between what Zygmunt Bauman (2000) calls fast- and slow-time. That said, we cannot simply say that the poor live in a 'slow' city and the educated and wealthy in a 'fast' one. The topologies of speed and slowness in the contemporary city are much more complicated.

Curiously, in *The Information Age,* Castells does not tell us much about the internal texture of everyday experience in this new hyper-connected city. Perhaps, then, it is useful to turn to another thinker who has. While there are many theorists of the post-industrial city, few are as bold or as acidic in their assessment of the changes wrought on the urban by acceleration, speed, and information technologies as French architect, planner, urbanist, and philosopher Paul Virilio. Virilio (1986, 1991, 1995, 2005) writes of an accelerated city of flows and bits; of an urban condition dominated by the condition and experience of speed. Like Schivelbusch, Virilio traces the emergence of the contemporary condition of speed back to the invention and proliferation of modern technologies like the train and the automobile. At the same time he is also interested in the alignment between such technologies and the imperatives and logistics of both information/communications systems and military technology (see also Mattelart, 1994; Luke and O'Tuathail, 2000). For Virilio, there is a certain entrainment and acceleration of the relative rhythms of these technologies – each operates according to the determining logic of speed. Thus, transport technologies have generated a 'shrinking' effect, which in turn has been amplified by an acceleration of information transmission, mediatization, and delivery.

Extract 17.2: From Virilio, P. (1991) *The Lost Dimension*, New York: Semiotext, pp. 9–27.

Thanks to the cathode-ray tube, spatial dimensions have become inseparable from their rate of transmission. As a unity of place without any unity of time, the City has disappeared into the heterogeneity of that regime composed of the temporality of advanced technologies. The urban figures are no longer designated by a dividing line that separates here from there. Instead it has become a computerized timetable. ...

Where once one necessarily entered the city by means of a physical gateway, now one passes through an audiovisual protocol in which the methods of audience and surveillance have transformed even the forms of public greeting and daily reception. Within this place of optical illusion, in which the people occupy transportation and transmission time instead of inhabiting space, inertia tends to renovate an old sedentariness, which results in the persistence of urban sites. With the new instantaneous communications media, arrival supplants departure: without necessarily leaving, everything 'arrives'.

After the spatial and temporal distances, *speed* distance obliterates the notion of physical dimension. Speed suddenly becomes a primal dimension that defies all temporal and physical measurements. This radical erasure is equivalent to a momentary inertia in the environment. The old agglomeration disappears in the intense acceleration of telecommunications, in order to give rise to a new type of concentration: the concentration of a domiciliation with domiciles, in which property boundaries, walls and fences no longer signify the permanent physical obstacle. Instead they now form an interruption of an emission or of an electronic shadow zone which repeats the play of daylight and the shadow of buildings. ...

Since the beginning of the twentieth century, the classical depth of field has been revitalized by the depth of time of advanced technologies. Both the film and aeronautics industries took off soon after the ground was broken for the grand boulevards. The parades on Haussmann Boulevard gave way to the Lumière brothers' accelerated motion picture inventions; the esplanades of Les Invalides gave way to the invalidation of the city plan. The screen abruptly became the city square, the crossroads of all mass-media. ...

Taken together these interlinked developments have facilitated the emergence of a kind *dromocracy* – in which relations between places, objects, and people are defined in terms of their relative speeds. As the accompanying extract indicates, Virilio's observations about the spatial dimensions of these changes amplify elements of the rather more sober arguments of Castells and others (see Sassen, 1991). Like Castells, Virilio depicts a post-industrial city fragmented and reorganized by the accelerating powers of information technology and transmission. In this context what matters is not so much physical distance from centres of power, but one's ability to access infrastructures of speed and connectivity.

For Virilio, the inevitable outcome of this new logic is the decline of the 'real' geographic city as a distinctive space, no longer structured along the familiar spatial reference points of urban and rural, centre or periphery. Indeed, for Virilio, the real city has become a mere spectral presence, a memory or residue of what it once was. If the 'real' city is disappearing, then emerging in its place is the city as a 'reality effect'. Key to understanding this is Virilio's interest in cinema as the distinctive logic of perception and the screen as the defining modern 'way of seeing'. As the accompanying extract demonstrates, Virilio points to an important association between the development of the screen and the experience of space. Here there are echoes of Schivelbusch's analysis of the experiential transformations associated with train travel. Similarly, in the present, the experience of the movement-space of aeroplanes has become closely linked to the logics of cinematic perception. Anyone who has taken a long-distance flight will recognize this in the simple fact that the cabin and seats of aircraft have come to function as entertainment centres. At the same time the airport has become 'nothing put a projector, a site of accelerated ejection' (Virilio, 2005: 98).

Virilio's vision of cities as sites of acceleration is often seductive, particularly because his writing style frequently conveys successfully a sense of movement, fragmentation, and acceleration. However, we need to treat this vision with a degree of caution, not least because Virilio is interested in mapping de-contexualized and trans-historical 'tendencies' rather than producing grounded empirical research (Luke and O'Tuathail, 2000; Thrift, 2005). Unsurprisingly, Virilio is too prone to hyperbole, too eager to assume the pervasiveness of processes he describes. Most hyperbolic is Virilio's frequent insistence on the annihilation of city space by the logics of accelerated temporality. Virilio's vision of the city of speed is also a profoundly dystopian and disenchanting one (Thrift, 2005). It presents the decline of the city under the tyranny of acceleration as a *fait accompli*. Furthermore, the sense of despair coursing through Virilio's writings about the city seems to spring from an implicit – if unelaborated – nostalgia for an older sense of urban life. Yet, at the same time, Virilio seems to offer little in the way of hope or alternative. If things are this bad, and if urban geo-politics – the politics of space and territory – has become everywhere out-paced by chrono-politics – the politics of speed – then what is to be done? How might we re-imagine life in other, less accelerated, less abstract, and less apocalyptic ways?

Slow cities, more sociality

Virilio heightens our awareness of the different ways in which the experience and organization of the urban can be understood and analyzed (with different degrees of nuance and rigour) through questions of speed. But he also accepts too easily that cities are becoming faster and this is eroding authentic social life. We need to be a little more cautious and considered. In the first place, acceleration is not always an inevitable consequence of technological change, a fact to which anyone stuck in traffic around a major urban centre will attest. In the second place, even if cities are faster – and we do not doubt they are – this is not a homogeneous process. They also contain spaces of 'slowness'. Or another way, we might better understand what is happening in cities in terms of a greater differentiation of the velocities of urban life. And, in turn, we also need to see this differentiation as generative of forms of sociality, of all sorts of ways of inhabiting cities.

One way to consider the importance of this differentiation is through questions of the body and embodied experience. As we have already seen, one of the most remarked upon elements of the introduction of new technologies is the way in which they seem to reconfigure embodied experience. So, as Schivelbusch documents, the railway journey seemed to confuse the familiar reference points of embodied experience, accelerating the individual while also rendering him or her immobile, a kind of sedentary sense of speed. A similar analysis can be made of the experience of other forms of travel, including the automobile and the aeroplane (see Baudrillard, 1988; Auge, 1995; Gottdeiner, 2000). Indeed, the logics of post-industrial city seem to shift the balance of experience even further away from the lived temporality and rhythm of the fleshy body (Figure 17.1). As we have already argued in the previous section, this kind of argument is best exemplified in the writing of Virilio. If we are to believe Virilio, the physical body is being left behind, transformed into code and bits, rendered redundant in the face of the speed of information transfer. And, as consequence of their enforced sedentariness, post-modern bodies need to find new ways of exercising (often alone, or with machines) in order to become fast, fit, and flexible enough to be actively responsible producers and consumers in a society of speed. To this one might add the growing imperative for urban dwellers to do a range of things 'on the go', as it were – make phone calls, work, eat energy bars, even wash their hair (Banham, 1973; Glieck, 1999; Honore, 2004).

But while we might well observe elements of these tendencies, we need to be careful not to use them to make sweeping diagnoses of the contemporary condition. Things are a lot more complicated than commentators such as Virilio and many other like-minded urban critics (e.g., Bauman, 2000; Dear, 2000; Soja, 2000) would have us believe. While a practice such as driving encourages a certain mobile sedentariness – to say nothing of obesity – it also needs to be remembered that driving itself remains a thoroughly embodied experience, one productive of distinctive affective attachments to the urban environment. It also generates spaces of experience which individuals find ways of using to their own idiosyncratic ends. John Katz, in his book *How Emotions Work* (1999), for example, examines how driving on the Los Angeles freeways is bound into a complex emotional economy of courtesy, respect, mutual obligation, and of course – sometimes – anger (see also Latham 2004; Latham and McCormack, 2004; Laurier, 2004; Sheller, 2004; Thrift, 2004). In addition, there are various efforts to deliberately work against the accelerating imperatives of a society in which everything, in the words of Virilio, seems to 'arrive'. Here one might point to the slow food movement, or indeed to the nascent sense among some consumers of the value of sourcing local food instead of assuming that anything can be sourced quickly at anytime in the year. Of course these are minor tendencies, but they point to the necessity of reminding ourselves of the different rhythms at which urban life can be actively experienced and potentially transformed (see Lefebvre, 2004).

And, as the extract below by Rebecca Solnit illustrates, eddies of slowness can emerge in the unlikeliest of spaces – in this case the strip in Las Vegas. Here Solnit

Figure 17.1 Cars and traffic define the rhythm and sense of speed of many contemporary cities

remarks upon the apparent rediscovery of an activity – walking – the pace of which many of the people take to define liveable cities. As Solnit documents in detail in the book from which this extract is taken, the relative slowness of walking as an embodied activity has been central to a number of attempts to apprehend the urban, typified by writers like Charles Baudelaire (Benjamin, 1973), Joseph Roth (2004), or Iain Sinclair (2002). For such commentators, the value of walking is precisely the way in which it allows one to engage in a kind of mobile contemplation, a slow but thorough immersion in the rhythms of everyday life. Clearly, there is a risk of romanticizing walking and one must not ignore the fact that only certain kinds of people tended to indulge in such activity – typically educated white males (although see Wilson, 1991). The point, however, is to reaffirm that such an activity – and its more leisurely, perhaps contemplative, pace – continues to exist alongside faster ways of being in the city.

Extract 17.3 From Solnit, R. (2001) *Wanderlust: A History of Walking*, London: Verso, pp. 254–5.

Las Vegas's downtown was built around the railway station: visitors were expected to get off the train and walk to the casinos and hotels of downtown's compact Glitter Gulch area around Fremont Street. As cars came to supersede trains for American travellers, the focus shifted: in 1941 the first casino-hotel complex went in along what was then the highway to Los Angeles, Highway 91, and is now the Las Vegas Strip. Long ago, after falling asleep in a car heading for the annual antinuclear gathering at the Nevada Test Site, I woke up when we came to a halt at a traffic light on the Strip to see a jungle of neon vines and flowers and words dancing, bubbling, exploding. I still remember the shock of that spectacle after the blackness of the desert, heavenly and hellish in equal measure. In the 1950s, cultural geographer J.B. Jackson described the then-new phenomenon of roadside strips as another world, a world built for strangers and motorists. 'The effectiveness of this architecture is finally a matter of what that other world is: whether it is one that you have been dreaming about or not. And it is here that you begin to discover the real vitality of this new other-directed architecture along our highways: it is creating a dream environment for our leisure that is totally unlike the dream environment of a generation ago. It is creating and at the same time reflecting a new public taste.' ...

In recent years, however, something wholly unexpected has happened on the Strip. Like those islands where an introduced species reproduces so successfully that its teeming hordes devastate their surroundings and starve *en masse*, the Strip has attracted so many cars that its eight lanes of traffic are in continual gridlock. Its fabulous neon signs were meant to be seen while driving past at a good clip, as are big signs fronting mediocre buildings on every commercial strip, but this Strip of Strips has instead in the last several years become a brand-new outpost of pedestrian life. The once-scattered casinos on the Strip have grown

(Continued)

together into a boulevard of fantasies and lures, and tourists can now store their car in one casino's behemoth parking lot and wander the strip on foot for days, and they do, by the millions – more than 30 million a year, upward of 200,000 at once on the busiest weekends. Even in August, when it was about 100 degrees Fahrenheit after dark, I have seen the throngs stream back and forth on the strip, slowly – although not much more slowly than the cars. ... It seemed to me that if walking could suddenly revive in this most inhospitable and unlikely place, it had some kind of future, and that by walking the strip I might find out what that future was.

Furthermore, and returning to Virilio's vision of the individual human body estranged from the material reality of the contemporary city, we can also see it is simply incorrect to reduce questions of leisure, sport, and fitness (including walking) to attempts to keep up to speed with an ever accelerating urban life. The fact is such activities work at a whole range of different paces. And they are also generative of different kinds of sociality, of various scales and intensities. Consider, for example, the case of running. While it can often be an individual affair, it is by no means exclusively so. Running also acts as the focus for all sorts of communities of interest – those of the running club, the athletics store, the informal Sunday running group, to name but three examples. What is more, running also organizes around itself all sorts of events of collective sociality in the city, including fun runs, 10k races, half marathons and marathons proper. These events can bring cities – or parts of cities – to a kind of halt, profoundly altering their rhythm even if only for short periods of time. The most recent New York marathon attracted two million spectators as well as 40,000 participants. Events such as these have become genuine folk festivals, where, along with the elite competitors, amateur runners, who are often anything but fast, are celebrated simply for taking part. Indeed, for the great majority of participants what matters is not the time taken to complete the marathon, but that they have made it to the end.

Summary

So we can say that cities are speeding up in certain ways and slowing down in others. Readers may find this a rather unsatisfactory conclusion. But to argue otherwise is to ignore the texture and detail of urban life and space. While both celebratory and hyper-critical accounts of the acceleration of urban life are seductive in their simplicity – we are either for or against speed – the real challenge for urban theory is to document the kinds of accelerations that are shaping our cities, and consider what this might mean for our understanding of cities as a collective enterprise (Hoffmann-Axthelm, 1993). This demands that we try to do four things.

First, following the exemplary work of Schivelbusch, we need to pay careful attention to the complex histories and geographies in which techniques and technologies of acceleration are implicated. What is most striking about the historical account presented by Schivelbusch is how the new technology of the railway did not simply make life 'faster'. Rather, the railway was involved in a reconfiguration of how society organized time-space that made some things slower just as it made others quicker. And while this reconfiguration was in no small part driven by the railway's remarkable capacity for overcoming distance, the ways this speed-up was experienced and put to use was equally bound up with parallel transformations in the organization of economic life. Second, and again informed by the work of social historians such as Schivelbusch, we need to avoid the temptation to exaggerate the accelerating tendencies of modern life, in the manner that a theorist like Virilio does. That is not to say that the kinds of account produced by Virilio have no value. But they need to be read in particular and limited ways. Rather than close analyses of how cities work or definitive mappings of the current state of the urban condition, Virilio's highly charged musings are better understood as speculative vehicles – ways of imagining possible futures. They allow us to race ahead and think about what might happen if certain processes continue in determinate directions. As such, they are distinctive kinds of techniques of thinking speed.

Third, if we are to use such speculative vehicles in negotiating the fine line between urban fact and urban futures, we need to avoid technological determinism, in which technology appears to work effortlessly to reshape urban space and experience. Technologies of urban acceleration, including information technologies, are embedded in complex assemblages of actors and devices. What appears inevitable in hindsight is the outcome of a multiplicity of negotiations and associations at many points of which things might have been otherwise. Indeed, one of the most striking elements of both Solnit and Schivelbusch's accounts is the degree to which the complex topologies of time-space they are exploring are emergent. They are not the product of singular, linear, or predictable processes of socio-economic change. Who would have expected that Las Vegas might be home to a rebirth of the pedestrian street? Or, that the remarkable speed of the train and airplane travel might generate a sense of timelessness?

So, finally, the speed of urban change is relative, only making sense in relation to notions of slowness. To claim that everything is accelerating is the same as claiming that nothing is. Ultimately, what we need to try to understand is the remarkable spatial-temporal flux within which contemporary cities are organized. As Robert Musil, writing about *fin-de-siècle* Vienna puts it:

> Like all big cities it was made up of irregularity, change, forward spurts, failures to keep step, collisions of objects and interests, punctuated by unfathomable silences; made up of pathways and untrodden ways, of one great rhythmic beat as well as the chronic discord and mutual displacement of all its contending rhythms (Musil, 1995: 4).

References

Auge, M. (1995) *Non-Place: An Introduction to the Anthropology of Super Modernity*, London: Verso.

Banham, R. (1973) *Los Angeles: The Architecture of Four Ecologies*, London: Penguin.

Baudrillard, J. (1988) *America*, London: Verso.

Bauman, Z. (2000) *Liquid Modernity*, Cambridge: Polity.

Benjamin, W. (1969 [1936]) *Illuminations*, London: Fontana.

Benjamin, W. (1973) *Charles Baudelaire*, London: Verso.

Castells, M. (1989) *The Informational City*, Oxford: Blackwell.

Castells, M. (1996) *The Rise of the Network Society*, Oxford: Blackwell.

Castells, M. (1997) *The Power of Identity: The Information Age Vol. II*, Oxford: Blackwell.

Castells, M. (1998) *End of Millennium: The Information Age Vol. III*, Oxford: Blackwell.

Coleman, B. (1973) *The Idea of the City in Nineteenth Century Britain*, London: Routledge.

Dear, M. (2000) *The Postmodern Urban Condition*, Oxford: Blackwell.

Fishman, R. (1987) *Bourgeois Utopias: The Rise and Fall of Suburbia*, New York: Basic Books.

Glieck, J. (1999) *Faster: The Acceleration of Just About Everything*, New York: Little Brown.

Gottdiener, M. (2000) *Life in the Air: The New Culture of Air Travel*, New York: Rowman & Littlefield.

Hall, P. (1969) *London 2000*, London: Macmillan.

Harvey, D. (1989) *The Urban Experience*, Oxford: Blackwell.

Hoffmann-Axthelm, D. (1993) *Die dritte Stadt: Bausteine eines neuen Gründungsvertrages*, Frankfurt am Main: Suhrkamp Verlag.

Honore, C. (2004) *In Praise of Slow: How a Worldwide Movement is Challenging the Cult of Speed*, London: Orion.

Jameson, F. (1991) *Postmodernism, or, The Cultural Logic of Late Capitalism*, London: Verso.

Katz, J. (1999) *How Emotions Work*, Chicago: The University of Chicago Press.

Kern, S. (2003) *The Culture of Time and Space, 1880–1918*, Cambridge, MA: Harvard University Press.

Latham, A. (2004) 'American dreams, American empires, American cities', *Urban Geography*, 25 (8): 788–91.

Latham, A. and McCormack, D. (2004) 'Moving cities: rethinking the materialities of urban geographies', *Progress in Human Geography*, 28 (6): 701–24.

Laurier, E. (2004) 'Doing office work on the motorway', *Theory, Culture, and Society*, 21 (4–5): 261–77.

Leach, W. (1994) *Land of Desire: Merchants, Power, and the Rise of a New American Culture*, New York: Vintage.

Lefebvre, H. (2004) *Rhythmanalysis: Space, Time and Everyday Life*, translated by S. Elden and G. Moore, London: Athlone.

Lehmann, U. (2000) *Tigersprung: Fashion in Modernity*, Cambridge, MA: MIT Press.

Luhmann, N. (1995) *Social Systems*, Palo Alto, CA: Stanford University Press.

Luke, T. and Ó'Tuathail, G. (2000) 'Thinking geopolitical space: the spatiality of war, speed and vision in the work of Paul Virilio', in M. Crang and N. Thrift (eds), *Thinking Space*, London: Routledge, pp. 360–79.

Matterlart, A. (1994) *Mapping World Communications: War, Progress, Culture*, translated by Susan Emanuel and James A. Cohen, Minneapolis: University of Minnesota Press.

Mitchell, T. (1995) *City of Bits: Space, Place and the Infobahn*, Cambridge, MA: MIT Press.

Mumford, L. (1938) *The Culture of Cities*, London: Secker & Warburg.

Musil, R. (1995) *The Man Without Qualities*, London: Picador.

Robins, K. (1997) *Into the Image: Culture and Politics in the Field of Vision*, London: Routledge.

Roth, J. (2004) *What I Saw: Reports from Berlin 1920–1933*, New York: W.W. Norton & Company.

Sassen, S. (1991) *The Global City*, Princeton, NJ: Princeton University Press.

Schivelbusch, W. (1986 [1977]) *The Railway Journey: The Industrialization of Time and Space in the 19th Century*, Berkeley: University of California Press.

Sheller, M. (2004) 'Automotive emotions: feeling the car', *Theory, Culture and Society*, 21 (4–5): 221–47.

Simmel, G. (1950 [1903]) 'The metropolis and mental life', in K. Wolff (ed.), *The Sociology of Georg Simmel*, New York: Free Press, pp. 409–24.

Sinclair, I. (2002) *London Orbital: A Walk around the M25*, London: Granta.

Soja, E. (2000) *Postmetropolis: Critical Studies of Cities and Regions*, Oxford: Blackwell.

Solnit, R. (2001) *Wanderlust: A History of Walking*, London: Verso.

Thrift, N. (2004) 'Driving in the city', *Theory, Culture and Society*, 21 (4–5): 41–59.

Thrift, N. (2005) 'Panicsville: Paul Virilio and the aesthetic of disaster', *Cultural Politics*, 1 (3): 337–48.

Tönnies, F. (1957 [1887]) *Community and Society*, New York: Harper & Row.

Virilio, P. (1986) *Speed and Politics*, New York: Semiotext.

Virilio, P. (1991) *The Lost Dimension*, New York: Semiotext.

Virilio, P. (1995) *The Art of the Motor*, Minneapolis: University Minnesota.

Virilio, P. (2005) *Negative Horizon: An Essay in Dromoscopy*, London: Continuum.

Wilson, E. (1991) *The Sphinx in the City: Urban Design, the Control of Disorder, and Women*, London: Virago.

Wollen, P. and Kerr, J. (2002) *Autopia: Cars and Culture*, London: Reaktion.

18 PLANNING AND CONFLICT

Malcolm Miles

This chapter

○ Considers planning as both process and practice, and the role of the planner as a professional

○ Notes a decline in the public's faith in the 'expert planner', arguing for a different (and more empowering) notion of planning

○ Suggests ways in which urban studies can contribute to a more pluralistic, active and engaging form of planning

Introduction

Planning has conventionally been seen as an application of technical expertise to competing claims for the city. The planner, a professional public servant, makes disinterested judgements in the public interest, reconciling demands for growth with the needs of a city's publics. Much day-to-day work in planning concerns infrastructure, but also involves attempts to resolve conflicting claims to space and to a city's public image. In extreme situations, such as the violence of Los Angeles in April 1992, it may seem that planning has broken down (though the failures may have been political, social, and cultural as well). Even in ordinary times, questions arise as to what constitutes the public good for groups with divergent interests; and whether the aim of a conflict-free city is more than fantasy. The assumptions which pertained when planning became a specialist profession a century ago may be outmoded. Cities are more complex, if not more violent, as difference is celebrated rather than assimilated. It can also be asked how much power a city, or even a state, can exercise today in face of global competition for inward investment. My aim is not to negate the case for planning expertise in mediating urban change. After outlining some challenges to planning, and testing a few assumptions, I draw attention to new planning practices which integrate the expertise of dwellers with that of planners. Yet I wonder if a conflict-free city can be realized, or would be desirable, while refusing a notion that cities are inevitably war, zones between classes, races, and genders.

Challenges to planning

I begin with some of the challenges faced by planning in the twenty-first century. Foremost in public debate is the globalization of capital and communications, and a shift of production to labour markets in the non-affluent world. Zygmunt Bauman (1998) considers the uneven benefits of globalization: while the rich travel and shop where they like, or in cyberspace, the poor, as economic migrants or asylum seekers, are restricted in movement; global capital appeals to bodies such as the World Trade Organization for selective deregulation but the power of governments and citizens to affect development decreases. A culture of deregulation affects planning as well as trade, especially in development zones. Jon Bird (1993) calls the outcome of deregulation in London's docklands a dystopia on the Thames. For corporate executives who work in such places as Canary Wharf, there are instabilities in the demise of long-term career patterns, while semi-skilled workers are de-skilled by automated processes of production (Sennett, 1998). As cities become de-industrialized, the potential for conflict increases between those who are dispossessed and those whose power and wealth increases. Yet while local and national governments lose influence, that of citizens' groups and campaigning organizations increases through global networking.

This leads me to another challenge: the rising assertion of differences of class, gender, ethnicity, age, mobility, and sexual orientation. If gay culture and gay parades are subsumed in city marketing (in Berlin and Brighton, for example) the assertion of difference still undermines the convention that planners serve a city as a whole. There may simply be no city-as-a-whole to serve, only its diverse elements. R.E. Pahl writes of the sociologist who is able to research how the aims of different social groups conflict, but unable to say what those aims should be: 'He [sic.] is a member of society as much as anyone else' (Pahl, 1975: 188). The perspectives of specific groups may not cohere, though new insights may arise when they collide. There is a nuance also in that difference moves the ground of planning from that of the abstract space of plans to the materialities of occupation. As Manuel Castells says, interviewed by Bob Catterall: 'Women's liberation, gay and lesbian liberation and ... the environmental movement are all about the politics of the body with the battleground being the body as a physical and cultural entity ... the conflict between this logic of networking, commodification, the extraordinary explosion of creativity and technology, and, on the other hand, meaning and identity, has reached the body' (Castells, 2001: 31).

Within this non-cohering matrix the neutrality of planning is questioned by groups who see themselves outside its cognizance: voices from the borderlands, as Leonie Sandercock calls them (1998: 110). Sandercock writes that 'planning theorists have so far shown almost no interest in the many debates around difference in social theory, feminist theory, and cultural studies' and that the discipline must therefore look outside its own parameters (1998: 109). There is much there. In the 1980s, for instance, feminist critics foregrounded the exclusion of women from the public realm (Hayden, 2003 [1981]; Wolff, 1989), and in the 1990s Doreen Massey took commentators such as Edward Soja, David Harvey, and Ernesto Laclau to task for their (to her) masculinist views of space and time (1994: 212–270). And Elizabeth Wilson argues that planning's task now is to revise its purpose: 'Town planning has too often been driven by the motor of capitalist profit and fuelled by the desire to police whole communities. ... The colonial imperative was close to the surface ... the city was essentially for the white, male bourgeoisie; all others were there on sufferance' (Wilson, 1991: 156; see also Extract 18.1). In place of a colonial attitude, Wilson calls for planning to regulate urban spaces for the use and excitement of all.

Extract 18.1: From Wilson, E. (1991) *The Sphinx in the City: Urban Life, the Control of Disorder, and Women*, Berkeley: University of California Press, pp. 157–8.

Women have fared especially badly in western visions of the metropolis because they have seemed to represent disorder. There is fear of the city as a realm of uncontrolled and chaotic sexual licence, and the rigid control of women in cities has been felt necessary to avert this danger. Urban civilization has come, in fact, to mean an authoritarian control of the wayward spontaneity of all human desires and aspirations. Women without men in the city symbolize the menace of disorder in all spheres once rigid patriarchal control is weakened. This is why women – perhaps unexpectedly – have represented the mob, the 'alien', the revolutionary.

It is therefore rather ironic that women have often appeared less daunted by city life than men. For example, most of the male modernist literary figures of the early twentieth century drew ... a threatening picture of the modern metropolis (an exception being James Joyce), but modernist women writers such as Virginia Wolff and Dorothy Richardson responded with joy and affirmation. ...

The urban crisis is a crisis on inequality, and of authoritarianism. Problems of overcrowding and population growth would always be difficult to solve, but they have been made much worse by the unequal competition for urban space and the ways in which the few have commandeered almost all the resources. I am not trying to reproduce the cliché of 'bright lights, big city' versus suburban respectability and drabness, to rely on a stereotype of urban chic. It is essential to acknowledge that city centres have become increasingly places of paradox: playgrounds for the rich, but dustbins for the very poor. Nor do I have a solution for the complex problems facing city dwellers and city governments. I am arguing that we will never solve the problems of cities unless we *like* the urban-ness of urban life. Cities aren't villages; they aren't machines; they aren't works of art; and they aren't telecommunications stations. They are spaces for face to face contact of amazing variety and richness. They are spectacle – and what's wrong with that?

Wilson is critical of anti-urbanism in the bias of British town-and-country planning towards a conflict-free city emulating the conditions of a village. Before commenting on the city–village dichotomy, which I also find inadequate, I want to say a little more about what I term the planning gaze. Briefly, a conventional city plan reduces the complexities of urban life to what can be represented visually and schematically on a blank ground. John Pickles (2004) notes that an expansion of map-making in sixteenth- and seventeenth-century Europe was linked to taxation and governance. Maps were tools of power as well as of military conquest. This gives rise to a cartographic gaze presupposing 'a parametric manifold within which nature and society can be thematized in terms of their spatial relations' (Pickles, 2004: 80). Massey argues similarly that modernism – the dominant cultural and intellectual framework of the twentieth century – 'privileged vision over the other senses and established a way of seeing from the point of view of an authoritative, privileged, and male, position' (Massey, 1994: 232). Forms of this argument are widely rehearsed; I would add only, following Foucault (1973), that the cartographic and planning gazes resembles that of a physician observing patients as abstract sites of anatomic malfunction. Castells' remarks on the body (above) are pertinent here. Equally pertinent is Catherine Belsey's observation that the development of linear perspective in art entails its undoing: 'But the new science of perspective ... came in due course to acknowledge the necessity of specifying the time and place of the act of seeing' (Belsey, 2005: 98). In other words, the view which is taken to be objective is, if taken far enough, subjective. The objective viewpoint of the city plan, likewise, summons a spectre of particular times, places, sounds, textures, smells, and acts which undermine its authority and dissolve its claim to disinterest: 'Detachment does not mean here disinterested' (Massey, 1994: 232). And as Audre Lorde argued, 'the master's tools will never dismantle the master's house' (Lorde, 2000 [1979]: 54).

Planning assumptions

The implication is that planning methods need to change if they are to address conflicts arising from difference, or from competing claims to the city (Lefebvre, 1996: 147–159). Indeed, planning has changed, and in the next section of the chapter I describe some of the ways it has done so. But I want first to test two assumptions which underpin the planning gaze: that cities are natural zones of competition leading to conflict, hence an imperative of control; and that planning can predict its outcomes.

Louis Wirth, in Chicago in the 1930s, saw the city as home to a new urban character, the mobile, self-interested urbanite who is indifferent to others' needs. He writes: 'Today all of us are men on the move and on the make, and all of us by transcending the bounds of our narrower society become to some extent marginal men' (Wirth, 1938, in Smith, 1980: 9; see also Wirth, 1964). He does not mention marginal women, but Chicago sociologists studied taxi-hall dancers (women who danced for money) as well as hobos (Anderson, 1923). Some people's marginality is still more comfortable than others', but Wirth's view, like that of Robert Park and Ernest Burgess, was influenced by early European sociologists Georg Simmel (1969 [1950], 1990 [1978]) and Fernand Tönnies. For Tönnies (1940, 1955), the city differed radically in its patterns of sociation from the village: rural life is characterized by continuity and ties of kin and land; but in cities, people from different backgrounds made more transient alliances, or 'communities of interest'. In his 1903 essay 'Metropoles and Mental Life' (Frisby and Featherstone, 1997: 174–185), Simmel writes of the smooth flow of rural life in contrast to an intensification of nervous (biological) stimulation in the city. The senses are divided against themselves, as it were, and the mind over-stimulated by constant signals with which it cannot fully cope. Hence the blasé metropolitan attitude.

Wirth's position extends Simmel's in the context of Chicago, and universalizes it. Similarly, Burgess (1925, in LeGates and Stout, 2003: 157–163) adopts a biological model for zones of transition produced by waves of migration. I am worried by this because the natural is outside history, except in the long cycles of natural selection, and hence outside human agency. The difficulty of planning as a response to a perceived inherent instability in the city is, too, that it implies control by means likely to be increasingly authoritarian. This is what happens in the case of Los Angeles as reported by Mike Davis (1990; see also Miles, 2000: 44–48). But I am sceptical of urban war stories, and of a model based on a supposed chaos for which the foil is village life. Pahl writes that the question as to whether a village is a stable community is raised 'more frequently than it is honestly answered' (Pahl, 1975: 41). He reads the idea of a village as a community made by ties of family and place as 'nearer to the middle-class image of a village' (ibid.) than to the reality of villages which, in the post-war years, became retreats for commuters and owners of weekend cottages. I wonder, too, if regulation works in its own terms of delivering predicted outcomes. A decline in the acceptance of the authority of professional expertise may follow a failure of the modernist project in planning and architecture to engineer a new society by design. This failure is evident in the destruction of tower blocks of social housing in London, Baltimore, and other cities, and in stories of the unease of living in 'concrete jungles'. The stories may be urban myths, and the estates of post-war re-housing were not all made of tower blocks (Chapter 4), but, as Edward Robbins argues (1996), they embodied a belief that poor people are unable to organize their lives and spaces for themselves and must therefore be given functionalized environments. The notion that design can produce new forms of behaviour, then, is flawed in that the kinds of behaviour produced by regulation which is disempowering may be as chaotic as those the new environment was designed to eliminate.

A further uncertainty is stated in complexity theory. From chaos theory, complexity theory (Cilliers, 1998) states that minor shifts in the conditions in which a given intervention is made may produce major shifts in outcomes. David Byrne (1997: 64–69) argues that this does not imply an abandonment of planning to market forces, nor evacuation of the concept of agency. He reads the post-war model's reliance on technocratic means to deliver democratic aims as contradictory but sees planning as a vehicle of human creativity, adding that a unity of opposites may deliver a fruitful synthesis which provides new tools for knowing the world and emphasize 'the capacity of agency in changing it' (Byrne, 1997: 51). Byrne argues for an ethical and democratic reconstruction of planning. His case is supported by evidence of self-organization among homeless people (Roschelle and Wright, 2003), and in informal settlements (Berg-Schlosser and Kersting, 2003), some of which have been legalized (Fernandes and Varley, 1998). Barbara Happe and Claus-Dieter König (2003: 119) conclude: 'The existence of numerous neighbourhood organizations in the shanty towns of South America confirms that the interests of the urban poor can, in principle, be effectively organized' (see Figure 18.1). Perhaps insights from work in the non-affluent world (Turner, 1976; Peattie, 1982; Hamdi, 1995; Serageldin, 1997) can be applied in the cities of global capital; and perhaps planning, as a profession of expertise, does accept the expertise of dwellers on dwelling as of equal value to its own. It needs also now to adapt to what Peter Marcuse (2002) terms a 'layered city', in which the spaces of business, power, production, and dwelling overlap with zones of luxury, gentrification, suburbanization, tenement-living, and abandonment: 'Each layer shows the entire space of the city, but no one layer shows the complete city.' (Marcuse, 2002: 106). And planning must face, too, the encroachment of privatization, not least on public urban spaces.

New attitudes to planning and conflict

Pahl, writing in *New Society*, observes 'a pathetic demand for a visionary, who can explain in one Sunday supplement how we should all live' (Pahl, 1975: 187). He observes, too, an underlying resentment of professional expertise: 'Planners are expected to make our life "better", but if they succeed they may be resented' (Pahl, 1975: 188). He cites sociologist and planner Herbert Gans (1962, 1967) as supporting user-oriented strategies which deliver most to those who have least (Pahl, 1975), implying that participation is key to dealing with such resentment. When Pahl wrote, in the 1960s, advocacy planning was a departure from the rational comprehensive model developed in Chicago and implemented most notoriously by Robert Moses in the planning of New York's freeways (Berman, 1982). Paul Davidoff (2003 [1965]: 2003: 389–398) argued that planners should represent the views of community groups rather than rely on quantitative data. Teaching in New York, Davidoff campaigned for the provision of racially integrated low-income housing. His position rests on a faith that 'the present can become an epoch in which the dreams of the past for an enlightened and just democracy are turned into a reality' (Davidoff, 2003 [1965]: 389). He continues that protest against discrimination has led to a range of welfare measures, and a Supreme Court ruling to guarantee equal protection under the law.

This may seem encapsulated in the 1960s, but Davidoff's idea of planning as a profession of articulation, working with specific groups to empower their voices can now be seen as forerunner to reconsiderations of planning in a multi-ethnic

Figure 18.1 Acts of resistance: political flyposters

society by Leonie Sandercock and John Forester. Sandercock sees advocacy planning as 'the first serious challenge to the rational comprehensive model' (Sandercock, 1998: 89). Both Pahl and Davidoff, nonetheless, retain the role of planning in the production of an orderly civic realm. They are reformers rather than revolutionaries. Pahl notes, though, that Gans doubts problems of poverty and segregation can in the end be solved by planning (Pahl, 1975: 190). Gans (1967) also argued that suburbanization, seen by some critics as a detraction from the urban, might have a value in housing a new middle class with coherent values, in sites such as Levittown (a private-sector mass housing development). Gans (1962) still sees suburban affluence as a gain at the expense of the poor and black, and here accords with Davidoff that planning should address the needs of specific interest groups. The question, now, is what balance is struck, and how mutably, between those interests (which are unlikely to agree with each other) and structures of governance. Sandercock suggests that Davidoff had a 'lawyer's faith in due process and enlightened plural democracy', that his model would 'serve to perfect both the rational model and pluralist democracy' as planners informed communities of the costs of development as well as representing their viewpoints (Sandercock, 1998). Pahl, Gans and Davidoff were, then, progressive in their efforts to re-vision planning as a means to address the contestation of urban space.

Urban conflict in the 1960s to 1980s was related to divisions of class and race, and the uses of space and concept of a public realm contested from a feminist position. Robert Beauregard also emphasizes the drain of population from North American cities through de-industrialization and suburbanization, in the context of a culture of decline. The armed response signs on the suburban lawns of Los Angeles on which Mike Davis based his essay 'Fortress L.A.' (Davis, 1990) denote a high point of this culture, but it was previously described by Richard Sennett in *The Uses of Disorder* (1970). While Davis wallows in disaster, and Beauregard (2003) cites renewal of central business districts as key to an urban upturn (though he also notes that for capital, cities are expendable), Sennett fuses insights from psychology and sociology. He describes the fear of outsiders in white suburbs as a fear of difference prolonging an adolescent mentality. Affirmation of white suburban homogeneity then becomes not only a mask for abuse in suburban interiors but also a necessary construction of coherence – because its lack is too terrible to bear. Sennett writes of the adolescent mind-set: 'In trying to enforce a vision of coherent order, the young person meets an immovable obstacle or social situation that is out of his [*sic.*] control. The disorderly world defeats the dreams of coherence' (Sennett, 1970: 115). He writes, too, that planning assumes a naïvely defined good city as its goal, to which conflict is 'conceived as a threat', a 'refusal to deal with the world in all its complexity and pain' (Sennett, 1970: 97). Sennett continues:

'The essence of the purification mechanism is a problem, planners think, best left to politicians and the like. Planners' sights are on that urban 'whole' instead; they are dreaming of a beautiful city where people fit together in peace and harmony (Sennett, 1970: 98).

The scale of urban failures in the 1990s put such dreams out of commission, putting new pressures on planning while shifting the profession's emphasis from mediation in the interests of a general public good to acceptance of difference and the imperfections of urban living. Beauregard, whom I read as fairly conservative, views urban decline as an inevitable corollary to growth, but also sees ambivalence towards cities as a product of a specific urban discourse in which cities are 'both fearful and alluring' (Beauregard, 2003: 245).

Sandercock notes that planning theory in the 1970s was informed by an emphasis on political economy in the work of David Harvey (1973) and Manuel Castells (1976, 1978). Her reservation is that an academic critique does not offer alternatives for 'what planners can do' (Sandercock, 1998: 92). Thomas Nagel, in *The View from Nowhere* (1978/86), writes of a related but not identical problem in moral philosophy: the difficulties of ethical living need to be dealt with 'not theoretically, but in life' (Nagel, 1989: 205). Nagel puts forward personal change and political change as ways towards a better world, without setting them in dualist opposition.

Extract 18.2: From Sanderock, L. (1998) *Towards Cosmopolis*, Chichester: Wiley, pp. 185–6.

If cultural imperialism and systemic violence are features of contemporary global urban and regional changes, then a politics of difference is a prerequisite for confronting these oppressions. A politics of difference is a politics based on the identity, needs, and rights of specific groups who are victims of any of the faces of oppression. ... The emergence of numerous social movements in the past two decades – feminism, gay liberation, Black power, indigenous rights – embodies the practice of a politics of difference, perhaps the most important aspect of which is a discursive politics. By that I mean the effort to reclaim and politicize the meaning of difference by asserting the positive qualities of the particular group and refusing to accept the dominant culture's definition of itself. In asserting gay pride, or Black is beautiful, there is a reversal of the devaluation of difference and an effort to overcome the internal colonization of selves by the dominant culture's definition.

The emergence of this identity politics has gone unchallenged, by either left or right, and a number of concerns do need to be aired. One is the assumption of group identity and homogeneity. ... Clearly none of the social movements which have asserted a positive group identity is in fact a unity. ... Another perceived problem ... is the alleged impossibility of working on broader agendas of social, economic, and environmental justice so long as oppressed groups insist on only fighting for their group-specific concerns. But while this may have been a reasonably accurate interpretation of identity politics when it first emerged, it is evident in the late 1990s that many of these social movements are now engaged in broader coalition politics. ... Further difficulties with identity politics may arise from the increasing 'hybridity' of global populations. ... The point here is that a politics of difference is not based in essentialist notions of identity but in situations, historical contexts, in which there are social relations of domination. Hybrid identities are just as vulnerable to stereotyping, vilification, and exclusion. A more serious charge against identity politics is that it, too, is oppressive, both to those inside the group ... and to those outside. ... Anyone who has been involved in social movement politics has experienced these practices; but so too have they experienced the internal struggles against them, as social groups strive to embody a more democratic politics.

This does not devalue critique, but signals a need to ground theory in practice just as the practice of dwelling is grounded in the contested values and uses of space. Discourse responds, as well, to extreme cases such as the failure of urban governance and planning in the Los Angeles insurrection of 1992: 'The city is burning. As the smoke and the glow from the fires begin to drift over the city, millions of horrified citizens huddle in front of TV sets which transmit images that confirm everyone's worst nightmares' (Sandercock, 1998: 13).

Los Angeles in 1992 is brought to mind by images of New Orleans in 2005: the poor (mainly black) marooned without water, food, or security in a city evacuated by those rich enough to leave in the days before a hurricane. Segregation is not absent in a Europe of guest-workers either, and debates on social exclusion in New Labour Britain reinforce a category which is not marked out by those said to belong to it. It emerges, though, that boundaries of otherness shift, opening a possibility for political action to reform social relations by redefining social categories. In this context, David Sibley (1995: 69) hopes 'the humanity of the rejected will be recognized and the images of defilement discarded'. A category such as the underclass remains, nonetheless, a source of fear in a process analogous to the self-coercion discerned by Foucault in the prison and asylum (1971, 1979).

David Sibley sees Foucault's work as informing discussion of social control, and I suggest a key area of reconsideration is the relation of means to ends. Sandercock charts successive departures from a conventional planning model, from equity planning (Krumholz, 1994) and communicative action (Habermas, 1984), linked to the work of Patsy Healey (1997) and Bent Flyvberg (1992), to radical planning. Drawing the chapter towards a close I want now to look at radical planning, and Iris Marion Young's work in political science. Among sources are Melvin Webber as precursor (1968, in LeGates and Stout, 2003: 470–474) and John Friedmann (1987, 1992). Peter Hall sums up Webber's position as a view of planning flatly denying 'the possibility of a stable predictable future or agreed goals' (Hall, in LeGates and Stout, 2003: 349). Radical planning operates from a recognition of inequality, and seeks to empower those regarded as outside the system: 'The identity of the radical planner ... is that of a person who has ... gone

AWOL from the profession, crossed over ... to work in opposition to the state and corporate economy' (Sandercock, 1998: 100). Time is spent in unstructured ways, hanging out with specific groups so as to be accepted by them and learn their mapping of the city. Partisan views and value judgements thus enter the planning process, which ruptures planning's claim to apply technical expertise for the general public good. The idea of a general public good may serve in any case to support the role of a dominant class, as argued by Antonio Gramsci. Susan Stokes, in context of a study of new social movements in Peru, characterizes Gramsci's understanding of hegemony as '[t]he ability of the dominant class to assume intellectual and moral leadership by claiming that their politics flows from their concern for the good of all' (Stokes, 1995: 122). Radical planning changes the power structure by beginning from an understanding of the viewpoints of those regarded as other.

There are gradations within radical planning. Sandercock relates that Allan Heskin (1991) and Jacqueline Leavitt (1994) align the radical planner with the community. Forester sees planners who deal only with data and regulations as failing 'to attend to the pressing emotional and communicative dimensions of local land-use conflicts' (Forester, 2003 [1987]: 385), but is concerned with creating dialogues between professionals and communities, not crossing to the community side, a strategy being to mediate for both sides: 'The planners may not be independent third parties who assist developers and neighbors in face-to-face meetings to reach development agreements – but they might still mediate such conflicts as "shuttle diplomats"' (Forester, 2003 [1987]: 381). For Friedmann, critical distance ensures the continuing viability of planning as a profession, which implies a paradoxical relation to sites of action – both distanced in order to be critical *and* embedded to be aware.

Artist Joseph Beuys said everyone is an artist (though not an art professor like himself); it could be argued that every dweller has a tacit knowledge of dwelling, and this constitutes a viable basis for grass-roots planning. Behind Beuys' remark is a point about the difference between representation in a social democracy and direct democracy in which citizens' acts are unmediated. Gans, in recent writing, proposes a citizens' democracy: not direct democracy nor the traditional town meeting, but more than present arrangements for involving electors in decision-making (Gans, 2003: 113). He argues, for instance, that the economy should be democratized through employee participation in company boards. While, then, capital encroaches on the public realm and decisions about the uses of space, Gans proposes a defence of public rights, or so it might seem, from within corporate bodies.

I see a connection to Young's critique of welfare capitalism. She writes 'insurgent social movements have questioned the welfare state's limitation of public debate to distribution, and sought to politicise the process of ownership and control' (Young, 1990: 66). This moves the matter out of the sphere of charity (for which gratitude is required) into that of power. It accepts power is not donated, and self-empowerment is possible. And it leads to the problem that the interests voiced in such a move will not cohere. Young states that impartiality in conventional legal models 'corresponds poorly to the social relations typical of family life and personal life, whose moral orientation requires not detachment from but engagement in and sympathy with the particular parties in a situation' (Young, 1990: 96). This challenges disinterested (Kantian) judgement as an inappropriate kind of judgement in situations of conflicting interests. Young argues, further, that the normative position taken as the outcome of disinterested judgement tends to equate with a majority position; and may not be in the interests of groups cast as minorities. Rather than being assimilated into a majority and its norms, it may be better for groups to develop distinct identities. Young (1990: 163) writes, 'Today and for the foreseeable future societies are certainly structured by groups' with emancipation viable when groups cultivate rather than submerge their sense of difference. She argues that 'from the assertion of positive difference the self-organization of oppressed groups follows' (Young, 1990: 167). But there is another question, as to how effective self-organized groups, or civic authorities, can be today in face of global capital, the fluidity of labour markets, and use of global media and branding to enforce the imperative to consume. As Bauman writes: 'The task is now to defend the vanishing public realm, or rather refurnish and repopulate the public space fast emptying' (Bauman, 2000: 39). And yet I wonder how inclusive the public realm ever was, in the slave economy of classical Athens or the beautiful cities of the nineteenth century, any more than in the arid wasteland of the twenty-first-century business park (or multi-storey car park). I see entirely the need for its defence, but equally for its repopulation to take a more radically democratic form. Activism has one response. Perhaps there are others (see Extract 18.3).

Extract 18.3: From Young, I.M. (2000) *Inclusion and Democracy*, Oxford: Oxford University Press, pp. 102–3.

Moral reason that seeks impartiality tries to reduce the plurality of moral subjects and situations to a unity by demanding that moral judgement be detached, dispassionate, and universal. But ... such an urge to totalization necessarily fails. Reducing differences to unity means bringing them under a universal category, which requires expelling those aspects of different things that do not fit into the category. Difference thus becomes a hierarchical opposition between what lies inside and what lies outside the category, valuing more what lies inside than what lies outside.

The strategy of philosophical discourse which Derrida calls deconstruction, and Adorno calls negative dialectics, exposes the failure of reason's claim to reduce difference to unity. ... The attempt to adopt an impartial and universal perspective on reality leaves behind the

(Continued)

particular perspectives from which it begins, and reconstructs them as mere appearances as opposed to the reality that objective reason apprehends. The experience of these appearances, however, is itself part of that reality. If reason seeks to know the whole of reality, then, it must apprehend all the particular perspectives from their particular points of view. The impartiality and therefore objectivity of reason, however, depends on its detaching itself from particulars and excluding them from its account of the truth. So reason cannot know the whole and cannot be unified.

Like other instances of the logic of identity, the desire to construct an impartial moral reason results not in unity, but in dichotomy. In everyday moral life, ... there are only situated contexts of action. ... The ideal of impartiality reconstructs this moral context into an opposition between its formally impartial aspects and those of its aspects that are *merely* partial and particular.

Impartial reason ... also generates a dichotomy between reason and feeling. Because of their particularity, feeling, inclination, needs, and desire are expelled from the universality of moral reason. Dispassion requires that one abstract from the personal pull of desire, commitment, care, in relation to a moral situation and regard it impersonally. ... This drive to unity fails, however. Feelings, desires, and commitments do not cease to exist and motivate just because they have been excluded from the definition of moral reason. They lurk as inarticulate shadows, belying the claim to comprehensiveness of universalist reason.

Summary

I began with reference to a view of urban planning as serving the public interest and seeing a city holistically. My argument is that the complexity of cities denies that option, but does not itself render conflict, particularly violence, inevitable. I reject Wirth's ethos of competitiveness, but equally the model of assimilation (which is always a dominant model). Recognition of difference, and self-empowerment by distinct publics, is one side of a coin, the other side of which is acceptance of continuous contestation. The alternative, often dressed in the garments of liberal reform (improvement) is repression. For Freud, the repressed has a way of returning to haunt us.

References

Anderson, N. (1923) *The Hobo,* Chicago: University of Chicago Press.
Bauman, Z. (1998) *Globalization: The Human Consequences,* Cambridge: Polity.
Bauman, Z. (2000) *Liquid Modernity,* Cambridge: Polity.
Beauregard, R.A. (2003) *Voices of Decline: The Postwar Fate of U.S. Cities* (2nd edn), London: Routledge.

Belsey, C. (2005) *Culture and the Real,* London: Routledge.

Berg-Schlosser, D. and Kersting, N. (eds) (2003) *Poverty and Democracy: Self-help and Political Participation in Third World Cities,* London: Zed Books.

Berman, M. (1982) *All That is Solid Melts into Air,* London: Verso.

Bird, J. (1993) 'Dystopia on the Thames', in J. Bird, B. Curtis, T. Putnam, G. Robertson and L. Tickner (eds), *Mapping the Futures: Local Cultures, Global Change,* London: Routledge.

Byrne, D. (1997) 'Chaotic Places or Complex Places? Cities in a Post-Industrial Era', in S. Westwood and J. Williams (eds), *Imagining Cities: Scripts, Signs, Memory,* London: Routledge.

Castells, M. (1976) *The Urban Question,* London: Edward Arnold.

Castells, M. (1978) *City, Class and Power,* London: Macmillan.

Castells, M. (2001) *The Making of the Network Society,* London: Institute of Contemporary Arts.

Cilliers, P. (1998) *Complexity and Postmodernism: Understanding Complex Systems,* London: Routledge.

Davidoff, P. (2003 [1965]) 'Advocacy and Pluralism in Planning', in R.T. LeGates and F. Stout (eds), *The City Reader* (3rd edn), London: Routledge.

Davis, M. (1990) *City of Quartz: Excavating the Future in Los Angeles,* London: Verso.

Fernandes, E. and Varley, A. (eds) (1998) *Illegal Cities: Law and Urban Change in Developing Countries,* London: Zed Books.

Flyvberg, B. (1992) 'Aristotle, Foucault, and Progressive Phronesis', *Planning Theory,* 7(8): 65–83.

Forester, J. (2003 [1987]) 'Planning in the Face of Conflict', in R.T. LeGates and F. Stout (eds), *The City Reader* (3rd edn), London: Routledge.

Foucault, M. (1971) *Madness and Civilization: A History of Insanity in the Age of Reason,* London: Tavistock. [First published as *Histoire de la Folie,* 1961, Paris: Libraire Plon.]

Foucault, M. (1973) *The Birth of the Clinic,* London: Tavistock. [First published as *Naissance de la clinique,* 1963, Paris: Presse Universitaires de France.]

Foucault, M. (1979) *Discipline and Punish: The Birth of the Prison,* Harmondsworth: Penguin. [First published as *Surveiller et punir,* 1975, Paris: Gallimard.]

Friedmann, J. (1987) *Planning in the Public Domain: From Knowledge to Action,* Princeton, NJ: Princeton University Press.

Friedmann, J. (1992) *Empowerment: The Politics of Alternative Development,* Oxford: Blackwell.

Gans, H.J. (1962) *The Urban Villagers,* Glencoe, IL: Free Press.

Gans, H.J. (1967) *The Levittowners,* Harmondsworth: Penguin.

Gans, H.J. (2003) *Democracy and the News,* Oxford: Oxford University Press.

Gramsci, A. (1971) *Selections from the Prison Notebooks,* New York: International Publishers.

Habermas, J. (1984) *The Theory of Communicative Action,* Boston: Beacon Press.

Hall, P. (2001) *Cities of Tomorrow: An Intellectual History of Urban Planning and Design in the Twentieth Century* (3rd edn), Oxford: Oxford University Press.

Hamdi, N. (1995) *Housing Without Houses: Participation, Flexibility, Enablement,* London: Intermediate Technology Publications.

Happe, B. and König, C.-D. (2003) 'Collective Interest Groups', in D. Berg-Schlosser and N. Kersting (eds), *Poverty and Democracy: Self-help and Political Participation in Third World Cities,* London: ZED Books, pp. 93–121.

Harvey, D. (1973) *Social Justice and the City,* London: Edward Arnold.

Hayden, D. (2003 [1981]) 'What Would a Non-sexist City be Like? Speculations on Housing, Urban Design, and Human Work', in R.T. LeGates and F. Stout (eds), *The City Reader* (3rd edn), London: Routledge.

Healey, P. (1997) *Collaborative Planning,* London: Macmillan.

Heskin, A. (1991) *The Struggle for Community,* London: Routledge.

Krumholz, N. (1994) 'Dilemmas of Equity Planning: A Personal Memoir', *Planning Theory,* 10 (11): 45–58.

Leavitt, J. (1994) 'Planning in an Age of Rebellion: Guidelines to Activist Research and Applied Planning', *Planning Theory,* 10 (11): 111–30.

Lefebvre, H. (1996) *Writings on Cities,* trans. E. Kofman and E. Lebas, Oxford: Blackwell.

LeGates, R.T. and Stout, F. (eds) (2003) *The City Reader* (3rd edn), London: Routledge.

Lorde, A. (2000) 'The Master's Tools will never dismantle the Master's House', in J. Rendell, B. Penner and I. Borden (eds), *Gender Space Architecture: An Interdisciplinary Introduction,* London: Routledge, pp. 53–55 [paper delivered in 1979 to the Second Sex Conference, New York, 29 September].

Marcuse, P. (2002) 'The Layered City', in P. Madsen and R. Plunz (eds), *The Urban Lifeworld: Formation, Perception, Representation,* London: Routledge.

Massey, D. (1994) *Space, Place and Gender,* Cambridge: Polity.

Miles, M. (2000) *The Uses of Decoration: Essays on the Architectural Everyday,* Chichester: Wiley.

Nagel, T. (1986) *The View from Nowhere,* Oxford: Oxford University Press.

Pahl, R.E. (1975) *Whose City? And Further Essays on Urban Society,* Harmondsworth: Penguin.

Peattie, L.R. (1982) 'Some Second Thoughts on Sites and Services', *Habitat International,* 6: 1–2.

Pickles, J. (2004) *A History of Spaces: Cartographic Reason, Mapping and the Geo-coded World,* London: Routledge.

Robbins, E. (1996) 'Thinking Space/Seeing Space: Thamesmead Revisited', *Urban Design International,* 1 (3): 283–91.

Roschelle, A.R. and Wright, T. (2003) 'Gentrification and Social Exclusion: Spatial Policing and Homeless Activist Responses in the San Francisco Bay Area', in M. Miles and T. Hall (eds), *Urban Futures: Critical Commentaries on Shaping the City,* London: Routledge.

Sandercock, L. (1998) *Towards Cosmopolis,* Chichester: Wiley.

Sennett, R. (1970) *The Uses of Disorder: Personal Identity and City Life,* New York: W.W. Norton and Co.

Sennett, R. (1998) *The Corrosion of Character: The Personal Consequences of Work in the New Capitalism,* New York: W.W. Norton and Co.

Serageldin, I. (1997) *The Architecture of Empowerment: People, Shelter and Liveable Cities,* London: Academy Editions.

Sibley, D. (1995) *Geographies of Exclusion,* London: Routledge.

Simmel, G. (1903) 'Metropoles and Mental Life', in D. Frisby and M. Featherstone, (eds), *Simmel on Culture,* London: Sage, pp. 174–185.

Simmel, G. (1969 [1950]) *The Sociology of Georg Simmel,* ed. K. H. Wolff, New York: Free Press.

Simmel, G. (1990 [1978]) *The Philosophy of Money* (2nd edn), ed. D. Frisby. London: Routledge.

Smith, M.P. (1980) *The City and Social Theory,* Oxford: Blackwell.

Stokes, S.C. (1995) *Cultures in Conflict: social movement and the state in Peru,* Berkeley: University of California Press.

Tönnies, F. (1940) *Fundamental Concepts of Sociology,* trans. C.P. Loomis, New York: American Book Company.

Tönnies, F. (1955) *Community and Association,* trans. C.P. Loomis, London: Routledge.

Turner, J.F.C. (1976) *Housing by People: Towards Autonomy in Building Environments,* London: Marian Bayars.

Wilson, E. (1991) *The Sphinx in the City: Urban Life, the Control of Disorder, and Women,* Berkeley: University of Chicago Press.

Wirth, L. (1938) 'Urbanism as a Way of Life', *American Journal of Sociology,* XLIV, 1 (July). Reprinted in R.T. LeGates and F. Stout (eds), *The City Reader* (3rd edn), London: Routledge.

Wirth, L. (1964) *On Cities and Social Life,* Chicago: University of Chicago Press.

Wolff, J. (1989) 'The Invisible *Flâneuse*: Women and the Literature of Modernity', in A. Benjamin (ed.), *The Problems of Modernity: Adorno and Benjamin,* London: Routledge

Young, I.M. (1990) *Justice and the Politics of Difference,* Princeton, NJ: Princeton University Press.

Young, I.M. (2000) *Inclusion and Democracy,* Oxford: Oxford University Press.

19   CRIME AND POLICING

Steve Herbert

This chapter

○ Offers different perspectives on why the city has been characterized by higher rates of criminality than the rural, and reviews literatures which focus on sites of urban criminality

○ Shows how the state and law seek to reduce and contain criminality through forms of spatial control that are becoming ever-more embedded in the fabric of everyday life

○ Highlights the contribution of urban scholarship to critical debates in criminology

---

Introduction

In 1751, Henry Fielding, a novelist and an early advocate of metropolitan policing, wrote the following about his home town, London:

> Whoever indeed considers the cities of London, and Westminster, with the late vast Addition of their Suburbs; the great irregularity of their Buildings, the immense Number of Lanes, Alleys, Courts and Bye-places; must think, that, had they been intended for the very purpose of Concealment, they could scarce have been better contrived. Upon such a view, the whole appears as a vast Wood or Forest, in which a Thief may harbour with as great Security, as wild Beasts do in the Deserts of Africa or Arabia (Quoted in Tobias, 1979: 10).

Nearly 250 years later, William Bennett, John DiIulio and John Walters (1996: 27) had this to say about America's inner cities:

> Here is what we believe: America is now home to thickening ranks of 'super-predators' – radically impulsive, brutally remorseless youngsters, who murder, assault, rape, rob, burglarize, deal deadly drugs, join gun-toting gangs, and create serious communal disorders. Many of these super-predators grow up in places that may best be called criminogenic communities – places where the social forces that create predatory criminals are far more numerous and stronger than the social forces that create decent, law-abiding citizens. At core, the problem is that most inner-city children grow up surrounded by teenagers and adults who are themselves deviant, delinquent, or criminal.

These excerpts make plain a longstanding reality of urban life: that the disorderly residents of cities spawn fear. As cauldrons of teeming diversity – along lines, most notably, of race, ethnicity and class – cities bring together the comfortable and the distressed, the established and the newly arrived, the powerful and the powerless. The gaps that divide urban denizens can be canyonesque; they can enable people to inhabit space in close proximity but to live worlds apart. Those who inspire the greatest fear – 'wild beasts', 'super-predators' – are often perceived to reside in abominable jungle-like neighbourhoods, radically different from mainstream society.

Given this yawning chasm of social difference, it is hardly surprising that fear of others is common among urbanites. Such fear often focuses upon the threat posed by those who are believed to be criminal. Indeed, the initial development of formal police departments, in the late 1700s and early 1800s, was directly tied to concerns about the threat posed by supposed criminals. These fears were especially exacerbated when the development of industrial capitalism created both concentrated wealth and an influx of immigrants into cities (Miller, 1977; Schneider, 1980). The police were the state's formal response to this fear. Authorized by law and regulated by bureaucratic rules, the police used the legitimacy accorded to the state to reinforce order. This connection between the power of law and societal notions of order has become evermore entrenched over time. Today the role of law in generating order is uncontested; urban areas are regulated by an immense raft of formal rules dictating how the built environment should be constructed, occupied and used. Police tactics have historically varied across time and space, but a formalized internal security unit is a commonplace among modern states, and is generally accepted as legitimate. To be sure, abuses of police power, most notably excessive force and corruption, regularly flare up in many places (Chevigny, 1995), but the police largely continue to be accorded high levels of legitimacy (Loader and Mulcahy, 2003), given that fear of crime is an ever-present condition for many urbanites.

Fear of crime can indeed be well-founded; cities are often dangerous places, especially in certain sectors (Bellair, 1997; Morenoff et al., 2001; Rohe and Greenberg, 1986; Sampson et al., 2002). Yet the distress that accompanies urban difference and disorder is a complicated matter, in terms of both its genesis and its implications. Understandings of the nature and consequences of crime and disorder vary significantly; they are the subject of vibrant debate in academic and policy circles. The primary goal of this chapter is to chart this debate in broad strokes, by contrasting two ways of understanding the spaces of urban criminality, what I term the 'instrumental' and the 'constitutive'. These approaches differ in their explanations of urban crime, their understandings of territorial action to repulse crime, their recommendations for government action, and their visions of an urban future. Regardless of these debates, the political significance of the crime issue demands that governments address it. Indeed, the formal social control apparatus – police, courts, corrections – is a robust component of modern states, a material and symbolic bulwark against the threats of disorder and decay. This apparatus does much of its work in urban environments, helping to manage the social anxiety that accompanies difference. And this work is of immense significance, particularly for those who are targeted for apprehension and punishment. This leads to the second goal of this chapter: to highlight just why and how efforts at urban social control matter so much, not just for their effects on crime and disorder, but also on the broader currents of urban social life.

Understanding the spaces of urban crime: competing paradigms

Crime persists as a morally and politically consequential component of urban society. The law and order apparatus developed to respond to crime says much about the society that erects it. That apparatus obviously affects deeply those who are brought into it because of suspected criminality. But it affects urban society more broadly, in ways that are not always interrogated fully. It can affect how we perceive urban space, how we want that space used and controlled, how we see and relate to the strangers we encounter. The significance of the crime issue becomes especially obvious through a contrast of two primary approaches to understanding and responding to the spaces of urban crime.

It is always a risky endeavour to categorize differing ways of understanding social phenomena. Such broad categories necessarily occlude various nuances, and thereby make matters a little too simple. However, it can be helpful to construct these categories nonetheless, to compare and contrast broadly understood schools of thought. This is certainly the case with urban criminality. Two approaches are distinctive here, what I label the instrumental and the constitutive. These approaches differ in how they explain crime, how they understand the meaning of territorial action to quell crime, and how they perceive the role of the formal social control apparatus.

The instrumental approach explains crime in terms of individual actors or the small-scale communities of which they are part; social dynamics that occur at wider geographic scales are largely bracketed from consideration. From this micro-orientation, territorial action to reduce crime is strongly encouraged, as a means to repulse and possibly banish those who are inclined to offend. Spatial control is thus an instrument to be used to defend against the individuals predisposed to criminal behaviour. Further, such defences are to be supported by a robust formal social control apparatus. The government should ensure that criminals are apprehended and punished. The public spaces of the city are thereby worthy of continual surveillance and visible police presence, the better to deter the criminogenic from misbehaving.

The constitutive approach contrasts in every respect. Here, crime is understood in more capacious terms, as a result of a wide array of economic and social phenomena. Individuals are not simply rational actors who choose to engage in crime. They are, instead, constituted by the societies in which they find themselves, and limited by the structurally-generated forces that confront them. Territorial action is also understood in terms of its relation to wider social processes; it is seen not just as a natural response to sensible fears, but as a means to protect social position and to enable economic betterment. Such suspicion of territorial action is extended to the practices of formal social control. State response to crime must similarly be interrogated for its connections to wider societal dynamics, to be questioned and potentially challenged for unfairly targeting certain groups or otherwise damaging the democratic fabric of urban life.

To make these differences more plain, I turn now to a more extensive elaboration of these two approaches.

The instrumental

As noted above, those who adopt what I term the instrumental approach see criminality as largely the function of individual action, heartily endorse territorial means to repulse it, and applaud robust formal social control. Its advocates see urban space in terms of its surface characteristics, namely whether its appearance invites or repulses crime. They thus seek to use space as an instrument to deter criminal behaviour. Urban areas need to be monitored and defended, with the active assistance of the police.

A useful way to elaborate this approach more fully is to illustrate it with two very popular contemporary approaches to explaining and responding to urban crime – 'broken windows' and 'situational crime prevention'. The theory of broken windows (see Kelling and Coles, 1996; Wilson and Kelling, 1997) suggests that crime will occur in places where informal social control is not exercised. The lack of informal social control is made obvious when broken windows and other symbols of disorder are not fixed. If a potential criminal recognizes a place is disorderly, he/she presumes it is ripe for the picking and will engage in crime there. A broken window today, the argument suggests, translates into a much more significant crime problem tomorrow.

Situational crime prevention (Clarke, 1983; Clarke and Mayhew, 1980; Crowe, 1991; Felson, 2002; see also Brantingham and Brantingham, 1981) is broadly similar to broken windows. It emphasizes the role the built environment plays in either inviting or repulsing crime. Places communicate messages about how susceptible they are to criminals. For instance, places that possess walls that repel outsiders or that enable close monitoring through bright lights and clear sight lines are 'defensible' (Newman 1972), and thus deter the possibility of crime. It follows that place managers need to construct the built environment to make its defences obvious.

One of the key figures advocating for situational crime prevention is Marcus Felson. The excerpt from his popular text, *Crime and Everyday Life* (2002; see Extract 19.1) captures well the sort of environmental modifications that might work to increase a space's defensibility against crime. Felson outlines a number of methods by which place managers can alter the environment to repel invaders, make surveillance more easy, and scare off those who wander too close.

Extract 19.1: From Felson, M. (2002) *Crime and Everyday Life*, Thousand Oaks, CA: Pine Forge Press, p. 152.

Crime can be prevented by at least four physical methods: target hardening, construction, strength in numbers, and noise. For example, universities harden targets when they bolt down computers, typewriters, television sets, projection equipment, and the like. As for construction, universities sometimes put up extra walls, fences, or other physical barriers to reduce unauthorized entry to university buildings, or simply to channel flows of people coming and going to make mild supervision possible. For example, the University of Southern California put a fence around the premises in preparation for the 1984 Olympic Games. The campus was still open at many gates, but offenders could no longer enter anywhere, attack anything or anyone, and exit anywhere afterward.

Strength in numbers is important for helping people to protect themselves and their property. One designer of a high-rise building for the elderly put the recreation room on the first floor with good lines of sight to the door. Together, the residents were able to keep people from wandering in without permission.

Noise is also important for crime. On the negative side, offenders use noise to determine whether you are present or absent, even banging on our door to make sure you are gone. On the positive side, noise can protect you. A good lock on your door is important not so much for preventing illegal entry as for making sure the offender makes enough noise to draw the attention of others. Alarm systems operate on the same principle. Noisy dogs can serve both to alert others and to scare off offenders directly. Noise also can be directed at offenders, as when subway station personnel with loudspeakers direct someone to cease an undesirable behavior. In either direction, noise reminds us that crime is a physical act and that our five senses are essential for both committing it and preventing it.

Figure 19.1   Gated community, Saskatoon, Canada

Each of these approaches situates the criminal act at the level of the individual. In each case, the criminal actor is assumed to be rational, that is to make decisions through a cost–benefit analysis (for a wider-ranging defence of rational choice theory in criminology, see Gottfredson and Hirschi, 1990). As rational actors, potential criminals assess urban neighbourhoods in terms of their defensive capacity. When they recognize an unprotected area, they act. The geography of urban crime, therefore, is determined largely by the symbolic messages communicated by the environment: crime is largely absent from well-defended places and concentrated instead in poorly defended ones.

From this emphasis on rational choice and on the importance of the messages that places communicate, it is a very short step to seeing territorial action as a necessary instrument to reduce crime. Indeed, communities are considered unhealthy to the extent that they do not fix broken windows or erect impenetrable defences. Urban neighbourhoods are thereby entirely justified in building resistance into their landscapes. This phenomenon finds its clearest expression in the gated communities that are now extremely common in the suburban USA and Canada (Blakely and Snyder, 1997; Low, 2003) (see Figure 19.1). In addition to creating defensibility, space can also be constructed to increase surveillance. Those who advocate situational crime prevention often tout the need to ensure spaces can be rendered visible to those who are responsible for them, the better to witness possible criminal behaviour. An even more pronounced manifestation of this trend is the spread of closed circuit television (CCTV) in much of the UK. The urban spaces of the UK are now largely monitored by CCTV cameras, to make public activities visible, and thereby to presumably reduce the propensity to commit crimes (Fyfe and Bannister, 1998).

Surveillance, of course, need not occur solely through the watchful eyes of the state *à la* CCTV. It can be exercised by local users of places, who can exert informal social control as occasions arise for its use (Jacobs, 1961; Rohe and Greenberg, 1986). Yet those who adopt the instrumental approach embrace warmly the use of the state's power to deter crime. In their elaboration of broken windows, Wilson and Kelling wax nostalgic for the days of the beat cop who knew his neighbourhood well and dispensed justice informally, even to the point of choosing to 'kick ass'. It is hardly surprising, then, it took little energy for some advocates of broken windows to move towards the concept of 'zero tolerance'. This means the police should employ their full enforcement powers against even the most minor of infractions. Recall that the logic of broken windows suggests that seemingly insignificant matters today can become more significant later. Better, then, to crackdown on misdemeanour offences to avert a snowball into more serious crime. The zero tolerance policy was employed most famously in New York City in the 1990s. Tough tactics were used to explain the drop in crime rates and thus to bolster the political fortunes of a mayor, Rudolph Guiliani, and a police chief, William Bratton (Bratton, 1998; Silverman, 1999).

The era of zero tolerance is also the era of 'problem-solving policing' (Goldstein, 1991), often oriented towards the 'hot spots' of criminality (Sherman et al., 1989). Problem-solving means that police officers seek to understand the underlying dynamics that cause crime to persist in particular locations, the so-called hot spots. In this endeavour, many police departments increasingly rely upon geographic information systems to help them understand where crime is concentrated (Turnbull et al., 2000). There is considerable debate about just how extensive these problem-solving operations actually are (see Cordner and Biebel, 2005; Herbert, 2005), but such operations work in accordance with the instrumental approach. Space is again viewed in terms of its surface characteristics, and is to be mapped, modified and controlled to rebuff the crime occurring there (Sparrow et al., 1990).

In sum, then, the instrumental approach views urban space in somewhat superficial terms. It concentrates on whether an area invites or repulses the individual criminal actor. Because deterrence is desired, urban neighbourhoods should communicate a clear defence; territoriality should be used to make potential criminals move along. And this power of deterrence can be enhanced by strong police tactics that eliminate even low levels of criminality and target hotspots so that persistent crime problems disappear. On each of these key tenets, the constitutive approach differs.

The constitutive

The constitutive approach to urban crime differs most noticeably by viewing matters from a different scalar perspective. Constitutive scholars do not focus on individual criminals, nor do they see urban space largely in terms of its surface characteristics. Rather, they view crime and space as emerging from a broad-range of social phenomena, many of which operate at scales beyond the local. Urban spaces and their inhabitants are constituted by multiple social forces, and so cannot be understood without cognizance of those forces.

This more expansive viewpoint translates into a different explanation for urban crime. Constitutive scholars cite such factors as economics, race, and government action in seeking to explain why crime is concentrated in particular urban neighbourhoods (Morenoff et al., 2001; Sampson et al., 2002; Smith, 1986). If a given neighbourhood lacks much by way of employment opportunities, its residents lack ready access to the mainstream economy. Crime may represent the best job option for many of these residents, or may be resorted to out of general frustration with the cramped circumstances of their lives (Smith, 1986; Sampson and Wilson, 2002). John Hagedorn (1994), for example, interviewed those engaged in the drug trade in an industrial US city, Milwaukee. He discovered that most did not like the work, and only pursued it because of a lack of opportunity in the formal economy. Such cramped societal circumstances may be fuelled by racial dynamics, as well. Residential segregation persists, especially in American cities (Chapter 12), which contributes to a lack of access to employment. And this segregation is buttressed by government policies which help determine how residential property is zoned and how access to homes and financing is allotted (Holloway and Wyley, 2001; Lipsitz, 1998; Orfield and Ashkinaze, 1991).

Just as actual patterns of criminality are seen, from this perspective, as produced by a wide range of social factors, so is the fear of crime. Although many urban residents possess a well-founded fear of crime (Pain, 2001), societal perceptions of crime do not always match susceptibility to crime. In other words, one's fear of crime does not always correlate with one's actual vulnerability. In the USA, for example, the evidence suggests that racial attitudes deeply inform fear of crime: urban residents' perceptions of a given neighbourhood as home to either crime or disorder are determined most significantly by their perceptions that young black males are visible in public space (Quillian and Pager, 2001; Sampson and Raudenbush, 2004). This fear of black males, in turn, is shaped by media reporting patterns, which tend to emphasize black perpetrators and white victims (Gilliam and Iyenger, 2000; Hancock, 2000).

Crime may be perpetrated by individual actors, but the constitutive approach suggests that these actors engage in this behaviour in particular spaces that are generated by wide-ranging forces. Absent a rich understanding of these geographic contexts, one's explanation of urban crime is limited. Territorial action to rebuff crime, for constitutive scholars, must similarly be situated within a multi-scalar perspective. Such territorial action is not viewed simply as a natural defensive gesture, but potentially as a means to further one's economic and social position. For instance, the cleansing of urban space that accompanied the zero tolerance policing inspired by the broken windows theory can work quite well to further urban gentrification (Body-Gendrot, 1999; Parenti, 2000; Smith, 1999); it helps to justify clearing urban spaces of street life that might deter gentrifiers from entering a previously blighted neighbourhood. Those who hide behind the walls of their suburban communities do so, in part, to preserve their property values, and to minimize their contact with unwanted outsiders whose social difference may be a source of discomfort (Low, 2002). This preservation of class position can also be buttressed through the employment of private security, another means by which the haves can better protect themselves from the have-nots (Davis, 1992; Newburn, 2001). Matters of class and race, then, may underlie much of the territorial action that instrumentalists seek to naturalize.

Constitutive scholars are similarly apprehensive of the work of formal social control to reduce crime. In the USA especially, the punishment apparatus has expanded monumentally; the rate of incarceration has quadrupled in the past twenty-five years (Mauer, 1999). In urban areas, this incarceration boom is notably assisted by the growth of zero tolerance policing and other mechanisms for arresting those considered 'undesirable', such as so-called 'civility codes' making it a crime to sit or sleep in public spaces (Mitchell, 2001). Constitutive scholars ask critical questions of this robust expansion of the formal social control. Some, as suggested above, connect it to gentrification and other forms of urban redevelopment. Such redevelopment can be plausibly understood as necessary in a globalized world in which capital is increasingly footloose. This prompts a competition between cities to attract capital and the consumption dollars of tourists. Concern about the aesthetics of public space is heightened, which prompts the rise of civility codes (Christopherson, 1994; Mitchell, 2003). Other constitutive theorists question what this expansion of the punishment apparatus means about the nature of the contemporary state (Herbert, 1999). It is notable that the rise of punishment coincided with the decline of the welfare state. Indeed, some speculate that this is not merely coincidental. While the poor and disenfranchised (and often minority) populations were considered deserving of support and assistance in the heyday of the welfare state, they are now viewed as dangerous and deserving of surveillance and punishment (Beckett, 1997). The decline of welfarism is a key component of the rise of neoliberalism, which generally seeks to redefine the responsibility of the state towards supporting market mechanisms for regulating social life (Brenner and Theodore, 2002; Peck and Tickell, 2002). From this perspective, it is short-sighted to see formal social control as simply a natural process to better defend urban space. Instead, it must be viewed as an important component in the restructuring of the economy, the redefinition of the modern state, and the ongoing marginalization of urban residents of colour (Wacquant, 2000).

In general, constitutive scholars seek to enlarge our understandings of urban criminality and the spaces that host it by situating crime – what causes it, how we should respond to it – in terms of economics, social and political relations. Criminals may be actors, but they act within certain spatial contexts constituted by a broad range of social forces. Societal fears about crime are not simply natural, nor is territorial action that seeks protection against crime. All of these must be arrayed against the various social forces that constitute cities and the relations between the strangers who populate them.

I contrast these two broadly understood schools of thought to chart the parameters of the debates about urban crime. These debates are interesting in their own right, but they are also extremely consequential. Significant implications flow from adopting one approach versus the other, not just for how we deal with crime, but for how we understand ourselves in relation to city life more generally. It is to a consideration of these implications that I now turn.

Fear, crime and urban divisions

Urban life can often be fear-inducing because city residents necessarily inhabit a 'world of strangers' (Lofland, 1973). This means they daily encounter those who are unknown, and who may therefore inspire some discomfort. Urban residents must negotiate on-street interactions to minimize that discomfort (Anderson, 1990; Merry, 1981). These fears of difference are often magnified when combined with concerns about crime; psychic discomfort intensifies into a sense of physical vulnerability. City dwellers understandably desire some degree of protection against such vulnerability. Because cities, and especially certain parts of cities, can indeed be dangerous places, these requests for greater security often seem commonsensical.

As we have seen, however, one can understand urban crime from significantly different perspectives; the instrumental and the constitutive approaches certainly apprehend urban criminality in starkly disparate ways. These contrasting explanations for criminal behaviour lead, in turn, to contrasting understandings of the meaning of territorial action to rebuff crime, and of the role of formal social control in responding to crime. Looked at even more broadly, these two perspectives contrast in terms of their implicit view of the desired urban society. The critical distinctions between these two approaches involve divisions – between differently-situated social groups, and between urban citizens and the state. Crime is a social and political issue particularly well-suited to fortifying such divisions.

If territorial action to ward off crime is understood as a natural and healthy behaviour, then urban neighbourhoods are justified in walling themselves off from those they fear. Surveillance in public space is perfectly legitimate, as is a robust police presence. These practices of surveillance and social control exist to ensure that some members of urban society are protected from others, that the fearful are shielded from the fear-inducing. When fears of crime become an ordinary fact of social life (Garland, 2001), and are magnified by news media coverage (Beckett, 1997), this line between the normal and pathological is fraught with anxiety and in need of vigilant protection.

But to fortify this division is, by definition, to rend urban society assunder, to weaken any sense of collective consciousness. One implication of this is a diminution of the vibrancy of public space. A robustly democratic conception of public space requires that inclusion trump exclusion. In this view, public space must be open to allow differing peoples to use it in differing ways (Mitchell, 2003). This inclusive view of public space necessarily means that the easily frightened will be forced to confront those they find distasteful. This is a prospect not easily countenanced. The popularity of the theory of broken windows, and zero tolerance policing, is undoubtedly underwritten by a deeply-felt desire to keep public space free of those considered to be nuisances. When such defensive measures become commonplace, an inclusive urban society withers. Those considered nuisances are deprived of the spaces they need to simply subsist, and of the political agency they need to press their claims (Feldman, 2004). Further, those who are frightened are not challenged; they avoid the embrace of difference by hiding behind fear (Sennett, 1970; Young, 1990).

One of the most influential critics of defensible space is Mike Davis. He sees these environmental modifications as a visible manifestation of class conflict, a means by which the comfortable can rebuff the impoverished. As the excerpt from his 'Fortress LA' demonstrates (Extract 19.2), he views situational crime prevention as inimical to democracy; class divisions are fortified, and a sense of public collectivity evaporates.

Extract 19.2: From 'Fortress Los Angeles: The Militarization of Urban Space', in M. Sorkin (ed.) (1992), *Variations on a Theme Park: The New American City and the End of Public Space*, New York: Hill and Wang, pp. 155–6.

The universal consequence of the crusade to secure the city is the destruction of any truly democratic urban space. The American city is being systematically turned inward. The 'public' spaces of the new megastructures and supermalls have supplanted traditional streets and disciplined their spontaneity. Inside malls, office centers, and cultural complexes, public activities are sorted into strictly functional compartments under the gaze of private police forces.

This architectural privatization of the physical public sphere, moreover, is complemented by a parallel restructuring of electronic space, as heavily guarded, pay-access databases and subscription cable services expropriate the invisible *agora*. In Los Angeles, for example, the ghetto is defined not only by its paucity of parks and public amenities, but also by the fact that it is not wired into any of the key information circuits. In contrast, the affluent Westside is plugged – often at public expense – into dense networks of educational and cultural media.

In either guise, architectural or electronic, this polarization marks the decline of urban liberalism, and with it the end of what might be called the Olmsteadian vision of public space in America. Frederick Law Olmsted, the father of Central Park, conceived public landscapes and parks as social safety-valves, *mixing* classes and ethnicities in common (bourgeois) recreations and pleasures: 'No one who has closely observed the conduct of people who visit Central Park,' he wrote, 'can doubt that it exercises a distinctly harmonizing and refining influence upon the most unfortunate and most lawless classes of the city – an influence favorable to courtesy, self-control, and temperance.'

This reformist ideal of public space as the emollient of class struggle is now as obsolete as Rooseveltian nostrums of full employment and an Economic Bill of Rights. As for the mixing of classes, contemporary urban America is more like Victorian England than the New York of Walt Whitman or Fiorella La Guardia. In Los Angeles – once a paradise of free beaches, luxurious parks, and 'cruising strips' – genuinely democratic space is virtually extinct. The pleasure domes of the elite Westside rely upon the social imprisonment of a third-world service proletariat in increasingly repressive ghettos and barrios. In a city of several million aspiring immigrants, public amenities are shrinking radically, libraries and playgrounds are closing, parks are falling derelict, and streets are growing ever more desolate and dangerous.

Here, as in other American cities, municipal policy has taken its lead from the security offensive and the middle-class demand for increased spatial and social insulation. Taxes previously targeted for traditional public spaces and recreational facilities have been redirected to support corporate redevelopment projects. A pliant city government – in the case of Los Angeles, one ironically professing to represent a liberal biracial coalition – has collaborated in privatizing public space and subsidizing new exclusive enclaves (benignly called 'urban villages'). The celebratory language used to describe contemporary Los Angeles – 'urban renaissance,' 'city of the future,' and so on – is only a triumphal gloss laid over the brutalization of its inner-city neighborhoods and the stark divisions of class and race represented in its built environment.

This sense of division is magnified, in contemporary times, by the decline of welfarist notions of a protective state. The spread of neoliberalism, and its advocacy of the need for individuals to secure their own well-being, helps reduce legitimacy for the welfare state. This can justify heightened punishment for those who engage in crime, and erode any sense of collective responsibility for the disadvantaged. Society's divides between the normal and the pathological, between those who 'choose' crime and those who avoid it, are again fortified. A sense of responsibility for the disadvantaged is replaced with a sense of fear; social welfare is replaced by social control.

So legitimated, the social control apparatus becomes hard to challenge politically. Because of the moral significance often attached to the issue of crime, state action done in the name of reducing crime often becomes sacrosanct and beyond critique. Crime is an issue susceptible to various 'moral panics' (Cohen, 1973), and a strong state response is thereby seen as essential and unquestionable. Witness the stunning popularity of the broken windows theory despite convincing refutations of its logic (Harcourt, 2001; Sampson and Raudenbush, 1999; Taylor, 2000). In short, the theory is wrong: disorder today does not necessarily lead to crime tomorrow. But yet the theory remains the lodestar for much police practice, and helps legitimate continuing arrests of those considered signs of disorder. The resilient popularity of broken windows is testimony to the moral significance of the crime issue, and how it can easily lead to an unquestioned use of police authority.

Or take the reform movement known as 'community policing'. Spawned in the 1970s, community policing rose to become an important model for organizing police departments by the mid-1980s (Greene and Mastrofski, 1988). Central to the community policing ideal is a strengthened relationship between officers and citizens. Urban neighbourhoods and the police officers with whom they work are to develop a strong partnership; together, they should define and seek to resolve problems of crime and disorder (Trojanowicz and Bucqueroux, 1994). Yet these partnerships rarely materialize in practice; police retain the dominant voice in encounters with community groups (Skogan and Hartnett, 1997), and persist with a strong crime-fighting orientation regardless of the desires of urban residents (Herbert, 2006). Further, even to hope to use community as a means to fight crime is necessarily a divisive exercise: it bolsters a sense of 'us against them', as members of one neighbourhood seek to protect themselves against the outsiders they fear (Klinenberg, 2001).

In short, the fact crime and crime-fighting are so politically and morally consequential means that urban citizens are increasingly divided from one another, in both social space and physical space. Barriers are erected and justified, supported where possible by the strong arm of law enforcement. These barriers may generate a sense of security for those able to erect them, even though this may be somewhat illusory. Residents of gated communities, for example, are not, in fact, any less prone to crime than similarly situated neighbourhoods (Blakely and Snyder, 1997). What is not illusory is the resultant *insecurity* experienced by those who are cast away. Those who live in poor and crime-ridden neighbourhoods face nearly insurmountable odds in their quest for the economic mainstream, especially if they find themselves captured by the net of formal social control. In the incarceration-happy USA, a criminal record nearly guarantees a lifetime of penury and disenfranchisement; economic activity is limited by the taint of the past (Pager, 2003; Petit and Western, 2004), political activity can be reduced by the inability to vote (Manza and Uggen, 2005). Those neighbourhoods where ex-convicts are concentrated feel the effects deeply, as potential jobholders are vacuumed into prison and returned as perpetually marginalized individuals (Fagan, 2004).

Indeed, the divisions cemented by ardent and politicized fears of crime find their starkest expression in the dire circumstances faced by such disadvantaged neighbourhoods. Ostracized as the home of rapacious fearsome predators, limited in their receipt of assistance from neoliberalized governments, such neighbourhoods subsist with little meaningful connection to wider society. This wide socio-spatial division is solidified by the potent discourse of crime. To bridge this division is necessarily an exercise in redirecting the discourse of crime towards a language of inclusiveness and collective responsibility.

Summary

Richard Sennett writes: 'A city isn't just a place to live, to shop, to go out and have kids play. It's a place that implicates how one derives one's ethics, how one develops a sense of justice, how one learns to talk with and learn from people who are unlike oneself, which is how a human being becomes human' (quoted in Lees, 2003: 3). Whatever resonance Sennett's ethical vision of the virtue of urban difference continues to possess, its legitimacy is threatened by the widespread fear of crime. No other issue is capable of creating greater psychic insecurity, and none is as effective in solidifying divisions between social groups and the urban neighbourhoods in which they live.

These divisions are actively supported by those who advocate the instrumental approach to urban crime. Instrumentalists see urban crime as an individual act against which neighbourhoods must protect themselves, in concert with a robust formal social control apparatus. In this way, unwittingly or not, instrumentalists fortify social and spatial distinctions while they seek to purify public space of unwanted elements. By contrast, those who approach crime from a constitutive perspective see things in diametrically opposite terms. Here, the emphasis is on the broad array of factors that motivate criminality and societal responses to it. Crime is a social act, as are the various efforts to lessen its incidence and impact. Those who engage in crime are not necessarily evil others against whom protection is necessary, but members of the collective in need of assistance. From this vantage point, the public is understood in broad terms, and energy is directed towards reducing rather than strengthening social and spatial divisions.

This latter vision possesses little political luster in these neoliberal times. Yet its unpopularity, as Sennett implicitly suggests, spells trouble for a rich version of city life. This vision is one that embraces difference, even if that difference occasionally frightens. When urban residents shrink from this difference and seek to banish those who inspire fear, they necessarily shrink from any connection to a broader public. This can reduce the vibrancy of public space, it can diminish a sense of urban life as a cauldron of social difference that is to be experienced and appreciated.

One can argue the virtues of either the instrumental or constitutive approaches. What is harder to argue is that the issue of crime, and the fear that it helps spawn, is one critical to the fabric of city life. Perceptions of crime influence how urban citizens react to one another in chance encounters, how they envision an ideal public space, how they seek to defend themselves, how they view formal efforts to reduce their vulnerability. In short, the discourses and practices of law and order are critical components of how city life is both envisioned and lived.

References

Anderson, E. (1990) *Streetwise: race, class and change in an urban community*, Chicago: University of Chicago Press.

Beckett, K. (1997) *Making Crime Pay: law and order in contemporary American politics*, New York: Oxford University Press.

Bellair, P. (1997) 'Social interaction and community crime: examining the importance of neighbor networks', *Criminology* 35 (4): 677–703.

Bennett, W. DiIulio, J. and Walters, J. (1996) *Body Count: moral poverty and how to win America's war against crime and drugs*, New York: Simon & Schuster.

Blakely, E.J. and Snyder, M.G. (1997) *Fortress America: gated communities in the US*, Washington, DC: Brookings Institution Press.

Body-Gendrot, S. (1999) *Social Control of Cities?* Oxford: Blackwell.

Brantingham, P. and Brantingham, P. (1981) *Environmental Criminology*, Beverly Hills, CA: Sage.

Bratton, W. (1998) *The Turnaround: how America's top cop reversed the crime epidemic*, New York: Random House.

Brenner, N. and Theodore, N. (2002) *Spaces of Neoliberalism: urban restructuring in North America and Western Europe*, Malden, MA: Blackwell.

Chevigny, P. (1995) *Edge of the Knife: police violence in the Americas*, New York: Free Press.

Christopherson, S. (1994) 'The fortress city: privatized spaces and consumer citizenship', in A. Amin (ed.), *Post-Fordism: a reader*, Oxford: Blackwell. pp. 409–27.

Clarke, R. (1983) 'Situational crime prevention: its theoretical basis and practical scope', in M. Tonry and N. Morris (eds), *Crime and Justice: an annual review of research*, Chicago: University of Chicago Press. pp. 225–56.

Clarke, R. and Mayhew, P. (1980) *Designing Out Crime*, London: HMSO.

Cohen, S. (1972) *Folk Devils and Moral Panics*, London: MacGibbon and Kee.

Cordner, G. and Biebel, E. (2005) 'Problem-oriented policing in practice', *Criminology and Public Policy* 4: 155–80.

Crowe, T. (1991) *Crime Prevention through Environmental Design: applications of architectural design and space management concepts*, Boston: Butterworth-Heinemann.

Davis, M. (1992) 'Fortress Los Angeles: the militarization of urban space', in M. Sorkin (ed.), *Variations on a Theme Park: the new american city and the end of public space*, New York: Hill and Wang. pp. 154–80.

Fagan, J. (2004) 'Crime, law and the community: dynamics of incarceration in New York City', in M. Tonry (ed.), *The Future of Imprisonment*, Oxford: Oxford University Press. pp. 27–59.

Feldman, L. (2004) *Citizens without Shelter: homelessness, democracy and political exclusion*, Ithaca, NY: Cornell University Press.

Felson, M. (2002) *Crime and Everyday Life*, Thousand Oaks, CA: Pine Forge Press.

Fyfe, N. and Bannister, J. (1998) 'The eyes upon the street: closed circuit television surveillance and the city', in N. Fyfe (ed.), *Images of the Street: planning, identity and control in public space*, London: Routledge. pp. 254–67.

Garland, D. (2001) *The Culture of Control*, Chicago: University of Chicago Press.

Gilliam, F.D. and Iyenger S. (2000) 'Prime suspects: the influence of local television news on the viewing public', *American Journal of Political Science* 44 (3): 560–73.

Goldstein, H. (1991) *Problem-Oriented Policing*, New York: McGraw-Hill.

Gottfredson, M. and Hirschi, T. (1990) *A General Theory of Crime*, Stanford, CA: Stanford University Press.

Greene, J. and Mastrofski, S. (1988) *Community Policing: rhetoric or reality?* New York: Praeger.

Hagedorn, J. (1994) 'Homeboys, dope friends, legits, and new jacks', *Criminology* 32 (2): 197–219.

Hancock, L. (2000) 'Framing children in the news: the face and color of youth crime in America', in V. Polakow (ed.), *The Public Assault on American's Children: poverty, violence and juvenile injustice*, New York: Teachers College Press.

Harcourt, B. (2001) *The Illusion of Order: the false promise of broken windows policing*, Cambridge, MA: Harvard University Press.

Herbert, S. (1999) 'The end of the territorially-sovereign state? The case of crime control in the US', *Political Geography* 18 (2): 149–72.

Herbert, S. (2005) 'POP in San Diego: a not-so-local story', *Criminology and Public Policy* 4 (2): 181–86.

Herbert, S. (2006) *Citizens, Cops and Power: recognizing the limits of community*, Chicago: University of Chicago Press.

Holloway, S. and Wyly, E. (2001) '"The color of money" expanded: geographically contingent mortgage lending in Atlanta', *Journal of Housing Research* 12 (1): 55–90.

Jacobs, J. (1961) *The Death and Life of Great American Cities: the failure of town planning*, Harmondsworth: Penguin.

Kelling, G. and Coles, C. (1996) *Fixing Broken Windows*, New York: Martin Kessler Books.

Klinenberg, E. (2001) 'Bowling alone, policing together', *Social Justice* 28 (3): 75–80.

Lees, L. (2003) 'The emancipatory city: urban (re)visions', in L. Lees (ed.), *The Emancipatory City: paradoxes and possibilities*, London: Sage.

Lipsitz, G. (1998) *The Possessive Investment in Whiteness: how white people profit from identity politics*, Philadelphia: Temple University Press.

Loader, I. and Mulcahy, A. (2003) *Policing and the Condition of England: memory, politics and culture*, Oxford: Oxford University Press.

Lofland, L. (1973) *A World of Strangers: order and action in urban public space*, New York: Basic Books.

Low, S. (2003) *Behind the Gates: life, security, and the pursuit of happiness in fortress America*, New York: Routledge.

Manza, J. and Uggen, C. (2005) *Locked Out: felon disenfranchisement and American democracy*, New York: Oxford University Press.

Mauer, M. (1999) *Race to Incarcerate*, New York: The New Press.

Merry, S. (1981) *Urban Danger: life in a neighborhood of strangers*, Philadelphia: Temple University Press.

Miller, W. (1977) *Cops and Bobbies: police authority in New York and London, 1830–1870*, Chicago: University of Chicago Press.

Mitchell, D. (2001) 'The annihilation of space by law: the roots and implications of anti-homeless laws in the US', in N. Blomley, D. Delaney and R.T. Ford (eds), *The Legal Geographies Reader*, Malden, MA: Blackwell.

Mitchell, D. (2003) *The Right to the City: social justice and the fight for public space,* New York: Guilford Press.

Morenoff, J., Sampson, R. and Raudenbush, S. (2001) 'Neighborhood inequality, collective efficacy, and the spatial dynamics of homicide', *Criminology* 39 (3): 517–60.

Newburn, T. (2001) 'The commodification of policing: security networks in the late modern city', *Urban Studies* 38 (5–6): 829–48.

Newman, O. (1972) *Defensible Space: crime prevention through urban design,* New York: Macmillan.

Orfield, G. and Ashkinaze, C. (1991) *The Closing Door: Conservative policy and black opportunity,* Chicago: University of Chicago Press.

Pager, D. (2003) 'The mark of a criminal record', *American Journal of Sociology* 108 (5): 937–75.

Pain, R. (2001) 'Gender, race, age and fear in the city', *Urban Studies* 38 (5–6): 899–913.

Parenti, C. (2000) *Lockdown America: police and prisons in the age of crisis,* London: Verso.

Peck, J. and Tickell, A. (2002) 'Neoliberalizing space', *Antipode* 34 (3): 380–404.

Petit, B. and Western, B. (2004) 'Mass imprisonment and the life course: race and class inequality in US incarceration', *American Sociological Review* 69 (2): 151–69.

Quillian, L. and Pager, D. (2001) 'Black neighbors, higher crime? The role of racial stereotypes in evaluations of neighborhood crime', *American Journal of Sociology* 107 (3): 717–67.

Rohe, W. and Greenberg, S. (1986) 'Informal social control and crime prevention in modern neighborhoods', in R. Taylor (ed.), *Urban Neighborhods: research and policy,* New York: Praeger.

Sampson, R. and Raudenbush, S. (1999) 'Systematic social observation of public spaces: a new look at disorder in urban neighborhoods', *American Journal of Sociology* 105 (3): 603–51.

Sampson, R. and Raudenbush, S. (2004) 'Seeing disorder: neighborhood stigma and the social construction of "broken windows"', *Social Psychology Quarterly* 67 (4): 319–42.

Sampson, R. and Wilson, W. (2002) 'A theory of race, crime, and urban inequality', in F. Cullen and R. Agnew (eds), *Criminological Theory: past to present,* Los Angeles: Roxbury. pp. 187–211.

Sampson, R., Morenoff, J. and Gannon-Rowley, T. (2002) 'Assessing "neighborhood effects": social processes and new directions in research', *Annual Review of Sociology* 28 (3): 443–78.

Sennett, R. (1970) *The Uses of Disorder: personal identity and city life,* New York: Knopf.

Schneider, J. (1980) *Detroit and the Problem of Order, 1830–1880,* Lincoln, NE: University of Nebraska Press.

Sherman, L., Gartin, P. and Buerger, M. (1989) 'Hot spots of predatory crime: routine activities and the criminology of place', *Criminology* 27 (1): 27–55.

Silverman, E. (1999) *NYPD Battles Crime: innovative strategies in policing,* Boston, MA: Northeastern University Press.

Skogan, W. and Hartnett, S. (1997) *Community Policing, Chicago Style,* New York and Oxford: Oxford University Press.

Smith, N. (1999) 'Which new urbaninsm? New York City and the revanchist 1990's', in R. Beauregard and S. Body-Gendrot (eds), *The Urban Moment: cosmopolitan essays on the late 20th-century city*, Thousand Oaks, CA: Sage.

Smith, S. (1986) *Crime, Space and Society*, Cambridge: Cambridge University Press.

Sparrow, M., Moore, M. and Kennedy, D. (1990) *Beyond 9/11: a new era for policing*, New York: Basic Books.

Taylor, R. (2000) *Breaking Away from Broken Windows*, Boulder, CO: Westview Press.

Tobias, J. (1979) *Crime and Police in England, 1700–1900*, Dublin: Gill and Macmillan.

Trojanowicz, R. and Bucqueroux, B. (1994) *Community Policing: how to get started*, Cincinnati: Anderson.

Turnbull, L., Hendrix, E. and Dent, B. (2000) *Atlas of Crime: mapping the criminal landscape*, Westport, CT: Greenwood.

Wacquant, L. (2000) 'The new "peculiar institution": on the prison as surrogate ghetto', *Theoretical Criminology* 4 (3): 377–89.

Wilson, J. and Kelling, G. (1997) 'Broken windows', in R. Dunham and G. Alpert (eds), *Critical Issues in Policing*, Prospect Heights, IL: Waveland Press. pp. 424–37.

Young, I. (1990) *Justice and the Politics of Difference*, Princeton, NJ: Princeton University Press.

20 TERROR AND SURVEILLANCE

Jon Coaffee and David Murakami Wood

This chapter

○ Argues that the city fulfils important defensive functions, protecting its citizens from particular threats, both internal and external

○ Focuses on recent threats to the city, suggesting that these have produced particular urban interventions and infrastructures designed to enhance urban surveillance

○ Concludes that despite the drive towards all-encompassing surveillance, fear can never be entirely removed; in fact, some forms of urban surveillance fuel fear

---

Introduction

This chapter considers the changing nature of urban socio-spatiality in response to fear of war, crime and terrorism. It will work on two interweaving themes: first, the form, and second, the scale of surveillance strategies. Its central argument is that forms of socio-spatial control have undergone scalar changes, moving outwards from the city through the nation-state to the world as a whole and, simultaneously, moving inwards through classes and groups within the city to the individual body, and, finally – through the fracturing of city, state and material body – to the displacement of the material as the object of control in favour of the virtual. However, this should not be regarded as a pure 'chronology'. City-wide strategies have far from vanished. Indeed, at all scales, socio-spatialities are constantly subject to reconfiguration in response to political economic trajectories and fears that (re)emerge. This is as much an age of 'urban resilience', or indeed 'urbicide', as it is of DNA-testing and 'data shadows'.

The development of the defensive city

Cities have always been characterized by material, psychological or metaphysical insecurity, whether these be of invasion by external forces, conflict between social classes and groups, crime, disorder and uprising. As Nan Ellin (1997) has remarked, 'Form follows fear'. Defence against external attack has been an ever-present pre-occupation. Archaeological records show that the early urban areas on the floodplains of great rivers such as the Nile, Tigris, Euphrates and Yangtze were often surrounded by walls, ditches and other defensive features to delimit the 'known' from the chaos and danger of the outside world. For example, Jericho was one of the first examples of a defensive city, which, in around 7000 BC, had a defensive wall and large bastion towers supported by a nine metre ditch which deprived attackers of a means by which they could approach the city walls. Other prominent examples include the Mesopotamian service towns of Uruk and Ur, which, through defensive strategies, created rich city-states and controlled territories (Atkins et al., 1998). These defensive features – a combination of a wall, tower and ditch – became the universal blueprint for the fortified city, a design which changed little between the building of Jericho and the introduction of gunpowder some 8,000 years later (Keegan, 1993).

What is clear is that where centralized modes of governance began to be established, construction of strategic defences around cities or regions became widespread. As Gold and Revill (1999: 230) noted, these defences served to secure 'the *interests* of an imperial power, serving to establish a presence and create an image of power that might impress an indigenous population of rival colonialists' (original emphasis). The development of such urban assemblages inside city walls meant the urbanization process could therefore be read as a process of social control. In particular, the threat of external attack meant the ruling class could justify the dense concentration of population into an easily regulated and surveilled space.

As cities developed, defensive systems became more complex to cope with the improving strategies of intruders. For example, Lanciani (1967) indicated that Ancient Rome was fortified seven times by different lines of walls between the fifth century BC and the third century AD. The castles and the walled towns of medieval Europe were also good examples of this trend. Here internal defences, as represented by the fortress which dominated the centre of the city, and external defence, in the form of a wall, were key features. Atkins et al. (1998) indicate that the developments in military technology at this time meant defensive technologies improved. Designs for stone-clad castles were imported into Europe from Arabia, which formed the centre of new settlements as urbanization spread within the safe confines of the city wall. Durham and Newcastle in northeast England, and the Italian cities of Florence, Venice, Milan and Rome provide good historic examples of such defended settlements. Likewise in the Far East, the castle-towns of feudal Japan, and the complicated spatial hierarchy that resulted from the edict of the ruling shogunate, separating cities into demarcated and often overtly surveilled areas for samurai, merchants and commoners (Sorensen, 2002). The city walls and castles served a defensive purpose but also became symbolic of social stratification within cities, and divisions between town and country. In time, the city wall and castle became less important, but still remained as symbols of wealth, privilege and power.

The history and spatial practice of these defensive and surveillant strategies is ineluctably bound up with other key social and technological developments in cities, in particular the history of measurement, and the establishment of bureaucratic forms of governance: a good example is the way in which the invention of the public clock transformed the ability of the rulers of cities in thirteenth-century Europe to more intensively regulate the spatio-temporal behaviour of the inhabitants (Crosby, 1997).

The modern period: classes and subjects

In time the city wall became less important as a symbol of wealth, privilege and safety, as technological advances – most notably the invention of gunpowder – made such traditional defences less effective. Cities, however, continued to be characterized by defensive features as new walled and gated spaces developed, increasingly inside city boundaries, as danger was increasingly seen to originate from within, rather than outside, the urban area. As such, by the mid-nineteenth century many Western cities were characterized by secure residential estates amid vast tracts of working-class housing, which were seen as *terra incognita*. The modern period saw massive spatial restructuring of major cities, moving away from the fortified, and impenetrable medieval city, to a model which would allow penetration deep into the urban fabric not by invading military forces, but by those of the nascent nation-state, in order to demonstrate the control of the means of violence and (potentially) to crush uprisings among the working class. Baron Haussmann's redesign of the city of Paris from 1853 to 1870, with wide boulevards radiating out from the centre of the city is the most famous and effective example of this process according to Walter Benjamin (1976), although recent work by Higonnet (2002) disputes the primarily military motivation. It is important to note here the connection with the management of colonial cities: such city redesign plans were often more effectively imposed upon subaltern populations.

The institutional measures designed to create such conditions of course served functions other than surveillance for the purposes of state control. For example, the massive effort in measurement that went into the Poor Laws in nineteenth-century Britain, the precursors of almost all modern health, sanitation and welfare systems, were inspired by philanthropic and patriarchal middle-class concerns over the health and hygiene of the working classes and not simply or even primarily by fear of revolution. Similarly, the invention of the passport (Torpey, 2000) allowed convenient movement through national borders and the development of trade, and not just the ability to trace the movements of suspicious individuals. Surveillance could be empowering (at least for some). However, this was also the age which developed the first 'biometric' identification technique, fingerprinting (Cole, 2001), and photography, which right from its conception was used, along with many other less long-lived techniques, as a tool to measure and 'scientifically' categorize suspicious people and criminal 'types' (Sekula, 1992), many of which were again pioneered in the colonies (e.g. Major, 1999).

Underlying the concern with the 'dangerous classes', the restructuring of urban space and institutions in response to moral concerns and the fear of revolution, a change was taking place in the conception of the person. The medieval, indeed ancient, conception of an all-seeing God, and his earthly representatives with the power of repression and torture at their disposal lost ground in the scientific and philosophical revolution of the Enlightenment, with the terrain of contestation shifting towards the body and the mind and the ability of individuals to regulate themselves: in other words, the development of a 'modern subject'. Michel Foucault (1975) argued that the development of this subject was the key purpose of modern institutions, such as the police service, the hospital, the asylum, and most importantly of all, the prison. In a key chapter in *Discipline and Punish* (1975), Foucault argues that this is encapsulated in spatial form by the diagram of the Panopticon, the reformatory and poor-house design developed at the end of the eighteenth century by Utilitarian philosopher and philanthropist, Jeremy Bentham (1791). This *dispositif* has been well described elsewhere: put simply, it consists of a circular arrangement of cells surrounding a central guard tower, from which a watcher may observe, unseen, the prisoners. The prisoners would not at any time know for certain that they were being watched, and thus, would choose to modify their own behaviour, to train their souls, and become docile bodies fitting the norms of the institution. Surveillance was no longer just about watching for the threat of an external enemy, but about watching the internal foe: whether inside society or inside oneself. Thus urban surveillance in the modern period was not simply a matter of large-scale spatial control, but of moral disciplinary measures at far smaller scales. And thus the most mundane institutions could provide settings for the operation of these forms of close supervision, for example, the public house (Kneale, 1999) or the new public libraries (Black, 2001).

War, intelligence and surveillance

The military continued to be a major agent of control, a primary source of surveillant practices, and an inventor, adopter and adaptator of existing technologies for purposes of surveillance and control. For example, one of the first suggested experimental uses of the hot-air balloon was in battlefield reconnaissance (Saatjan, 2002) and the speedy production of rifles was the original purpose of the modern production line (Hounshell, 1984), which provided for the machinic control of industrial workers within urban factories, and this interrelationship between the social, the industrial and the military continued throughout the twentieth century. In particular, the field of espionage and political policing grew enormously (Porter, 1992). Espionage has traditionally been known as the 'second oldest profession', but towards the end of the nineteenth century and particularly in the early part of the twentieth century, states began to create more effective systems of monitoring other nations and internal threats.

The internal 'threats' came from the increasing organization of the dissatisfied urban working class. These intensified with the success of the Bolshevik Party in overturning the Tsarist regime in Russia in 1919. But in colonialist nations, the development of internal security, like many urban redesign policies, were also strongly influenced by the experience of policing overseas possessions, for example, for the UK, Ireland and India (Bunyan, 1977). Political police thereafter proliferated both in nations opposed to communism and those seeking to control 'counter-revolutionary' activity in nascent communist countries (see, for example, Mazower, 1997). Examples include MI5 (the Secret Service) in the UK, the Federal Bureau of Investigation (FBI) in the USA (which was also the national-level serious crime agency), the Special Higher Police (Tokkō) in Japan and the Gestapo in Nazi Germany. While some of these agencies (the Gestapo and Tokkō in particular) were disbanded with the defeat of the Axis regimes during the Second World War, others (such as the FBI) became stronger during the post-Second World War 'Cold War' (see below), and were responsible for public 'purging' of political opponents. New organizations evolved in new post-colonial states, for example Mossad in Israel. It is arguable that the apogee of this ongoing era of internal security was with the Stasi in the former German Democratic Republic, which at its peak was the largest national employer, held a detailed file on every citizen and listed almost one-sixth of the population as informers (Gieseke, 2001).

External intelligence agencies as they exist today are also a product of the twentieth century (Richelson, 1995). For example, in Britain, the Special Intelligence Service (SiS, or MI6) and General Communications Headquarters (GCHQ) both evolved from the Admiralty's naval intelligence organizations in the First World War. In the USA, agencies proliferated and flourished in the Cold War, the most important being the Central Intelligence Agency (CIA) which focuses on Human Intelligence (HUMINT), the National Security Agency (NSA) which deals with Signals Intelligence (SIGINT), and the National Reconnaissance Organization (NRO) (satellite operations) which grew to be the world's most expensive intelligence agency. The Cold War saw massive resources devoted to espionage with the increased exploitation of flight, in particular spaceflight and the development of satellites (Richelson, 1999), but also of computing. Computers were developed for two main purposes during the Second World War: the cracking of codes and the control of atomic weapons. The Cold War saw massive technological development in telecommunications monitoring (interception or 'tapping'), filtering and eventually automated word, phrase and voice recognition systems such as the NSA's ECHELON system (Campbell, 1997) and the combination of computers with telecommunications in ARPANET, the emergency distributed communications system that became the Internet. The combination of satellite surveillance, telecommunications and computing formed the basis of massive attempts by rival powers to surveil the entire planet, a strategy which Edwards (1996) has called the creation of a 'closed world'.

One of the key features of this period for urban areas, which had begun with the massive destruction of cities through bombing in the Second World War (London, Coventry, Dresden, Tokyo, Hiroshima, Nagasaki, etc.) was the shift from the city as defensible space to the city as target: strategies shifted from trying to prevent land attack to an 'anxious urbanism' of flight from the cities, retreat, abandonment and 'acceptable losses'. Farish (2003) has pointed out how the US interstate highway system was particularly designed with this in mind, and the Cold War saw the proliferation of 'secret underground cities' (McCamley, 1998): networks of fortified spaces below and beyond major cities that would enable a small elite portion of society, particularly government, to survive the destruction of cities and provide secure command and control centres. The latter, largely removed from urban centres, exist not only in the 'homeland' of the nation-state, but in the case of imperial powers, distributed throughout the world: there is a whole 'hidden geography' of such bases and bunkers (Wood, 2001), the most extensive network being that of the USA.

Crime, terrorism and the city in the late twentieth century

The geopolitical urban strategies of nation-states were largely invisible to ordinary people and even to most involved in the everyday governance of cities, who had their own problems of control. In the late 1960s and early 1970s, defensive architecture and urban design were increasingly used in American cities as a direct response to the urban riots which swept many US cities in the late 1960s, as well as the perceived problems associated with the physical design of the modernist high rise blocks, which were seen as breeding grounds for criminal activity (Jacobs, 1961). This was also a result of research which indicated a relationship between certain types of environmental design and reduced levels of violence (Gold, 1970), concerns that enhanced urban fortifications were socially and economically destructive (in terms of economic decline of the city centre and social polarization) and that the provision of security was becoming increasingly privatized as individuals, having lost faith with the public authorities to provide a safe environment, increasingly sought to defend themselves.

As a result of such concerns, American urban planners and designers looked for strategies to reduce the opportunity for urban crime. These came initially through an 1971 approach called *Crime Prevention through Environmental Design*

(CPTED). However, it was the publication of Oscar Newman's *Defensible Space: Crime Prevention through Urban Design* (1972) that stimulated the most intense debates on the relationship between crime and the built environment. In his studies Newman did not rule out the use of security fences or electronic surveillance technologies, but relying on these measures was seen as a last resort if more subtle design solutions were unsuccessful. Newman's work on housing estates in New York and St Louis led to the concept of *defensible space*, which he saw as a 'range of mechanisms – real and symbolic barriers ... [and] improved opportunities for surveillance – that combine to bring the environment under the control of its residents' (Newman, 1972: 3). Defensible space was seen as the physical expression of a social fabric that could defend itself and could arguably be achieved by the manipulation of architectural and design elements.

Newman's ideas were inexorably linked to the late 1960s and early 1970s and reflected the enhanced interest of the architects, planners and urban designers in linkages between environment and behaviour and especially ideas of territoriality. Poyner (1983: 8) further highlighted that 'defensible space' was considered attractive at this time because the 'emphasis was on the use of the environment to promote residential control and therefore somehow return to a more human and less threatening environment'. In short, defensible space offered an alternative to the target-hardening measures being introduced to new residential communities at this time in America and subsequently in other Western countries, most notably in the UK (Coleman, 1985).

In the British context, defensible space ideas were to have wider adaptations than the residential context. For example, Boal (1975) argued that, in relation to the need for anti-terrorist security in Northern Ireland, the ultimate level of security provision in a city is defensible space with its emphasis on territoriality, existing alongside physical barriers. In particular, Belfast in the 1970s could be seen as a laboratory for radical experiments on the fortification of urban space with a number of distinct defended territories created along sectarian lines to give the occupants of a defined area, or individual buildings, enhanced security. This was most noticeable in Belfast city centre where, following a series of car bomb attacks, a security cordon was enacted in 1972 in an attempt to protect retail premises (Brown, 1985). The drastic security measures were taken due to the unsuccessful attempts by the authorities to tackle the security problem and can be seen as a radical example of territoriality. By 1974 the barbed wire fences encircling the central area had been replaced by a series of tall steel gates which became known locally as the 'ring of steel'. All shoppers were searched upon entering this pedestrianized cordon. Subsequently, as the risk of terrorist attack subsided during the 1980s, the cordon contracted in size and urban planners have sought to re-image this 'pariah city' in an attempt to attract businesses back (Neill et al., 1995) (see Extract 20.1). Indeed, any decrease in security was offset by a centralized CCTV scheme, which became operational in December 1995.

Extract 20.1: From Boal, F.W. (1995) *Shaping a City: Belfast in the Late Twentieth Century*, Belfast: Institute of Irish Studies, pp. 81–91.

As the 1960s progressed it was beginning to be evident that the City Centre was threatened by two trends – the growth in road traffic and the first signs of suburban shopping development. … Traffic congestion and suburban shopping centres were nothing unique to Belfast. The third threat – urban terrorism – was another matter altogether. Here, from 1969 onwards, violence rapidly emerged and in many ways the City Centre was literally at the focus of the action.

The City Centre of the early 1990s is a far cry from that of the early 1970s. Then the bombing campaign of the Provisional I.R.A., which appeared to be a concerted attempt to cripple the city's commercial life, led to the destruction of some 300 retail outlets and resulted in a loss of almost one quarter of the total retail floor space. Security measures introduced in response to the bombing campaign, together with the campaign itself, combined to make shoppers reluctant to patronise centrally located retail establishments. … The 1980s and the early 1990s have witnessed a remarkable recovery in the fortunes of the City Centre. A decrease (and now cessation) of the Provisional I.R.A. bombing campaign, together with a related relaxation of City Centre security arrangements, led to renewed confidence among consumers and investors alike. City Centre accessibility was improved … while the environment of the centre was enhanced by extensive pedestrianization (itself made easier by the security measures taken to exclude car bombers) and widespread tree planting. A net retail floor-space of 1.3 million square feet in 1967, having declined to 1.1 million by 1975, recovered to 1.3 million again by 1985. This means that much of the evident growth in trade and investment since 1970 has actually been constituted by a recovery to a former position.

Fortress cities: strategies of city-wide urban defence

In the 1990s new defensible space approaches to the operationalization of pro-security discourses once again served to influence the design and management of the urban landscape. The response of urban authorities to insecurity in some cases was dramatic, especially in North America, and in particular in Los Angeles (LA), where it is argued that the implementation of crime displacement measures and the surveillance of particular spaces has been taken to an extreme. In LA the social and physical fragmentation of the city is often shown to be very pronounced, and which, according to certain commentators, could set a precedent for 'postmodern urbanism' (Dear and Flusty, 1998).

During the 1990s LA assumed a theoretical primacy within urban studies with an overemphasis on its militarization, portraying the city as an urban laboratory for anti-crime measures. Fortress urbanism was highlighted as the order of the day, as an obsession with security became manifested in the urban landscape with 'the physical form of the city … divided into fortified cells of affluence and places of terror where police battle the criminalized poor' (Dear and Flusty, 1998: 57). For example, it was reported that, in 1991, 16 per cent of Los Angelians were living in 'some form of secured access environment' (Blakely and Snyder, 1997: 1).

Mike Davis is perhaps the most cited author on 'Fortress LA'. Davis depicts how in recent years the authorities and private citizen groups in LA have responded to the increased fear of crime by 'militarizing' the urban landscape. His dystopian portrayal of LA in *City of Quartz* (1990) provided an alarming indictment of how increasing crime trends could theoretically affect the development and functioning of the future city through the radicalizing of territorial defensive measures with the Los Angeles Police Department (LAPD) becoming a key player in the development process. As the boundaries between the two traditional methods of crime prevention – law enforcement and fortification – have become blurred, defensible space and technological surveillance, once used at a micro-scale level, are being used at a meso and macro level to protect an ever-increasing number of city properties and residences. In *Beyond Blade Runner: Urban Control, the Ecology of Fear*, Davis (1992; see also Davis, 1998) extrapolated current social, economic and political trends to create a vision for the future city in the year 2019, which in this account had become technologically and physically segregated into zones of protection and surveillance such as high security financial districts and segregated gated communities. In this vision, economic disparities have created an urban landscape of cages and wasteland (Extract 20.2).

Extract 20.2: From Davis, M. (1998) *Ecology of Fear: Los Angeles and the Imagination of Disaster*, New York: Metropolitan Books, pp. 359–61.

Every American city boasts an official insignia and slogan. Some have municipal mascots, colors, songs, birds, trees, even rocks. But Los Angeles alone has adopted an official nightmare. In 1998, after three years of a debate, a galaxy of corporate and civic celebrities submitted to Mayor Bradley a detailed strategic plan for Southern California's future. Although most of the *L.A. 2000: A City for the Future* is devoted to hyperbolic rhetoric about Los Angeles's rise as a 'world crossroads' comparable to imperial Rome or LaGuardian New York, a section on the epilogue, written by historian Kevin Starr, considered what might happen if the city failed to create a new 'dominant establishment' to manage its extraordinary ethnic diversity. 'There is of course the *Blade Runner* scenario: the fusion of individual culture into a demotic polyglotism ominous with unresolved hostilities.'

Blade Runner – Los Angeles's dystopic alter ego. Take the Grayline tour in 2019: the mile-high neo-Mayan pyramid of the Tyrell Corporation drips acid rain on the mongrel masses in the teeming ginza far below. Enormous neon images float like clouds above the fetid, hyperviolent, while a voice intones advertisements for extraterrestrial suburban living in 'Off World'...

With Warner Brothers' release of a more hardboiled 'director's cut' a few months after the Rodney King riots, Ridely Scott's 1982 film ... reasserted its sway over our increasingly troubled sleep. Ruminations about the future of Los Angeles now take for granted the dark imagery of *Blade Runner* as a possible, if not inevitable, terminal point for the Land of Sunrise. ...

Events since the 1992 riots – including a four-year-long recession, a sharp decline in factory jobs, deep cuts in welfare and public employment, a backlash against immigrant workers, the failure of police reform and an unprecedented exodus of middle class families – have only reinforced spatial apartheid in greater Los Angeles. As the endless summer comes to an end, it seems that L.A. 2019 might well stand in a dystopian relationship to the most traditional ideals of a democratic metropolis.

But what kind of dystopian cityscape, if not *Blade Runner*'s, might the unchecked evolution of inequality, crime, and social despair ultimately produce? Instead of following the grain of traditional clichés and seeing the future merely as grotesque ... would it not be more fruitful to project existing trends along their current downward-sloping trajectories?

Davis's work has been elaborated on by many subsequent authors (see Chapter 19). For example, Flusty (1994) provided a categorization of the different types of fortress urbanism which, he argued, had thrown a blanket of fortified and surveillance security over the entire city. He referred to the spaces of security as 'interdictory space' which are designed to exclude by their function and 'cognitive sensibilities'. A typology of such spaces is shown in Table 20.1.

Table 20.1   Typology of interdictory space

Stealthy space	Passively aggressive with space concealed by intervening objects
Slippery space	Space that can only be reached by means of interrupted approaches
Crusty space	Confrontational space surrounded by walls and checkpoints
Prickly space	Areas or objects designed to exclude the unwanted such as unsittable benches in areas with no shade
Jittery space	Space saturated with surveillance devices

Source: Adapted from Flusty (1994)

Flusty highlighted how such defended spaces, alone or in combination, have pervaded all aspects of urban life, leading to an ever-increasing number of highly secure gated communities, bunker architectures and highly policed ghettos in the disadvantaged poor areas of the city. He also noted how the commercial privatization of space is taken to an extreme as a strong fear of the public realm leads to highly inclusive business facilities either in isolation or in self-contained agglomerations.

Although 'Fortress LA' in the 1990s became a powerful vision for the city, it is important to realize that there are many other ways in which urbanism in LA may be viewed. For example, critics of Davis have argued that he is portraying a very dystopian image of the city as one shackled with terror, fear and anxiety and under the constant gaze of surveillance cameras.

Hard and soft boundaries: rings of concrete and rings of surveillance

During the late 1980s attempts to design out or reduce the impact of terrorism at specific targets were often crude and rudimentary, but nonetheless, high profile (see Figure 20.1). By the early 1990s, when fortress urbanism was a popular practice among planners and urban designers, there was also a noticeable targeting of global cities, and in particular their economic infrastructure, by terrorist organizations in order to attract global media publicity and cause severe insurance losses and significant disruptions in trade. This was perhaps most noticeable in London, where the Provisional IRA successfully attacked a number of key economic targets with large bombs exploding in the City of London (the Square Mile) in April 1992 and April 1993 and the London Docklands in 1992 (unexploded) and 1996. These bombings and the subsequent reaction of urban authorities and the police served to highlight the use made of both territorial and technological approaches to counter-terrorist security, creating a series of interlocking of hard and soft boundaries to counter the terrorist threat (Coaffee, 2003, 2004). Such an approach can be seen as an enhancement of pre-existing methods already employed in many cities as a result of the increase in the fear of crime.

Following the first major bombing of the City of London in April 1992, there were calls for an impenetrable security cordon to be constructed, although at this time such radical security was dismissed as a propaganda gift to the bombers. However, in the aftermath of the 1993 bomb in the City, what was referred to in the media as a Belfast-style 'ring of steel' was activated in the City, securing all entrances to the central financial zone. Locally, the ring of steel was referred to as the 'ring of plastic' as access restrictions were based primarily on the funnelling of traffic through rows of plastic traffic cones.

The territorial approaches to security were backed up by the retrofitting of ever-advanced CCTV in both private and public spheres. The police, through an innovative partnership scheme known as 'CameraWatch', encouraged private companies to install CCTV in liaison with neighbouring businesses, while at the entrances of the ring of steel as well as at strategic points around the Square Mile the most technologically advanced CCTV cameras available were installed (in 1997) in the form of 24-hour Automatic Number Plate Recording cameras

Figure 20.1 Downing Street entrance

(ANPR), linked to police databases. These digital cameras were capable of processing the information and giving feedback to the operator within four seconds. In the space of a decade, where terrorism had been considered a serious threat, the City of London was transformed into the most surveilled space in the UK (and perhaps the world) with over 1,500 surveillance cameras operating, many of which are linked to the ANPR system (Coaffee, 2003).

In a similar way to the City, the London Docklands, containing the Canary Wharf complex, was also the focus for counter-terrorist planning through the 1990s. This area was subject to a failed bombing in 1992 as well as a devastating explosion in the southern part of the area in 1996. Following the 1992 Canary Wharf attack, managers initiated their own 'mini-*ring of steel*' essentially shutting down access to 'their' private estate within the Docklands complex (Coaffee, 2000; Graham and Marvin, 2001). Security barriers were thrown across the road into the complex, no-parking zones were implemented, a plethora of private CCTV cameras were installed and identity card schemes were initiated. After the 1996 bomb in the southern part of the Docklands the business community successfully lobbied the police to set up an anti-terrorist security cordon to cover the whole of the Docklands – the so-called *Iron Collar* modelled on the City of London's approach (see Figure 20.2) – amidst fears that high-profile businesses might be tempted to relocate away from the Docklands.

Figure 20.2   Entrance through the Docklands' Iron Collar

Over time, such securitization against certain 'at risk' sites from terrorism has lead to the inevitable dislocation of London into zones of differential risk and security. That said, advancements in technology have subsequently allowed a more expansive security blanket over central London. The ANPR technology that was developed throughout the City's attempts to deter terrorists has now been 'rolled out' across central London for use in traffic 'congestion charging'. This system became operational in February 2003 and uses 450 cameras in 230 different positions. In essence, central London has been circled by digital cameras creating a dedicated 'surveillance ring' affording London's police forces vast surveillance gathering capabilities for tracking the movement of traffic and people, and by inference highlighting potential terrorist threats. This became a key priority after the attacks in New York and Washington DC on 11 September 2001.

From CCTV to dataveillance

Closed circuit television (CCTV) has played a significant role in these various strategies of spatial control. Attempts had been made to utilize broadcast television cameras for crime control from very early in the history of television: examples can be found in Nazi Germany, and in post-war Britain, where for the coronation of Queen Elizabeth II, the police requested the British Broadcasting

Corporation to provide them with access to footage for purposes of crowd control, but were refused (Williams, 2003). Yet CCTV's origins are neither exclusively in urban policing nor military surveillance. A significantly underplayed strand in the history of surveillance is that of consumption and leisure, and it is here where CCTV first flourished. It was in the 1960s casinos in the USA, largely to prevent fraud, and thereafter, particularly with the development of videotape in the massive suburban shopping malls that CCTV developed (Norris et al., 2004). Theme parks are also crucial here: the parks owned by the Disney Corporation became important sites for the experimental utilization of private security (in the form of undercover security) and also CCTV (Shearing and Stenning, 1985). This is important because it is the semi-public, closed and controlled world of malls and theme parks that is increasingly seen as providing a model for neo-liberal urban renewal initiatives (Sorkin, 1992).

However, for urban spatial control, the UK was the pioneer. It has been estimated that there are 4 million CCTV cameras in the UK (Norris et al., 2004) and over 85% of local authorities now have at least one system in place (Webster, 2004). CCTV was first introduced in public space in Bournemouth, a resort town on the south coast of Britain in 1985 (for a detailed account of the history of CCTV in Britain, see Norris and Armstrong, 1999). The year before had seen the Provisional IRA mount a devastating bomb attack on the ruling Conservative Party conference in Brighton, almost killing the Prime Minister, Margaret Thatcher. Bournemouth was to be venue for the next conference. That CCTV soon penetrated British towns and cities so thoroughly was not entirely due to the fear of terrorism. The neo-liberal relaxation of planning laws and the expansion of out-of-town shopping had seen traditional space of consumption in town centres decline. Other fears were also used to justify the installation of CCTV systems: particularly football hooliganism, and high-profile crimes against children, especially the kidnapping and murder of James Bulger in 1993. Crucial to this process was the role of the state in providing funding and limiting regulation. Thus while CCTV was expanding, it was also able to be normalized as an expected feature of public space, or even a 'fifth utility' (Graham et al., 1996). In the wake of September 11, CCTV, along with other advanced technologies of urban surveillance, have 'surged' forward (Wood et al., 2003), and the UK has come to be seen as a 'model' for the implementation of urban security by other nation-states.

The relationship between surveillance and subjectivity observed by Foucault in the modern period has continued to evolve. First, biometric identification (based on quantified bodily traces) has moved far beyond fingerprinting to the more intimate. Identification has penetrated the surface of the body and involves analysis of samples like hair, urine and blood, and at more fundamental levels, such as with DNA fingerprinting (Nelkin and Andrews, 1999). However, forms of surveillance also involve less intrusive biometrics, such as facial and iris recognition, some of which can be linked into digital CCTV (Introna and Wood, 2004). Current fears exist within a far more complex technologically-dominated polity than in the early modern period of Foucault's *dispositif panoptique* (Barry, 2001). This techno-logical politics crosses all domains from the mechanical to the biological. For example new genetics metaphors are bound up in the notions of militarization, security and resilience (Dillon, 2002), and police and urban planners are adopting neo-Victorian notions of threat from 'genetically dangerous' classes of people, instead of criminality as individual 'deviancy' (Rose, 2000).

At the same time, the surveillance of the body itself has been supplemented by associated data. Mark Poster (1990), Gilles Deleuze (1992), and Oscar Gandy (1993) posit movement away from internalized soul-training to surveillance through searchable databases of information, sometimes called 'dataveillance' (Clarke, 1988). This dispersed and fragmentary information could constitute a separation or division of the self, creating multiple 'dividuals' (Deleuze, 1992) or 'data subjects' (Gandy, 1993), as (or more) important than the embodied subject. This is a step-change from the Stasi paper file: computer databases allow greater integration and automated algorithmic operations to be performed effectively in real time, without the bodily subject knowing (Graham and Wood, 2003).

Urban surveillance beyond 9/11: militarization, automation and pervasiveness

Since the devastating attacks targeted at the symbolic urban centre of neo-liberal Western capitalism of 11 September 2001, there have been growing tendencies towards a situation where 'military and geopolitical security now penetrate utterly into practices surrounding governance, design and planning of cities and region' (Graham, 2002: 589). The 'war on terrorism' has already served as a 'prism being used to conflate and further legitimize dynamics that already were militarizing urban space' (Warren, 2002: 614). In the immediate aftermath, many commentators also (incorrectly) predicted the demise of the skyscraper and the changing functionality of urban centres. Others highlighted the potential for terrorism to lead to a new counter-urbanization trend among business and wealthier citizens in search of 'space and security' (Vidler, 2001), or for the increased fragmentation of urban space to con-tinue through 'concentrated decentralization' (Marcuse, 2002).

Extract 20.3: From Lyon, D. (2003) *Surveillance after September 11*, Cambridge: Polity, pp. 4–5.

We can understand 9/11 – the events and their aftermath – in two ways: 9/11 may be viewed as both revealing and actually constituting major social change. The attack brought to the surface a number of surveillance trends that had been developing quietly, and largely unnoticed for the previous decade and earlier. ... In other words, the establishment of 'surveillance societies' that affects the lives of ordinary people was already underway long before 9/11. The aftermath of the attacks helps us to see more clearly what is already happening.

At the same time, the 9/11 event may also be read as an opportunity – to some, even a golden opportunity – that gave some already existing ideas, policies, and technologies their chance. In this way it helped to constitute merging social and political realities. ... The desire of several governments to hold on to some semblance of social control, which some felt had been slipping away from them in a globalizing word, now found an outlet in 'anti-terrorist' legislation.

Technologically, the US administration was fairly quick to come up with the astonishingly comprehensive 'Total Information Awareness' scheme at the Pentagon. The data-mining technologies had been available for some time in commercial settings, but until 9/11 no plausible reason existed for deploying them – and the customer data that they analyse – within a national security apparatus. The drive towards large-scale, integrated systems for identifying and checking persons in places such as airports and at borders, urged for years by technology companies, received its rationale as the twin towers tumbled.

Such accounts of the post-9/11 city also tend to present bleak portrayals and worst case scenario options. The concern is that anti-terrorist defences and heightened surveillance, if constructed, could mean the virtual death of the urban areas as functioning entities. For example, in London, the unprecedented events of 9/11 led to an instant counter-response from London police forces focused on digitalized tracking technologies as well as the overt fortressing of 'at risk' sites. For example, certain prominent landmark buildings were crudely fortified against vehicle-borne bombs, such as the US embassy in central London, which has become a virtual citadel, separated from the rest of London by fencing, waist-high 'concrete blockers', armed guards and mandatory ID cards. Furthermore, in May 2003, in response to a heightened state of alert, a vast number of waist-high concrete slabs were placed outside the Houses of Parliament to stop car bombers. This so-called 'ring of concrete' was later painted black to make it more 'aesthetically pleasing' (Coaffee, 2004).

Anti-terrorist security at key sites in major cities is both visible (as in overt fortressing and defensible space measures at key target sites) and invisible as surveillance activity. The latter forms what Lianos and Douglas (2000) refer to as Automated Socio-Technical Environments (ASTEs): normative notions of good behaviour and transgression and, increasingly, stipulations and punishments (for example, electronic tagging) are encoded using software into the space–time fabrics of cities. Right-wing commentators in the USA, such as Huber and Mills (2002), went so far as to demand a war between 'our silicon' and 'their sons', and argued that a pervasive automated surveillance apparatus of micro-sensors 'dispersed along roadsides, hills, and trails ... will report just about anything that may interest us – the passage of vehicles, the odor of explosives, the conversations of pedestrians, the look, sound, weight, temperature, even the smell, of almost anything'. This surveillant adaptation of 'pervasive', 'ubiquitous' or 'ambient' computing (Cuff, 2003) has the potential to enable what Thrift and French (2002) have called the 'automatic production of space'.

With some arguing that 'fear and urbanism are at war' (Swanstrom, 2002), urban policy-makers now have to think carefully when balancing security with mobility and risk with recklessness. Since 9/11, issues of trust, risk and danger in cities have increasingly come to the fore, with 'trust [being] replaced with mistrust and as such "the terrorist threat" triggers a self-multiplication of risks by the de-bounding of risk perceptions and fantasies' (Beck, 2002: 44) which are over-exposed in the global media and uniquely concentrated in the global city. This had led to areas becoming disconnected, physically and technologically, from the rest of the city through the development of securitized 'rings of confidence' (Coaffee, 2003), contributing to a fractured or 'splintered' urbanism (Graham and Marvin, 2001). The response of urban authorities and public and private security agencies to this threat poses serious consequences for urbanity and the civic realm, and in particular for freedom of movement. This is particularly serious when militarized security perspectives are bound up with neo-liberal agendas on urban regeneration (Raco, 2003). Worryingly, such processes can often be seen as selective or exclusionary. As Coaffee (2004: 209) notes, 'the policy processes which are leading to the ever-increasingly automatic control and militarization of urban space have ultimately lacked transparency and scrutiny and have often been promoted in terms of traffic management or crime inhibiting measures. As such, this points inevitably to the splintering potential of such rings of security and rings of confidence, which are slowly but surely becoming 'rings of exclusion'.

Summary

Terrorism is the most recent in a long line of threats which have shaped the fabric, social relations and governance of the city. This chapter has traced the changes in the nature and scale of response from external fortification through modernization, bureaucratization and self-surveillance to a city of increasingly pervasive surveillance conducted by a wide range of institutions, groups and individuals with an expanding range of technologies. At the beginning of the twenty-first century we believe that several key areas form the focus for future research. The first concerns the complex connections between the military and the civil. The second involves the diffusion of the new forms of surveillance practices and technologies to the cities of the global south, and hence the need for cross-cultural and comparative studies of surveillance. Third, the concept of pervasiveness, ubiquity or ambience in surveillance demands attention for its capacity to infiltrate buildings and infrastructure, consumer products, clothing and even bodies. Finally, and most immediately, there is the growing concentration on what is being commonly termed 'urban resilience' (Coaffee, 2006) – a critique of which includes the serious and potentially dangerous impacts that surveillance might have for equity and conviviality in cities. One of the key roles that academics can play is in questioning the underlying 'fears' and 'threats' and to make sure that the variety of potential outcomes follow hope, not fear.

References

Atkins, P.J., Simmons, I. and Roberts, B. (1998) *People, Land and Time*, London: Arnold.

Barry, A. (2001) *Political Machines: Governing a Technological Society*, London: Athlone.

Beck, U. (2002) 'The terrorist threat: world risk society revisited', *Theory, Culture and Society*, 19 (4): 39–55.

Benjamin, W. (1976) 'The Paris of the second empire in Baudelaire', in W. Benjamin, *Charles Baudelaire: A Lyric Poet in the Era of High Capitalism*, London: Verso. pp. 9–106.

Bentham, J. (1791) *Panopticon: or The Inspection-House* (2 vols), London: T. Payne.

Black, A. (2001) 'The Victorian information society: surveillance, bureaucracy, and public librarianship in 19th-century Britain', *Information Society*, 17 (1): 63–80.

Blakely, E.J. and Synder, M.G. (1997) *Fortress America: Gated Communities in the US*, Washington, DC: The Brookings Institution.

Boal, F.W. (1975) 'Belfast 1980: a segregated city?', *Graticule*, Belfast: Department of Geography, Queens University of Belfast.

Boal, F.W. (1995) *Shaping a City: Belfast in the Late Twentieth Century*, Belfast: Institute of Irish Studies.

Brown, S. (1985) 'Central Belfast's security segment: an urban phenomenon', *Area*, 17 (1): 1–8.

Bunyan, T. (1977) *The History and Practice of the Political Police in Britain* (revised edition), London: Quartet Books.

Campbell, D. (1999) *Interception Capabilities 2000*, Luxembourg: European Parliament, STOA Programme.

Clarke, R.A. (1988) 'Information technology and dataveillance', *Communications of the ACM*, 31 (5): 498–512.

Coaffee, J. (2000) 'Fortification, fragmentation and the threat of terrorism in the City of London in the 1990s', in J.R. Gold and G. Revill (eds), *Landscapes of Defence*, London: Prentice Hall.

Coaffee, J. (2003) *Terrorism, Risk and the City*, Aldershot: Ashgate.

Coaffee, J. (2004) 'Rings of steel, rings of concrete and rings of confidence: designing our terrorism in central London pre and post 9/11', *International Journal of Urban and Regional Research*, 28 (1): 201–11.

Coaffee, J. (2006) 'From counter-terrorism to resilience', *European Legacy – Journal of the International Society for the Study of European Ideas* (ISSEI), 11 (4): 389–403.

Cole, S. (2001) *Suspect Identities: A History of Criminal Identification and Fingerprinting*, Cambridge, MA: Harvard University Press.

Coleman, A. (1985) *Utopia on Trial: Vision and Reality in Planned Housing*, London: Hilary Shipman.

Crosby, A.W. (1997) *The Measure of Reality: Quantification and Western Society, 1250–1600*, Cambridge: Cambridge University Press.

Cuff, D. (2003) 'Immanent domain: pervasive computing and the public realm', *Journal of Architectural Education*, 57 (1): 43–9.

Davis, M. (1990) *City of Quartz: Excavating the Future in Los Angeles*, London: Verso.

Davis, M. (1992) *Beyond Blade Runner: Urban Control, the Ecology of Fear*, Open Magazine Pamphlet Series, Westfield, NJ.

Davis, M. (1998) *Ecology of Fear: Los Angeles and the Imagination of Disaster*, New York: Metropolitan Books.

Dear, M. and Flusty, S. (1998) 'Postmodern urbanism', *Annals of the Association of American Geographers*, 88 (1): 50–72.

Deleuze, G. (1992) 'Postscript on the societies of control', *October*, 59 (Winter): 3–7.

Dillon, M. (2002) 'Network society, network-centric warfare and the state of emergency', *Theory, Culture and Society*, 19 (4): 71–9.

Edwards, P.N. (1997) *The Closed World: Computers and the Politics of Discourse in Cold War America*, Cambridge, MA: MIT Press.

Ellin, N. (1997) *Architecture of Fear*, Princeton, NJ: Princeton Architectural Press.

Farish, M. (2003) 'Disaster and decentralization: American cities and the Cold War', *Cultural Geographies*, 10 (2): 125–48.

Flusty, S. (1994) *Building Paranoia: The Proliferation of Interdictory Space and the Erosion of Spatial Justice*, Los Angeles: Los Angeles Forum for Architecture and Urban Design, 11.

Foucault, M. (1975) *Discipline and Punish: The Birth of the Prison*, New York: Vintage.

Gandy, O.H. Jr. (1993) *The Panoptic Sort: A Political Economy of Personal Information*, Boulder, CO: Westview Press.

Gieseke, J. (2001) *Mielke-Konzern: Die Geschichte der Stasi 1945–1990*, Stuttgart: Deutsche Verlags-Anstalt.

Gold, R. (1970) 'Urban violence and contemporary defensive cities', *Journal of the American Institute of Planning*, 36 (3): 146–59.

Gold, J.R. and Revill, G. (1999) 'Landscapes of defence', *Landscape Research*, 24 (3): 229–39.

Graham, S. (2002) 'Special collection: reflections on cities, September 11 and the "War on Terrorism" – one year on', *International Journal of Urban and Regional Research*, 26 (3): 589–90.

Graham, S. (ed.) (2004) *Cities, War, and Terrorism: Towards an Urban Geopolitics*, Oxford: Blackwell.

Graham, S., Brooks, J. and Heery, D. (1996) 'Towns on the television: closed circuit TV systems in British towns and cities', *Local Government Studies*, 22 (3): 3–27.

Graham, S. and Marvin, S. (2001) *Splintering Urbanism: Networked Infrastructures, Technological Mobilities and Urban Condition*, London: Routledge.

Graham, S. and Wood, D. (2003) 'Digitising surveillance: categorisation, space, inequality', *Critical Social Policy*, 23 (2): 227–48.

Higonnet, P. (2002) *Paris: Capital of the World*, Cambridge, MA: Harvard University Press.

Hounshell, D.A. (1984) *From the American System to Mass Production, 1800–1932*, Baltimore, MD: Johns Hopkins University Press.

Huber, P. and Mills, M.P. (2002) 'How technology will defeat terrorism', *City Journal*, 12 (1) (no page numbers), http://www.city-journal.org/html/12_1_how_tech.html.

Introna, L. and Wood, D. (2004) 'Picturing algorithmic surveillance: the politics of facial recognition systems', *Surveillance and Society*, 2 (2/3): 177–98, http://www.surveillance-and-society.org/articles2(2)/algorithmic.pdf.

Jacobs, J. (1961) *The Death and Life of Great American Cities*, London: Peregrine.

Jeffery, C.R. (1971) *Crime Prevention through Environmental Design*, Thousand Oaks, CA: Sage.

Keegan, J. (1993) *A History of Warfare*, London: Hutchinson.

Kneale, J. (1999) 'A problem of supervision: moral geographies of the nineteenth-century British public house', *Journal of Historical Geography*, 25 (3): 333–48.

Lanciani, R. (1967) *The Ruins and Excavations of Ancient Rome*, New York: Benjamin Bloom Publisher.

Lianos, M. and Douglas, M. (2000) 'Dangerization and the end of deviance: the institutional environment', *British Journal of Criminology*, 40 (2), 261–78.

Lyon, D. (2003) *Surveillance after September 11*, Cambridge: Polity.

McCamley N. (1998) *Secret Underground Cities*, Barnsley: Leo Cooper.

Major, A.J. (1999) 'State and criminal tribes in colonial Punjab: surveillance, control and reclamation of the "dangerous classes"', *Modern Asian Studies*, 33 (3): 657–88.

Marcuse, P. (2002) 'Urban from and globalization after september 11: the view from New York', *International Journal of Urban and Regional Research*, 26 (3): 596–606.

Mazower, M. (ed.) (1997) *The Policing of Politics in the Twentieth Century: Historical Perspectives*, Providence, RI: Berghahn Books.

Neill, W.J.V., Fitzsimons, D.S. and Murtagh, B. (1995) *Reimaging the Pariah City: Urban Development in Belfast and Detroit*. Aldershot: Avebury.

Nelkin, D. and Andrews, L. (1999) 'DNA identification and surveillance creep', *Sociology of Health & Illness*, 21 (5): 689–706.

Newman, O. (1972) *Defensible Space: Crime Prevention Through Urban Design*, New York: Macmillan.

Norris, C. and Armstrong, G. (1999) *The Maximum Surveillance Society: The Rise of CCTV*, Oxford: Berg.

Norris, C., McCahill, M. and Wood, D. (2004) 'Editorial. The growth of CCTV: a global perspective on the international diffusion of video surveillance in publicly accessible space', *Surveillance and Society*, 2 (2/3): 110–35, http://www.surveillance-and-society.org/articles2(2)/editorial.pdf.

Porter, B. (1992) *Plots and Paranoia: A History of Political Espionage in Britain, 1790–1988*, London: Routledge.

Poster, M. (1990) *The Mode of Information: Poststructuralism and Social Context*, Cambridge: Polity Press.

Poyner, B. (1983) *Design against Crime: Beyond Defensible Space*, London: Butterworths.

Raco, M. (2003) 'Remaking place and securitising space: urban regeneration and the strategies, tactics and practices of policing in the UK', *Urban Studies*, 40 (9): 1869–87.

Richelson, J. (1995) *A Century of Spies: Intelligence in the Twentieth Century*, New York: Oxford University Press.

Richelson, J. (1999) *America's Space Sentinels: DSP Satellites and National Security*, Kansas: University Press of Kansas.

Rose, N. (2000) 'The biology of culpability: pathological identity and crime control in a biological culture', *Theoretical Criminology*, 4 (1): 5–34.

Saatjan, M. (2002) 'Balloon flight and the invention of airspace (1783–1870)', paper given at 'Transforming Spaces: The Topological Turn in Technology Studies', Technological University of Darmstadt, 22–24 March.

Sekula, A. (1992) 'The body and the archive', in R. Bolton (ed.), *The Contest of Meaning: Critical Histories of Photography*, Cambridge, MA: MIT Press. pp. 343–88.

Shearing, C.D. and Stenning, P. (1985) 'From the panopticon to Disneyworld: the development of discipline', in A.N. Doob and E.L. Greenspan (eds), *Perspectives in Criminal Law*, Aurora: Canada Law Books. pp. 335–49.

Sorensen, A. (2002) *The Making of Urban Japan: Cities and Planning from Edo to the Twenty-first Century*, London and New York: Routledge.

Sorkin, M. (1992) *Variations on a Theme Park: The New American City and the End of Public Space*, New York: Hill and Wang.

Swanstrom, T. (2002) 'Are fear and urbanism at war?' *Urban Affairs Review*, 38 (1): 135–40.

Thrift, N. and French, S. (2002) 'The automatic production of space', *Transactions of the Institute of British Geographers* (NS), 27 (4): 309–35.

Torpey, J. (2000) *The Invention of the Passport: Surveillance, Citizenship and the State*, Cambridge: Cambridge University Press.

Vidler, A. (2001) 'Aftermath; a city transformed: designing "defensible space"', *New York Times*, September 23, 2001. http://query.nytimes.com/gst/fullpage. html?res=9502E4DB163AF930A1575 ACOA9679C8B63 [accessed 10/10/2007].

Warren, R. (2002) 'Situating the city and September 11: military urban doctrine, "pop up" armies and spatial chess', *International Journal of Urban and Regional Research*, 26 (3): 614–19.

Webster, W.R. (2004) 'The diffusion, regulation and governance of closed-circuit television in the UK', *Surveillance and Society*, 2 (2/3): 230–50, http://www.surveillance-and-society.org/articles2(2)/diffusion.pdf.

Williams, C.A. (2003) 'Police surveillance and the emergence of CCTV in the 1960s', in M. Gill (ed.), *CCTV*, Leicester: Perpetuity Press.

Wood, D. (2001) 'The hidden geography of transnational surveillance: social and technical networks around signals intelligence sites', unpublished PhD thesis, University of Newcastle upon Tyne, UK.

Wood, D., Konvitz, E. and Ball, K. (2003) 'The constant state of emergency: surveillance after 9/11', in K. Ball and F. Webster (eds), *The Intensification of Surveillance: Crime, Terror and Warfare in the Information Era*, London: Pluto Press.

21 DREAMS AND NIGHTMARES

Stuart C. Aitken

This chapter

- ○ Considers the relationship between urban life and mental life

- ○ Uses examples from film to demonstrate that the real and fantasized city are inseparable

- ○ Demonstrates the potential contribution of post-structural and psychoanalytic theory to urban studies

Walk down the right back alley in Sin City and you can find anything (Marv, *Sin City*, 2005)

What if I'm wrong? I've got a condition. I get confused sometimes. What if I've imagined all this? What if I've finally turned into what they've always said I would turn into? A maniac. A psycho killer. (Marv, *Sin City*, 2005)

Introduction

Marv walks out of Frank Miller's spectacularly violent comic book series and into *Sin City* (2005), a movie directed by Robert Rodriguez and Quentin Tarantino. His noir world is infested with criminals, crooked police and *femme fatales*; some searching for revenge, some for salvation and others for both. It is a city of social pathology, anomie, angst, and schizophrenia. The film incorporates storylines from three of Miller's graphic novels, including *Sin City*, which launched the long-running, critically acclaimed series, as well as *That Yellow Bastard* and *The Big Fat Kill*. The series and the movie articulate an archetypically shadowed Jungian city where fantastic sex and violence reverberate through men and women anti-heroes of a wholly likable kind pitted against pure evil (emanating very specifically and twistedly in *Sin City* from the powerful clergyman, Cardinal Roark). In the movie, Mickey Rourke plays Marv with a fantastically sympathetic obsession. Marv is an outcast misanthrope with a psychic 'condition' requiring constant medication. He is on a mission to avenge the death of a hooker who was nice to him. I am drawn to this brutish character who murders the bad guys with a penchant for sadism: 'I love hitmen. No matter what you do to them, you don't feel bad.' Marv's attractiveness comes from three main sources: first, he is constructed as a classic vigilante, with a hugely appropriate moral streak; second, despite his quoted doubts in the epigram above, he is probably not delusional or even 'conditional' – he is just (and justly) upset; third, his sense of vigilante justice is in a city that is structured around enormously evil institutional power. The city is the delusion through which Marv thrives.

In what follows I look at the way cities are related to collective and individual psychoses. To do so, I focus on popular movies and art. Raymond Williams (1981) emphasized the point that a critical analysis of the content of popular media is necessary to understand contemporary urban culture. Representations of the city and their less representable affects map the material landscape by engaging viewers in the construction of new geographies that display the social and material world. The use of media imagery to construct and shape meanings of place is important, as is the role of images in the re-coding of space for new cultural orders (Short et al., 1993). What is less representable, and what becomes the fulcrum for my arguments, are the affects of these re-codings on a variety of dreams and nightmares and vice versa. In this chapter, I briefly look at desires, fantasies and fetishes in general before turning more specifically to agoraphobia, vertigo and schizophrenia. The chapter begins with a short discussion of the ways various social theorists have dealt with the relations between urban space and mental wellness.

Warping spaces

Understandings of cities, over the last one hundred years or so, have become a focus for coming to terms with quite a number of different things about ourselves, including our dreams, fantasies and psychoses. The 'city' has come to embrace symbols and symptoms of almost every social, mental and cultural process: at the very least, cities are concentrations and distillations of those processes; at their largest, cities are embodiments and producers of those processes (Bell and Haddour, 2000). Through this, Anthony Vidler (2000) argues that space is warped in two ways: first, as a psychological space, it is full of disturbing forms, including those of architecture; and second, as a produced space, when artists and architects break the boundaries of a genre to depict space in new ways.

Today, discussions of warping spaces are best understood from the writings of post-structural theorists. However, this 'spatial warping' and its connections to our mental health was first discussed by social theorists at the end of the nineteenth century. In terms of psycho-social processes, agoraphobia and claustrophobia began to be attributed to aspects of urban living in the late nineteenth century with the work of influential writers such as Emile Durkheim (1893/1932), Max Weber (1904/1946) and Georg Simmel (1903/1950) (see Extract 21.1). Durkheim's notion of anomie, for example, refers to breakdowns of social norms in modern urban society to the extent that norms no longer control activities. He argues that individuals cannot find their place in society without clear rules, and changing conditions as well as adjustment to urban life leads to conflict, deviance and social pathology. Durkheim observed that social periods of disruption (with rapid urbanization, for instance) brought about greater anomie and higher rates of dissatisfaction, crime and suicide. Among other things, Weber focused on the problematic dehumanizing contexts of urban bureaucracy:

> When fully developed, bureaucracy stands ... under the principle of *sine ira ac studio* (without scorn and bias). Its specific nature which is welcomed by capitalism develops the more perfectly the more bureaucracy is 'dehumanized', the more completely it succeeds in eliminating from official business love, hatred, and all purely personal, irrational and emotional elements which escape calculation. This is the specific nature of bureaucracy and it is appraised as its special virtue (Weber, 1946/1958: 215–16).

The influence of Durkheim and Weber led Simmel to postulate that a huge burden of urban society was its focus on organization and rationality over emotion and irrationality, and homogenization over differentiation and heterogeneity. These burdens played out in the psyche primarily because urban individuals increasingly give up their hearts to react with their heads, an '... organ which is least sensitive and quite remote from the depth of the personality'.

Extract 21.1: From Simmel, G. (1903/1950) 'The Metropolis and Mental Life', in K.H. Wolff (ed.), *The Sociology of Georg Simmel*, London and New York: Tree Press, pp. 409–24.

The deepest problems of modern life derive from the claim of the individual to preserve the autonomy and individuality of his existence in the face of overwhelming social forces, of historical heritage, of external culture, and of the technique of life. ... Nietzsche sees the full

(Continued)

development of the individual conditioned by the most ruthless struggle of individuals; socialism believes in the suppression of all competition for the same reason. Be that as it may, in all these positions the same basic motive is at work: the person resists to being leveled down and worn out by a social-technological mechanism. An inquiry into the inner meaning of specifically modern life and its products, into the soul of the cultural body, so to speak, must seek to solve the equation which structures like the metropolis set up between the individual and the super-individual contents of life. Such an inquiry must answer the question of how the personality accommodates itself in the adjustments to external forces. ... The psychological basis of the metropolitan type of individuality consists in the intensification of nervous stimulation which results from the swift and uninterrupted change of outer and inner stimuli. Man is a differentiating creature. His mind is stimulated by the difference between a momentary impression and the one which preceded it. Lasting impressions, impressions which differ only slightly from one another, impressions which take a regular and habitual course and show regular and habitual contrasts – all these use up, so to speak, less consciousness than does the rapid crowding of changing images, the sharp discontinuity in the grasp of a single glance, and the unexpectedness of onrushing impressions. These are the psychological conditions which the metropolis creates. ... Thus the metropolitan type of man – which, of course, exists in a thousand individual variants – develops an organ protecting him against the threatening currents and discrepancies of his external environment which would uproot him. He reacts with his head instead of his heart. In this an increased awareness assumes the psychic prerogative. Metropolitan life, thus, underlies a heightened awareness and a predominance of intelligence in metropolitan man. The reaction to metropolitan phenomena is shifted to that organ which is least sensitive and quite remote from the depth of the personality.

With the destruction and rebuilding of European cities after the First World War, phobias and anxieties came to be seen by many as the mental condition of modern metropolitan life. Sensations of anxiety and shame, whose centre could not be located and, therefore, could not be placated, found form in representations of the city and its architecture. Two notable examples, Fritz Lang's movie *Metropolis* (1926) and Franz Kafka's novel, *The Castle* (1921), accentuated anomie, paranoia and mass hysteria. After the Second World War, existentialist writers such as Jean Paul Sartre and Albert Camus sought further understanding of the hopelessness, despair and loneliness that accompanied modern living and dystopian futurists such as George Orwell extrapolated the psychic horrors of paternalistic patriarchy and societal compliance.

Towards the end of the twentieth century, postmodern writers such as Fredric Jameson, David Harvey, and Jean Baudrillard spoke to how seeming interior psychic conditions and larger social structures – for example, planned depthlessness (Baudrillard, 1988) and political unconsciousness (Jameson, 1991) – intertwined with cultural products, in particular the spatial arts of architecture, urbanism and film. Feminist writers such Marilyn French (1977) and Elizabeth Wilson (1991) countered male-dominated urban and architectural strictures, offering shrewd insights into what makes cities magical and fun, despite their vulgarity, rabble, vice and empty corporate plazas. French was one of the first feminists to recognize that women's penchant for depression and suicide might stem from suburban spatial entrapment; Wilson suggested planners repeatedly attempt to control and regulate women with grandiose, utopian plans that all but destroy urban culture. She argued many urban experiences offer women (and men) freedom:

> The failure of many well-intentioned planners was to allow the horrors of city life to blind them to its virtues. Those who believe that spiritual values can develop only in an atmosphere of calm and orderliness will always dislike and fear the city. For others it will represent the possibility of the highest levels of spirituality in its excesses and extremes: the city as the ultimate sublime (Wilson, 1991: 158).

Some scholars, attempting to understand a seeming increasingly chaotic and foreboding urbanness at the end of the twentieth century, embraced the sensibilities that developed from postmodernism's careless attitude to structure (Dear and Flutsy, 1998; Soja, 2000). Others elaborated psychoanalytic and therapeutic geographies deriving from Freudian and post-Freudian theorists (Pile, 1996; Aitken, 2002; Callard, 2003b).

In this chapter, I try to move beyond interiority and a sculpted psychoanalytic understanding of the city deriving from Freud and the object relations theorists who followed him by leaning more fully on the post-structural appraisals of Lacan and, in particular, Deleuze and Guattari's (1983, 1987) experimental, affective and playful schizoanalysis. Deleuze and Guattari's schizoanalysis uses a multitude of concepts to express the propensity of material to mutate, transform, and thus give rise to a plurality of urban spatial formations. Their post-structural urban space is folded and warped, containing affective and irrational 'pre-verbal intelligible content' that is not about a universal language of images, or some existential or Lacanian 'lack' in the viewer. Affect alludes to the motion part of emotion that sloshes back and forth between our perceptions of cities (the ways they are rendered as cultural products) and our actions (what we do). It is a movement of expression that carries stories between different levels of articulation, for example, between the visceral and the moral (Aitken, 2006). With these suggestions, Deleuze and Guattari reinsert in wholly appropriate ways the seemingly lost affective dimensions to urban culture mourned by Durkheim, Weber and Simmel. In what follows, I try to elaborate more fully the importance of this mobile perspective before noting its relations to a variety of phobias.

Moving through, over and beneath the city

Like Foucault, I find it more productive to think of the geography of these dreams and nightmares – and the struggles to interject them into the order and chaos of the city – as sites of self-making rather than as sites of Freudian repression. Like Lacan, I want to elaborate a more post-structural, mobile appraisal of Freud's stories of desire and eroticism. Lacan brings a sense of mobility lacking in Freudian analyses of interiority. Mobility in this sense refers to a dynamic interrelation between urban residents and their occupancy of city space. I suggest elsewhere that a Lacanian analysis of urban science fiction films can be used to draw out useful aspects of this particular topology, in that the screen portrays images from which the viewer apprehends the on-screen world as a reflective plane that offers a sense of 'wholeness', that is, a feeling of being complete and secure in one's identity (Aitken, 2002). As Crang (1997) notes, the attenuation of other senses within the darkened interiors of theatres is an especial practice of viewing which sets up the possibility of the illusory eye/I following the camera (see Doel and Clarke, 1997). For Cresswell and Dixon (2003), mobility can be thought of in an even broader sense as a certain attitude, at times openly radical and at times quietly critical, towards fixed notions of people and cities as they may appear in film, art and architecture. Accordingly, an emphasis on mobility suggests a certain scepticism in regard to stability, rootedness, surety and order, and yet it is not entirely about disorder either.

But Lacan does not go far enough. Durkheim and Weber created concepts such as 'anomie' and 'the iron cage' to reveal the ways capitalism and bureaucracy create detachment from the actual urban realm. This larger social structure is missing from Lacanian analysis. In an important sense, psychotic detachment from the actual is not just about interior processes but is produced by determinable social functionings within a real network of power relations:

> In order to grasp the conditions of existence of these phenomena it is necessary to reattach them to their obscure 'vertical content' in all its fractal glory. This is precisely what Lacanians omit to do in their treatment of the unconscious as a metonymic-metaphorical deep structure (Massumi, 2002: 45).

To understand associations with larger power structures, Deleuze and Guattari (1983, 1987) suggest that urban space is not only mobile, but schizophrenic. At times and in different places, it is striated or girded, and movement within it is confined to a horizontal plane, and limited by the order of that plane to preset paths. It is also, at other times and in other, different places, smooth or open-ended. The urban dweller occupies a nomadic existence, rising up at any point to move to another existence and then, at other times and other places, being constrained by or complicit with corporate capitalism. To understand this more fully, take Guillana Bruno's (1987: 69) proposition that the android replicants in Ridley Scott's *Blade Runner* (1982) are caught in a 'schizophrenic vertigo' as they search for their creator and a sense of unknown self in the corporate towers and abandoned buildings of 2019 Los Angeles. They are ultimately unsuccessful because their attempts to undo the vertigo by killing their creator in his corporate tower (a smoothing) is insufficient to change their pre-programmed terminations (a striation). In another movie whose urban realm resembles

Scott's future Los Angeles, human societal nightmares are constructed and compressed by alien striations. Protagonist John Murdock (Rufus Sewell) is caught up in the perpetual nightmare of Alex Proyas' *Dark City* (1998) until he is able to move upward, outward and through to another realm, and his fantasy of the seaside resort Shell Beach. Murdock's search is contrived through alleyways and across rooftops that strongly resemble the city setting of Lang's *Metropolis* and Scott's *Blade Runner*, except these settings continually change as the aliens (existentially labelled 'Strangers') construct and reconstruct the stories of the individuals who live in Dark City. There is no day, but at the stroke of midnight the humans in Dark City fall asleep and urban space literally warps: skyscrapers erupt and buildings twist into new forms as the city is remade and its characters are reworked to fit new narratives (see Aitken, 2002). So, like Michel de Certeau's (1984) spatial stories, individual narratives comprise the structure of Dark City's urban space.

These narratives do not just appear. Stewart (2002) notes that erotic dreams and nightmares are inflected by various historical power structures such as Christian asceticism, medicine or philosophy. A reason for this tenacity, he argues, is the ease with which the affective sensations of the fantastic and the erotic, such as terror and arousal, have jumped between genres such as monastic handbooks, medieval folk-tales and gothic fiction. As one of the Strangers in *Dark City* explains to Murdock: 'We fashioned the city on stolen memories: different eras, different pasts.' Societal memories 'localize' and 'punctuate' people's activities in space and time in a classic Lefebvrian sense. Stewart's work demonstrates the importance of an historical perspective in identifying and understanding culturally the power of urban nightmares contextualized by paranoia and indifference, and dreams contextualized by fantasies and fetishes.

Dreams and desires: fantasies and fetishes

It is appropriate not only to think about structures of fear and perversion suggested by movies such as *Sin City*, *Blade Runner* and *Dark City*, but also about structures of fantasy and its relations to identity. Linda Williams (1991: 11) insists that fantasies 'are not ... wish-fulfilling linear narratives of mastery and control leading to closure and the attainment of desire. They are marked, rather, by the prolongation of desire, and by the lack of fixed position with respect to the objects and events fantasized.' Fantasy is a place where conscious and unconscious – part and whole – may meet. It is a place resembling what Kaja Silverman (1992) refers to as a mobius-strip reflecting relations between interiority and exteriority, where each interchanges with the other imperceptibly as one move along the strip. In their now classic essay 'Fantasy and the Origins of Sexuality', Laplanche and Pontalis (1986/1964) argue that fantasy is the staging of desire, its *mis-en-scène* rather than its object. In their later book on psychoanalysis, Laplanche and Pontalis (1973: 277) note that people live their lives through a spatial relationship to the city that is 'the entire complex outcome of a particular organisation of personality, of an apprehension of objects that is to some extent or other phanta-sized'. In this sense, it is difficult to untangle fantasies from urban spaces.

Williams (1991: 11) notes further that Laplanche and Pontalis's (1986/1964) understanding of fantasies is linked to 'myths of origins' which moves between two discrepancies: an irrecoverable original experience and the uncertainty of its imaginary revival. In this sense, fantasies move us 'mobiusly' between interiority and exteriority, consciousness and unconsciousness, the real and the imagined. The important point that Williams raises is that the juncture of this irrecoverable real event and the totally imaginary event has no fixed temporal or spatial exis-tence: it is entirely mobile. Her argument melds well with how contemporary Freudian and object relations theorists, pulling from Melanie Klein and Julia Kristeva, understand 'the origin of the subject' as unfixed, unstable and transi-tional. Fantasies, then, are about mythic origins and their power derives from con-temporary urban reality as well as the unconscious. What needs further com-menting upon are the ways fantasies relate to phobias, and how this relation turns on an understanding of urban social space. Here, again, I gain insight from Lacan.

Lacan (1978) suggests that with desire we are trying to assert a sense of being, a way to mark our existence. If, in this sense, desire is conceptualized as a positive source of new beginnings, then urban space gets to mark these beginnings and pro-pel them forward. Alternatively, as a seeming illness emanating from, say, patri-archy, then men are taught to occupy urban space in ways that connote strength potency and assertiveness in the oppression of others. The male body is translated into an objective, physical project, subject to the motivation and will of its owner, a view that leads to the achievement-oriented, impervious, self-sufficient urban masculinity of corporate offices as well as bordellos and commercial pornography. This is a brutal, and often phallic, body politic. Jim Craine and I argue that it is this problematic detachment of desire and threat that artists and popular commentators on the urban condition, such as musician Matt Johnson, eschew for a recognition of body, soul and desire as one (Aitken and Craine, 2002).

For Matt Johnson, desire is drawn out of and embodied in urban realms. Musical images are joined with urban art to travel through Johnson's self-images that reside in notoriously corporate offices and salubrious spaces. Part of the appeal of Johnson's art and, it could be argued, many of the negative representations of urban space suggested by rap and hip-hop artists such as Snoop Dogg and Arrested Development, are their tendency to evoke tensions that always turn on the listener's ability to relate to those feelings. Many of Johnson's lyrics, for exam-ple, place the underside of lust and despair in quixotic parodies of urban angst and anomie (Aitken and Craine, 2002: 103). They also relate in important ways to how angst and anomie are related to globalized commodity fetishes.

Commodification is perhaps the most important theme of Johnson's *Nakedself* album, which weaves some of his long-term concerns with lust, loneliness and desire and his concerns with corporate capitalism in an eerie and apocalyptic soundscape of despair, where I am 'mobilized, globalized, hypnotized and homog-enized' in a 'Kentucky Fried Genocide' (Global Eyes, *Nakedself*, 2001). Whereas elsewhere Jonhson tilts against the sexual, spiritual and political malaise of urban society, *Nakedself* conjoins existential and sexual themes about commodified global consumerism. The music and its lyrics are a potent plea for global social activism that no longer denies the addictive power of sex and material goods, but redirects its energy against patriarchy and global capitalism (Aitken and Craine, 2002). While listening to and viewing Jonhson's art, it is difficult to disregard the famous admonition of Guy Debord (1983/2000) that the current spectacle *is* the transformation of desire and fantasy into the reality of the commodity occupying the totality of urban life (Extract 21.2).

Extract 21.2: From Debord, G. (1983/2000) *Society of the Spectacle*, Detroit: Black and Red, p. 42 (original emphasis).

The spectacle is the moment when the commodity has attained the *total occupation* of social life. Not only is the relation to the commodity visible but it is all one sees: the world: the world one sees is its world. Modern economic production extends its dictatorship extensively and intensively. In the least industrialized places, its reign is already attested by a few star commodities and by the imperialist domination imposed by regions which are ahead in the development of productivity. In advanced regions, social space is invaded by a continuous superimposition of geological layers of commodities. At this point in the 'second industrial revolution,' alienated consumption becomes for the masses a duty supplementary to alienated production. It is *all* the *sold labor* of a society which globally becomes the *total commodity* for which the cycle must be continued. For this to be done, the total commodity has to return as a fragment to the fragmented individual, absolutely separated from the productive forces operating as a whole. Thus it is here that the specialized science of domination must in turn specialize: it fragments itself into sociology, psychotechnics, cybernetics, semiology, etc. watching over the self-regulation of every level of the process.

This is also, clearly, the world of David Fincher's film *Fight Club* (2000), based on the critically acclaimed novel by Chuck Palahniuk of the same name. Although touted and popularized as a film about urban violence (a form of commodification in and of itself), *Fight Club* is also about a cathartic escape from the clutches of consumerism and the ways material possessions construct monotony and isolation (Craine and Aitken, 2004). For example, protagonist Jack (Ed Norton) has a sense of isolation that is reinforced by his condominium – on the fifteenth floor of a glass and steel tower – a space filled with Ikea products and other brand name consumer items that standardize and therefore define his existence. Viewers become a part of this space through a wonderfully crafted scene near the beginning of the movie as Jack moves through his apartment as if it were a page from an Ikea catalogue.

The mobile nature of Jack's job as a coordinator of product-recall campaigns for a leading automobile company further buttresses his belief that he is nothing more than a component part of a stifling, life-taking capitalist machine. Although potentially a nomad, he is constructed by his job and his creativity is constricted by its formulae. Jack travels around the country visiting accident sites, using a mathematical formula to figure out whether the cost of recalling defective cars exceeds that of paying out-of-court settlements to accident victims and their families. This is Palahniuk's equivalent of Philip Dick's (1968) Viogt-Kampff test (see discussion below). Jack's formulae are normalizing agents, and his job suggests the system is monstrous and inhuman – people and their sorrow are measured in units of profitability and loss. In a prophetic moment, Jack observes that he is nothing more than a 'single-serving' friend, just like the servings of butter and sugar and nuts offered by the air-hostess (Craine and Aitken, 2004).

Towards the end of *Fight Club*, violence becomes both terrifying and liberating – it is indeed debatable as to whether the exquisitely detailed 'fights' in the movie are real or just nihilistic fantasies. Either way, Jack exists in, and we have been watching, a dream – a dream that serves as a long prologue to an awakening. As he watches corporate towers blow up and fall in the last scenes of the movie while holding hands with his girlfriend, there is a sense that a liberating transformation is in place fomented by the urban devastation they are witnessing. As each tower falls, so too falls the grand plans of corporate America. There is a sense that Jack is moving from a ridiculously violent city to Elizabeth Wilson's (1991) sublime city (or perhaps it is simply capitalism changing its face).

Nightmares and fears: agoraphobia, schizophrenia and vertigo

Desire, in a Lacanian formulation, constructs city space as an integral part of a creative and productive voyage of discovery that relates wholly to the transformation of minds, bodies and souls. As such, this city of desire is supremely sublime. Countering this, and Wilson's assuredly up-beat proclamations about women in the city, Felicity Callard (2003a) notes the phenomena of fear and anxiety require that we take seriously the difficulty of concluding a joyful articulation between the sphere of urban social space and that of the psyche. Callard (2003a) highlights how fear (and its corollary desire) infects social relations through 'pathological' experiences of anxiety and how fear can emerge in particular social and spatial settings to beset those who otherwise appear 'sane'. If the affective elements of desire and fantasy tie into fears and phobias, she argues, then both sets of relations are about disruption of an individual's connection to a symbolic/spatial hegemonic norm. Madness, in the form of syndromes such as agoraphobia and vertigo, and also in the form of conditions such as schizophrenia, suggest, in Deleuzian (1994) terms, no-madism. The mobile spirit of the urban no-mad folds, warps and smooths the repetitious striations of hegemonic norms.

In the case of agoraphobia taken by Callard, space that is open foments anxiety because it suggests disorder. And yet it may be possible for an individual to reorder urban space with 'props'. For example, skirting an open plaza enables the agoraphobe to remain close to buildings or perhaps the crossing is possible when accompanied by a child in a buggy. In movies like *Sin City* and *Dark City*, the agoraphobe is treated to continual retreats to strangely familiar alleyways and back-door entrances. Callard notes that in the late nineteenth century, commentators circled around two problematics: the difficulty of adjudicating what it was that the agoraphobe, in his/her fear of the agora, actually feared; and the dilemma posed by the fact that 'the will' seemed utterly inadequate for overcoming the disorder. The question that Callard raises relates to what models of the social and the spatial underlie these two problematics, because both rely on the presumption that causes external to the individual precipitate fear when, in actuality, the condition mixes the internal and external in novel ways.

A schizophrenic, in the clinical sense, is someone who attempts an escape from self-identity, and who is thwarted by society or otherwise fails. In *Dark City*, schizophrenia is raised for Murdock when he yells at his wife 'I'm living someone else's nightmare'. That the people in the dark city live out other people's lives under the direction of the Strangers is a motif of the 'schizophrenic vertigo' alluded to earlier (Bruno, 1987: 69). Schizophrenia and vertigo are conflated with castration fears in *Dark City*, *Blade Runner* and *Sin City*. Indeed, in the latter, castration fears are realized spectacularly by several 'victims' of the anti-heroes' justice. The whole leitmotif of 'yellow bastard' is powered by his castration at the hands of the incorruptible street-savvy detective played by Bruce Willis.

The urban reality of many schizophrenics is about times and spaces that are discontinuous. This is clearly the case in *Dark City*, *Sin City* and *Blade Runner*'s Los Angeles, which are places of perpetual night-time and rain. Eyes and time-pieces are central icon of all three movies. Doel and Clarke (1997) posit further that, for the schizophrenic, time is not only discontinuous but the 'perpetual present' is lived with more intensity. For the inhabitants of Dark City and Sin City, and the replicants in *Blade Runner*, the present is always part of a larger experiment controlled by the 'law of the father' (the Strangers, Cardinal Roark, and Tyrell respectively). Doel and Clarke outline in detail how *Blade Runner* positions the replicants on Oedipal journeys that are their attempt to accede into the symbolic order, and how Roy Batty fails spectacularly to bend to the 'law of the father'. But his patricidal actions are also the beginnings of his 'subsequent acceptance of terminal breakdown' (Doel and Clarke, 1997: 149). Indeed, in a wonderfully crafted irony, the demise of Cardinal Roark (Rutger Hauer) and the hand of Marv in *Sin City* is remarkably similar to the demise of Tyrell at the hands of replicant Roy Batty (Rutger Hauer) in *Blade Runner*.

Schizophrenia is also about the erasure of difference. Caroline Knowles (2000) points out that the condition is really about an interpretation of particular forms of social divergence from urban norms and their political regimes, which are assumed to hold in place the behaviour of the 'no-mad', as well as being a description of the behavioural manifestations of particular forms of private terror and anxiety. Knowles points out that schizophrenic people are typecast and politically hidden in the streets of cities. These non-people/no-mads are designated as such because they deviate from standards set by society. As Doel and Clarke (1997: 157) point out for *Blade Runner*, the Viogt-Kampff test is a representation of a statistical norm around which deviations may be distributed. It is a quantitative standard, a curve, a statistical marker of normalcy that can, seemingly, identify the no-mad. How people, places and processes are represented and marked determines who are a city's villains, victims and salvationists. These constructions of normalcy are the conduit through which planning and policy interventions are understood and advanced, laying out city 'truths' and 'facts' in seamless everyday spaces. Larry Knopp (1998: 150–1) sums up this evolving perspective by suggesting that for urban theorists today, the city may be viewed as a representation of social relations and meanings in space, at densities and scales that are at once sufficiently large and complex as to feel overwhelming and incomprehensible, while at the same time remaining navigable and meaningful from the vantage point of those who control the norms of identity.

Mary Anne Doane (1982: 76) notes this is an illusory, symbolic depth inhabited and controlled by (white, affluent) men. The vertical exaggeration of cities is not only a symbolic but also a literal construction. The vertical space of power also bears meaning for the Tyrell Cooperation's Aztec-like pyramid and in *Metropolis*'s elite garden of the bourgeoisie ('where there is the sun and life'), and Cardinal Roarke's towered citadel in *Sin City*. Scott Bukatman (1993: 128) argues further that although the oppressive scale of cities is emphasized in many films, it is also presented as 'a weightless space, an area of suspension and vertical boundlessness'. This contrived depth and weightlessness constructs a sanctuary against the anxieties of vertigo. It transcends the seventy years between *Metropolis* and *Sin City*: elevators transport the workers of Metropolis to the subterranean 'workers' city' while the secret stairways are known only to John Fredersen, the corporate father of Metropolis; hover cars in *Blade Runner* are seen to be used only by city officials; only the Strangers are able to fly through Dark City; the anti-heroes of Sin City are able to jump through widows and doors at the top of tall buildings and land on the street unscathed.

What work on the fantasies and phobias of urban places calls into question is the assumption of, first, a discrete individual and an external threatening object or condition and, second, a majoritarian standard around which normality and humanness are judged. And so we can understand fear of open spaces or heights in the same way as we understand fear of snakes and blood because each is constructed through a fixed symbolic/spatial hegemonic norm. A central concern, then, revolves around ways that certain phobias – vertigo, agoraphobia and schizophrenia – are *embodied* in representations of urban spaces (filmic and other) because these speak to changing identities and changing social/spatial norms. For Deleuze and Guattari (1983), these phobias are not maladies; they are processes of becoming. They break away into the unstable equilibrium of continuing self-invention.

Becoming other urban embodiments

For Lefebvre (1991: 227), cities are a 'homogeneous matrix of capitalist space' and, as such, for individuals they narrate the intersection of particular forms of power with a specific form of political economy. Buildings contain activities in socially controlled spaces and sites equipped for particular kinds of production and reproduction. The power of capital to assert its requirements against the onslaught of dreams and nightmares dominate this cityscape but, as Pile (1996: 213) points out, Lefebvre is also unequivocal in his assessment that buildings and monuments conceal both a 'phallic realm of (supposed) virility' and a repressed space of panoptic surveillance and voyeurism (see Extract 21.3).

Extract 21.3: From Pile, S. (1996) *The Body and the City: Psychoanalysis, Space and Subjectivity*, London and New York: Routledge, pp. 213 and 224.

Monuments and monumental buildings present and re-present the phallus. They both make visible and 'mirror' back to the 'walker on the street' their place in the world, geographically, historically and socially; they reproduce repressive spaces which, while ostensibly acting as celebrations of events and people, have both feet in terror and violence; and they repeat not just people's experiences of themselves and their relations to others but also modalities of power. Symbols on the royal road to the 'unconscious' of urban life may not be that hard to find; indeed, it is possible to analyze urban spaces as if they had been dreamt. ...

The Tower is the production of space as a script, a spectacle, a sex aid. A vision, a sight and a site, but none of this means that it has an identity. Its identity is constantly shifting – Phallus, Ego, Face – and it thereby remains a veil, a chameleon, a hybrid of non-essences, deterritorialized and reterritorialized by flows of desire and power.

Rather than focus on the phallic symbolism of urban space, Meagan Morris (1992) uses Deleuze and Guattari's (1987) thesis on 'faciality' to suggest embodiments of a different kind (see also Deleuze, 1986). Deleuze argues that most often it is the face or its equivalent that gathers and expresses affect in complex, elusive ways. In cinema, close-ups make the face the pure building block of affect, what Deleuze (1986: 103) calls its *hylé*, from the Greek word meaning 'matter' or 'content'. This embodiment of urban content, argues Morris, is anchored in the city in the form of façades and picture-postcard renderings of urban space. Flows of capital and desire, for Delueze and Guattari (1987), are etched/striated on the face of the earth, continually deterritorializing and reterritorializing the body and the city. From this perspective, argue Morris and Pile (1996: 223), the shift from phallus/penis to face/faciality marks the deterritorialization of one masculine space with the reterritorialization of another. With a collection of postcard scenes of urban skylines the brutish body politic of the phallic urban tower is softened by a more acceptable face of capitalism.

Marv has a particularly ugly, scarred face. His face is all about resistance. Marv's resistance takes the form of systematically killing his way to the top. Many gruesome murders later he climbs the spiral staircase to the bed-chamber of Cardinal Roark. Roark, confronting Marv, speaks to the beauty of killing and eating people so that he could also eat their souls. The man of the church feels completely justified and comfortable with his deeds. This is a root of the structure upon which Sin City is built. Marv confronting Roark, kills him:

CARDINAL ROARK: Will that bring you satisfaction, my son? Killing a helpless, old, fart.
MARV: Killing? No. No satisfaction. Everything up until the killing, will be a gas.

Marv is a creation of Sin City ('I am not going to leave, I like it here') in the same way as Roy Batty is a creation of the Tyrell Corporation. So what comes next? With *Sin City* I expect that there will be other Cardinal Roarks and other Marvs. These kinds of movie sell well.

Summary

In representing urban space as a product of resistance and its relation to identity, it is important not to over-generalize or homogenize the process of dissent itself as a psychological imperative. Nor is it appropriate, as I may have suggested here, to imply that there is homogeneity at the roots of agoraphobia, vertigo and schizophrenia; that all these syndromes and phobias are about fighting, with Marv, the law of the father. To suggest that this is so would also deny difference. Rather, resistance is fundamentally related to the constitution of bodies in space, and thus is highly individuated and individuating. Individuation is exhibited by, and expressed through, forms of resistance that elaborate many kinds of spaces, encompassing numerous temporalities. The kinds of dreams and nightmares outlined here are movements as a form of resistance and resilience that includes not only broad horizontal shifts across land, but also vertical shifts up and down towers. There are distinctions, and there are moments of conformity. Indeed, conformity is an important spatial aspect of Deleuzian theory, for to focus exclusively on resistance is an act of generalization and purification, which does not effectively loosen and make more mobile the connections between dreams and nightmares on the one hand, and urban becomings on the other. In each of the syndromes and conditions outlined in this chapter, a journey is embarked upon from which a final destination remains elusive.

References

Aitken, S.C. (2002) 'Tuning the self: city space and SF horror movies', in R. Kitchen and J. Kneale (eds), *Lost in Space: Geographies of Science Fiction.* London and New York: Continuum Press.

Aitken, S.C. (2006) 'Leading men to violence and creating spaces for their emotions', *Gender, Place and Culture*, 13 (5): 491–507.

Aitken, S.C. and Craine, J. (2002) 'The pornography of despair: lust, desire and the music of Matt Johnson', *ACME, An International E-Journal for Critical Geographers*, 1 (1): 91–116.

Baudrillard, J. (1988) *America*, London: Verso.

Bell, D. and Haddour, A. (2000) 'What we talk about when we talk about the city', in D. Bell and A. Haddour (eds), *City Visions*, Harlow: Prentice Hall. pp. 1–11.

Bruno, G. (1987) 'Ramble City: postmodernism and *Blade Runner*', *October*, 41 (1): 61–74.

Bukatman, S. (1993) *Terminal Identity*, Durham, NC: Duke University Press.

Callard, F. (2003a) 'Conceptualisations of agoraphobia: implications for mental health promotion', *Journal of Mental Health Promotion*, 2 (4): 34–42.

Callard, F. (2003b) 'The taming of psychoanalysis in geography', *Social and Cultural Geography*, 4 (3): 285–312.

Craine, J. and Aitken, S.C. (2004) 'Street fighting: placing the crisis of masculinity in David Fincher's *Fight Club*', *Geo Journal*, 59: 289–96.

Crang, M. (1997) 'Watching the city: video, surveillance and resistance', *Environment and Planning A*, 28 (12): 2099–104.

Creswell, T. and Dixon, D. (2003) *Engaging Film: Geographies of Mobility and Identity.* Lanham, MD: Rowman & Littlefield.

Dear, M. and Flutsy, S. (1998) 'Postmodern urbanism', *Annals of the Association of American Geographers*, 88 (1): 50–72.

Debord, G. (1983/2000) *Society of the Spectacle*, Detroit: Black and Red.

De Certeau, M. (1984) *The Practice of Everyday Life*, Berkeley: University of California Press.

Deleuze, G. (1986) *Cinema 1: The Movement-Image*, trans. Hugh Tomlinson and Barbara Habberjam, London: The Athlone Press.

Deleuze, G. (1994) *Difference and Repetition*, New York: Columbia University Press.

Deleuze, G. and Guattari, F. (1983) *Anti-Oedipus: Capitalism and Schizophrenia*, Minneapolis: University of Minnesota Press.

Deleuze, G. and Guattari, F. (1987) *A Thousand Plateaus: Capitalism and Schizophrenia*, London: The Athlone Press.

Dick, P.K. (1968) *Do Androids Dream of Electric Sheep?* New York: Ballantine Books.

Doane, M.A. (1982) 'Film and the masquerade: theorizing the female spectator', *Screen*, 23 (3–4): 74–87.

Doel, M.A. and Clarke, D.B. (1997) 'From Ramble City to the screening of the eye: *Blade Runner*, death and symbolic exchange', in D.B. Clarke (ed.), *The Cinematic City*, London and New York: Routledge.

Durkheim, E. (1893/1932) *The Division of Labor in Society/De la Division du Travail Social*, Glencoe, IL: Free Press.

French, M. (1977) *The Women's Room*, New York: Jove Publications.

Jameson, F. (1991) *Postmodernism or, The Cultural Logic of Late Capitalism*, London and New York: Verso.

Knopp, L. (1998) 'Sexuality and urban space: gay male identity politics in the US, the UK and Australia', in R. Fincher and J. Jacobs (eds), *Cities of Difference*, New York: Guilford Press.

Knowles, C. (2000) *Bedlam on the Streets*, London and New York: Routledge.

Lacan, J. (1978) *The Four Fundamental Concepts in Psychoanalysis*, trans. Alan Sheridan, New York: W.W. Norton.

Laplanche, J. and Pontalis, J. (1973) *The Language of Psycho-Analysis*, trans. Donald Nicholson-Smith, New York: W.W. Norton.

Laplanche, J. and Pontalis, J. (1986/1964) 'Fantasy and the origins of sexuality', in V. Burgin, J. Donald and C. Kaplan (eds), *Formations of Fantasy*, London and New York: Routledge. pp. 5–34.

LeFebvre, H. (1991) *The Production of Space*, Oxford: Blackwell.

Massumi, B. (2002) *Parables for the Virtual: Movement, Affect, Sensation*, Durham, NC, and London: Duke University Press.

Morris, M. (1992) 'Great moments in social climbing: King Kong and the human fly', in B. Colomina (ed.), *Sexuality and Space*, Princeton, NJ: Princeton Architectural Press.

Pile, S. (1996) *The Body and the City: Psychoanalysis, Space and Subjectivity*, London and New York: Routledge.

Short, J., Benton, L., Luce, B. and Walton, J. (1993) 'Reconstructing the image of an industrial city', *Annals of the Association of American Geographers*, 83 (2): 207–24.

Silverman, K. (1992) *Male Subjectivity at the Margins*, London and New York: Routledge.

Simmel, G. (1903/1950) 'The metropolis and mental life', in K.H. Wolff (ed.), *The Sociology of Georg Simmel*, London and New York: The Free Press.

Soja, E. (2000) *Postmetropolis: Critical Studies of Cities and Regions*, Oxford: Blackwell.

Stewart, C. (2002) 'Erotic dreams and nightmares from antiquity to the present', *The Journal of the Royal Anthropological Insititute*, 8 (2): 279–309.

Vidler, A. (2000) *Warped Space: Art, Architecture, and Anxiety in Modern Culture*, Cambridge, MA: MIT Press.

Weber, M. (1904/1946) *The Protestant Ethic and the Spirit of Capitalism*, trans. T. Parsons, New York: Scribner.

Weber, M. (1946/1958) *From Max Weber*, trans. and ed. H.H. Gerth and C. Wright Mills, New York: Galaxy.

Williams, L. (1991) 'Film bodies: gender, genre and excess', *Film Quarterly*, 44 (4): 2–13.

Williams, R. (1981) *Culture*, London: Fontana Press.

Wilson, E. (1991) *The Sphinx in the City: Urban Life, the Control of Disorder and Women*, Berkeley: University of California Press.

INDEX

Note: Page numbers in italics refer to illustrations.